T0296779

Handbook of Obesity in Obstetrics and Gynecology

Handbook of Obesity in Obstetrics and Gynecology

Edited by

Tahir Mahmood

Department of Obstetrics & Gynaecology, Victoria Hospital, Kirkcaldy, United Kingdom

Chu Chin Lim

Department of Obstetrics & Gynaecology, Victoria Hospital, Kirkcaldy, United Kingdom

ELSEVIER

ACADEMIC PRESS
An imprint of Elsevier

Academic Press is an imprint of Elsevier
125 London Wall, London EC2Y 5AS, United Kingdom
525 B Street, Suite 1650, San Diego, CA 92101, United States
50 Hampshire Street, 5th Floor, Cambridge, MA 02139, United States
The Boulevard, Langford Lane, Kidlington, Oxford OX5 1GB, United Kingdom

Notices
Knowledge and best practice in this field are constantly changing. As new research and experience broaden our understanding, changes in research methods, professional practices, or medical treatment may become necessary.

Practitioners and researchers must always rely on their own experience and knowledge in evaluating and using any information, methods, compounds, or experiments described herein. In using such information or methods they should be mindful of their own safety and the safety of others, including parties for whom they have a professional responsibility.

To the fullest extent of the law, neither the Publisher nor the authors, contributors, or editors, assume any liability for any injury and/or damage to persons or property as a matter of products liability, negligence or otherwise, or from any use or operation of any methods, products, instructions, or ideas contained in the material herein.

ISBN: 978-0-323-89904-8

For Information on all Academic Press publications
visit our website at https://www.elsevier.com/books-and-journals

Publisher: Stacy Masucci
Acquisitions Editor: Patricia M. Osborn
Editorial Project Manager: Sara Pianavilla
Production Project Manager: Niranjan Bhaskaran
Cover Designer: Mark Rogers

Typeset by MPS Limited, Chennai, India

Working together
to grow libraries in
developing countries

www.elsevier.com • www.bookaid.org

Dedication

To Aasia and Berenice

For their support and endless love during our professional lives and their tolerance during the editing of this volume, which we hope will help our younger colleagues to provide effective reproductive healthcare.

Contents

Section 1 Gynaecology

Section 2   Obstetrics

List of contributors

Ibrahim Alsharaydeh Department of Obstetrics and Gynaecology, Raigmore Hospital, Inverness, Scotland

Diogo Ayres-de-Campos Department of Obstetrics and Gynecology, Santa Maria University Hospital, Lisbon, Portugal; Obstetrics Department, North Lisbon University Hospital Center, Lisbon Academic Medical Center, Lisbon, Portugal

Dominique Baker Department of Obstetrics & Gynaecology, Victoria Hospital, Kirkcaldy, United Kingdom

Rashda Bano Zita West Assisted Fertility Clinic, London, United Kingdom

Swetha Bhaskar Department of Obstetrics and Gynaecology, Victoria Hospital, NHS Fife, Scotland

Caroline Brewster Department of Obstetrics and Gynaecology, Royal infirmary of Edinburgh, Edinburgh, Scotland

F.P. Dunne College of Medicine Nursing and Health Sciences, National University of Ireland Galway, Galway, Ireland

A.M. Egan Division of Endocrinology, Diabetes and Metabolism, Mayo Clinic, Rochester, MN, United States

Mohamed ElMoursi Consultant of Obstetrics and Gynaecology, Victoria Hospital, NHS Fife, Kirkcaldy, United Kingdom

Andreia Fonseca Obstetrics Department, North Lisbon University Hospital Center, Lisbon Academic Medical Center, Lisbon, Portugal

Asma Gharaibeh Royal Infirmary of Edinburgh, Edinburgh, Scotland

Ailie Grzybek Ninewells Hospital, Dundee, Scotland

Alasdair Hardie Department of Obstetrics and Gynaecology, Royal Infirmary of Edinburgh, Edinburgh, Scotland

Kahyee Hor Department of Obstetrics and Gynaecology, Royal Infirmary of Edinburgh and University of Edinburgh, Scotland

Asma Khalil Fetal Medicine Unit, Department of Obstetrics and Gynaecology, St. George's University Hospitals NHS Foundation Trust, London, United Kingdom; Vascular Biology Research Centre, Molecular and Clinical Sciences Research Institute, St George's University of London, London, United Kingdom; Twins Trust Centre for Research and Clinical Excellence, St George's University Hospital, London, United Kingdom

Suresh Kini Assisted Conception Unit, Department of Obstetrics and Gynaecology, Ninewells Hospital and University of Dundee, Dundee, Scotland

Chu Chin Lim Department of Obstetrics & Gynaecology, Victoria Hospital, Kirkcaldy, United Kingdom

Tahir Mahmood Department of Obstetrics & Gynaecology, Victoria Hospital, Kirkcaldy, United Kingdom

Inês Martins Department of Obstetrics and Gynecology, Medical School, University of Lisbon, Portugal

Nirmala Mary NHS Lothian, Edinburgh, Scotland; Royal Infirmary Edinburgh, Edinburgh, United Kingdom

C. Newman College of Medicine Nursing and Health Sciences, National University of Ireland Galway, Galway, Ireland

Katrine Orr Ninewells Hospital, Dundee, Scotland

Nithiya Palaniappan Department of Obstetrics and Gynaecology, Victoria Hospital, Kirkcaldy, Scotland

Smriti Prasad Fetal Medicine Unit, St George's Hospital, St George's University of London, London, United Kingdom; Vascular Biology Research Centre, Molecular and Clinical Sciences Research Institute, St George's University of London, London, United Kingdom

Mythili Ramalingam Assisted Conception Unit, Department of Obstetrics and Gynaecology, Ninewells Hospital and University of Dundee, Dundee, Scotland

Gamal Sayed Women's Wellness and Research Centre and Clinical Department, College of Medicine, Qatar University, Doha, Qatar

Rabia Sherjil Royal Infirmary of Edinburgh, Edinburgh, United Kingdom

Laura Stirrat Royal Infirmary of Edinburgh, Edinburgh, Scotland

Omar Thanoon Department of Obstetrics & Gynaecology, Victoria Hospital, Kirkcaldy, United Kingdom

Hannah Waite Ninewells Hospital, Dundee, Scotland

Preface

Obesity has reached epidemic proportions globally. The impact of obesity on women's reproductive health and well-being is significant. The consequences of obesity start preconceptually in utero and its effect is transgenerational on reproductive health. In short, it will affect the well-being of women from 'Cradle to Grave'. The editors have assembled comprehensive reviews on various aspects of obstetrics and gynaecologic care that are impacted by this increasingly prevalent condition.

In the obstetrics section of this handbook, an overview of the genetic, molecular, and psychological bases of obesity is explored first. Preconceptual care and weight optimisation strategies have been discussed prior to looking at the specific details of the pregnancy journey. The antepartum care section includes elderly obese mothers, and the important role of ultrasound in early and late pregnancy is explained. A fresh look at the role of emerging viruses including zika and corona virus in pregnant women is also included. Management of specific complications and comorbidities among obese women, including hypertension, preeclampsia, venous thromboembolism, hyperglycaemia, and metabolic syndrome in pregnancy, is discussed. There are significant challenges in the provision of intrapartum and postpartum care of the obese pregnant women. All these have also been looked at along with practical solutions.

In the gynaecology section of this handbook, discussion begins at adolescent age. The management of conditions including polycystic ovarian syndrome, sexual behaviour, contraception, and hirsutism in obese women are presented. This is followed by chapters discussing infertility issues including recurrent pregnancy loss, assisted reproduction, sexual dysfunction, male obesity and its effect on semen quality, as well as potential therapies. Contraception can be challenging for obese women, and with the increasing popularity of bariatric surgery, a section has been dedicated for contraceptive care for women post bariatric surgery. The effect of obesity in myriad disorders is presented including menstrual disorders, urinary and faecal incontinence, pelvic organ prolapse, gynaecologic and breast cancer, osteoporosis, menopause, chronic pelvic pain, and psychosexual disorders. The management of the above-mentioned conditions has been discussed with particular consideration of obese women.

The purpose of this handbook is to compile a handy handbook enabling physicians to access the latest useful information on obesity and reproductive health. The book is specifically written in concise short paragraphs and often bullet point format to enable the reader to access the information more rapidly. This is especially useful as a bedside or office reference in a busy clinical setting. It is also valuable for revision purposes in preparation for postgraduate or board examinations.

Tahir Mahmood
Chu Chin Lim

Section 1

Gynaecology

Obesity, the onset of adolescence and menstrual disorders

Tahir Mahmood
Department of Obstetrics & Gynaecology, Victoria Hospital, Kirkcaldy, United Kingdom

1. Obesity among children, adolescents, and adults is set to be one of the most important public health concerns of the 21st century.
2. Over the last three decades, the incidence of obesity in childhood and adolescence has been a growing epidemic, with a rise by more than a half of overweight and a doubling of obesity.
3. All around the world, 1 in 10 young people aged 5−17 years are overweight or obese, and most of them live in developing countries,
4. More than 60% of children who are overweight before puberty will become overweight young adults.
5. Globally, prevalence of childhood overweight and obesity increased from 4.25 in 1990 to 6.7% in 2010 with 8.5% in Africa and 4.9% in Asia.
6. Data collected by WHO Europe show a prevalence of overweight/obesity ranging from 5% to more than 25%, with great variability among countries and a still growing incidence in more than half of them.
7. In general, a greater proportion of overweight/obesity was found in boys than in girls, and more so in western and southern Europe as compared to Northern European countries.
8. The prevalence rates are approximately double in Mediterranean nations than those of Northern European countries
9. 30% of North American children and adolescents are overweight or obese, with the highest rates among minorities and low-income families.
10. Obesity rates of both genders are the highest in Mexican Americans (31%), followed by non-Hispanic blacks (20%), non-Hispanic whites (15%), and Asian Americans (11%).

1.1 Role of BMI charts

1. In children and adolescents, obesity has not been as well defined as in adults, and therefore it is not a perfect measurement, but BMI is still considered to be a gold standard for diagnosis of overweight and obesity.
2. Even by using a BMI percentile chart, different definitions of being overweight and obese have been described in different systems.
3. Table 1.1 compares definitions for childhood and adolescent obesity as used by different organisations.

1.2 Aetiology

1. Obesity is a multifactorial condition and involves both genetic and nongenetic factors but lack of physical activity and unhealthy eating habits are key determinants.
2. Gungor[1] has published a comprehensive list of possible aetiological factors of obesity in childhood and adolescence.

Handbook of Obesity in Obstetrics and Gynecology. DOI: https://doi.org/10.1016/B978-0-323-89904-8.00007-X

Table 1.1 Various cut-off points of BMI used to describe childhood and adolescent obesity.

Definitions	CDC	WHO	IOTF	NCMP	SIGN
Overweight	85th−95th	85th−97th	91st	>85th	91st
Obesity	>95th	>97th	99th	95th	98th
Severe obesity					>99.6th

Key: *CDC*, Centre for Disease Control; *IOTF*, International Obesity Task Force; *NCMP*, National Child Measurement Program; *SIGN*, Scottish Intercollegiate Guidelines Network; *WHO*, World Health Organisation.

Genetic variations:

Rare genetic defects of leptin secretion, and more frequent genetic syndromes causing obesity, such as Prader-Willi syndrome.

Epigenetics:

In utero environments acting on DNA methylation which induce heritable changes in obesity expression (Table 1.1).

Endocrine disorders:

Hypothyroidism, growth hormone (GH) deficiency, and excess cortisol.

CNS diseases:

Congenital or acquired hypothalamic pathologies (Infiltrative diseases, tumours, or after treatment sequelae) that alter the hypothalamic regions in charge of hunger and satiety.

Intrauterine exposures:

Intrauterine exposures to gestational diabetes or extreme maternal adiposity (macrosomic and small for gestational age babies are at risk of childhood obesity).

BMI rebound:

An early postinfancy increase in weight before the age of 5.5 years is a risk factor for the development of obesity at later ages.

Diet:

1. High energy intake food in infancy
2. Excessive consumption of sweetened soft drinks in childhood
3. Poor eating habits, such as inadequate intake of vegetables and fruits
4. Skipping breakfast
5. Eating out frequently
6. Comfort eating
7. Fast food with high calorie content

Low-energy expenditure:

1. Poor physical activity
2. Excessive time spent in sedentary activities (Television or other screen viewing activities)

Sleep pattern:

Shorter sleep duration in infancy and childhood

Infections:

1. Microbial infections
2. Composition of the gut flora

Iatrogenic:

1. Cranial irradiation or surgery induced hypothalamic injury
2. Psychotropic medication (Olanzapine and Risperidone)
3. Chemotherapeutics
4. Hormonal contraception—medroxy progesterone acetate

Ethnic origin:
More frequent in Hispanic and South Asian children and adolescents.
Country of birth:
Children from developing countries, who are born underweight, are at higher risk.
Residence in urban versus rural areas:
There is higher incidence of obese children in urban area globally.
Socioeconomic level:
Children of the lowest socioeconomic groups living in high income countries.

There is a complex interplay between an obesogenic environment and the individuals' predisposition to adiposity.

It involves a number of appetite stimulating hormones, such as ghrelin, the anorexigenic peptide YY, the pancreatic polypeptide, glucagon, and others.
Onset of puberty:

1. An adequate nutritional status is a prerequisite for the central onset of puberty
2. At puberty there is an increase in BMI and subcutaneous adiposity.
3. Fat acts as metabolic trigger for initiation of puberty.
4. Obesity may lead to premature activation of the gonadotrophin-releasing hormone pulse generation.
5. Obese children frequently show a tall stature for their age, associated with an accelerated epiphyseal growth plate maturation and early puberty in both sexes, but mainly in girls.
6. Chronic malnutrition delays the onset of puberty.

Puberty is a developmental process during which a child becomes a young adult, characterised by the secretion of gonadal hormones and the development of secondary sexual characteristics that lead to sexual maturation and reproductive capability.

The age of onset of puberty ranges from 8 to 13 years old (average age of 10 years in White Americans and at 8.9 years in African-Americans).
Main physiological events include:
Although the physical sequelae of gonadarche and adrenarche occur concomitantly, a discordance of the two processes may also occur in normal development.
Adrenarche:

1. Involves the increased production of androgens by the adrenal cortex.
2. Typically precedes gonadarche.
3. There is increased secretion of dehydroepiandrosterone, dehydroepiandrosterone sulfate (DHEAS), and androstenedione.
4. Leads to the appearance of sexual hair (pubarche).
5. Absence of adrenarche does not prevent Gonadarche or the attainment of fertility.

Pubarche: is the appearance of pubic hair, primarily due to the effects of the androgens from the adrenal gland. It also refers to the appearance of axillary hair

Gonadarche:

1. It comprises growth and maturation of the gonads with increased secretion of gonadal sex steroids by the pituitary hormones FSH and LH,
2. Gonadarche leads to thelarche and menarche.

Thelarche: is the appearance of breast bud under the areola under the influence of oestrogen,

Menarche:

1. Is the onset of the menstrual cycle, caused by the action of oestradiol on the endometrium and is usually not associated with ovulation;
2. Menarche arrives on average at age 12.5 years, regardless of ethnicity, generally occurs 2−3 years after the onset of breast development;
3. Following thelarche on average by 2.5 years (range 0.5−3 years);
4. Menarche occurs in Tanner stage three or four;
5. Peak height velocity occurs at Tanner stage 2−3.

Precocious puberty:
When signs of puberty develop before 8 years of age in girls.

Delayed puberty:
The absence of thelarche or menarche by age 13 and 16 years, respectively.

Marshall and Tanner Stages of Breast and Pubic Hair Development.

Pubic hair scale

1. Stage 1: No hair
2. Stage 2: Downy hair
3. Stage 3: Scant terminal hair
4. Stage 4: Terminal hair that fills the entire triangle overlying the pubic region
5. Stage 5: Terminal hair that extends beyond the inguinal crease onto the thigh

Female breast development scale

1. Stage 1: No glandular breast tissue palpable
2. Stage 2: Breast bud palpable under areola (1st pubertal sign in females)
3. Stage 3: Breast tissue palpable outside areola; no areolar development
4. Stage 4: Areola elevated above contour of the breast, forming "double scoop" appearance
5. Stage 5: Areolar mound recedes back into single breast contour with areolar hyperpigmentation, papillae development, and nipple protrusion

(Continued)

(cont'd)

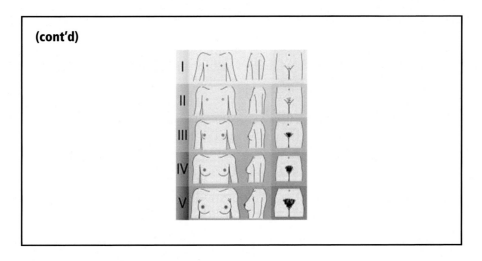

https://healthjade.net/tanner-scale/

Age of menarche:

A major determinant of pubertal timing is genetic. Other factors that influence pubertal development are race, general health, nutrition, and environmental effects.

There has been a decline in the age of menarche in Europe and North America from 16–17 years in the early 19th century, to 13 years of age in the latter half of 20th century.

Breast and pubic hair development are appearing earlier than they were 50 years ago.

In a study of American girls, mean ages for menarche were 12.1, 12.2, and 12.7 years for American-African, Mexican-American, and White-American girls, respectively.

A critical body weight or per cent body fat is the primary determinant of the development of secondary sexual characteristics.

It has been proposed that the onset of the growth spurt requires a critical weight of 47.8 kg, and an increased body fat (BMI excess by 25%–30%) can lead to an early height growth spurt and menarche age in puberty,

Puberty starts as a result of increase in the pulsatile secretion of *Gonadotropin-releasing hormone* (GnRH) from the hypothalamus, and the suppression of inhibitors of GnRH secretion.

Gonadotropin-releasing hormone (GnRH)

Around the age of 8, hypothalamus increases production of nocturnal GnRH pulse.

Slow GnRH pulsatility initially stimulates the release of FSH, resulting in follicular development and oestrogen production.

With the acceleration of GnRH pulsatility, LH is produced and this is preceded without any external manifestation of puberty by several years.

Kisspeptin, a hypothalamic neuropeptide, is now recognised as a key regulator of pulsatile GnRH secretion.

The irregular nocturnal pattern of LH secretion gradually increases until regular LH pulses are established, occurring every 90 minutes irrespective of time of day.

The production of gonadal steroids contributes to the development of external sexual characteristics (thelarche).

Adrenarche reflects the maturation of the hypothalamic- pituitary-adrenal axis.

During puberty, the sensitivity of the hypothalamus and the pituitary increase, thus allowing the production of sex steroids that stimulate secondary sexual characteristics (breast development, hip broadens, pubic hair growth).

Gonadotropins:

LH:

1. Stimulates androgen production by theca cells of the ovarian follicles; and
2. progesterone production from luteinised granulosa cells of the corpus luteum,

FSH:

1. Is important for follicular recruitment and selection;
2. in granulosa cells of the developing follicle, FSH induces expression of aromatase, which is responsible for aromatisation of androgens secreted by the theca cells; and
3. FSH also induces LH-R (receptor) in granulosa cells of the dominant follicle, which selectively amplifies the effect of declining FSH concentrations on the dominant follicle.

Inhibin:

1. There is increased secretion of gonadal inhibin A and B during Gonadarche;
2. inhibin B begins to rise with the onset of puberty, as does Inhibin A in breast stages 3 and 4,
3. adult levels are attained at approximately 14 to 15 years of age; and
4. during the menstrual cycles, inhibin B predominates in the circulation of follicular phase, and Inhibin A levels are elevated in the luteal phase.

Leptin:

1. Mainly secreted by white adipose tissue;
2. leptin appears to be one of the several factors that influence the activity of the GnRH pulse generator;
3. it also stimulate both GH secretion and GH response to GH-releasing hormone;
4. GH causes increase of linear growth via acting directly on skeletal growth centres by inducing both proliferation and differentiation of chondrocytes;
5. leptin is secreted in pulses, acting directly and dose dependently stimulating GnRH secretion in the arcuate hypothalamic neurons fostering the release of LH and FSH from the anterior pituitary;
6. leptin dose-dependently stimulates adrenal 17 alpha- hydroxylase and 17, 20-lyase, with higher adrenal androgen levels that are involved in accelerated growth of obese girls. Marked weight gain and obesity may therefore induce premature adrenarche;
7. higher Leptin levels in girls are associated with increased body fat, and an earlier onset of puberty; and
8. leptin levels decrease with increasing Tanner stages of puberty, and there is an increased sensitivity to Leptin.

Androgens:

Obesity is often accompanied with increased synthesis of androgens, with both peripheral and central action on the hypothalamic-pituitary axis, enhancing pubertal rise of GnRH secretion.

1. BMI positively correlates with free Testosterone Index−;
2. total Testosterone is fourfold elevated in pre-pubertal 7−19 years old obese girls;

3. testosterone is 1.75 fold elevated in pubertal 10- to 12 years old obese girls;
4. mean Free Testosterone levels in Tanner stage-matched obese girls have two- to ninefold elevated levels, compared with normal weight controls; and
5. obesity is also associated with approximately 40% elevation of DHEAS levels.

Reduced Sex Hormone Binding Globulin (SHBG) Levels:
It increases the availability of sex steroids, oestradiol, which can induce premature thelarche.

Increased adipose tissue aromatisation of androgens:
Increased aromatisation of androgens into oestrogens contributes to both accelerated growth rate and precocious puberty.

Higher volume adipose tissue acts as an endocrine organ and releases higher levels of sex steroids and adipokines and contributes to linear growth.

Increased levels of insulin:
Due to increased insulin resistance in obesity, it stimulates the insulin growth factor-1 (IGF-1) receptor, and higher levels of IGF-1 bioavailability.

Hyperinsulinaemia, with obesity-associated insulin resistance, may also stimulate the onset and progression of puberty by accelerating weight gain and growth.

Hyperinsulinaemia also increase LH-stimulated ovarian and adrenal steroidogenesis.

Puberty changes and obesity:
Obesity contributes to early breast development as the first sign of puberty, in comparison with girls who have pubic hair development first with normal BMI, and an earlier age of menarche as well (before the age of 12).

Health consequences of obesity in adolescence

Clinical issues relevant to gynaecological practice	General health issues
Earlier sexual maturation and reproductive dysfunction	Abnormal glucose metabolism
Alterations in menstruation	Hypertension
Chronic anovulation and PCOS	Dyslipidaemia
Hyperandrogenism associated symptoms	Hepatic steatosis
Heavy irregular periods and dysmenorrhea	Sleep apnoea
Risky sexual behaviour and STI risk	Psychological issues/emotional disorders
Inefficient use of contraception and higher risk of adolescent pregnancy	Eating disorders
Macromastia (large breasts) can be extremely distressing	Depression, chronic fatigue

General assessment of an obese adolescent:
History:

1. Age of onset of obesity (Gradual or Rapid)
2. Dietary habits
3. Activity history
4. Medication
5. A review of the systems to rule out: hypothyroidism, Cushing's syndrome, Albright' hereditary osteodystrophy syndrome, hypothalamic congenital or acquired diseases (previous surgery, autonomic dysfunction, etc.)

6. Family history of cardiovascular disease and diabetes
7. Psychosocial issues: mood changes, sleep disturbance, sense of loneliness, teasing at school
8. Smoking, drug abuse
9. Menstrual history
10. Relationship issues, concerns, risks

Clinical examination:

1. Dysmorphic features (short stature, fat distribution over the body especially around abdomen, interscapular area, on the face and around neck
2. Height and weight for BMI calculation, and waist/hip measurement
3. Hirsutism, acanthosis nigricans (may be suggestive of hyperandrogenism/hyperinsulinaemia)
4. Blood pressure
5. Any sign of developmental delay

Pelvic examination:

1. In sexually active adolescent and STI screening following sensitive counselling only
2. Ultrasound examination if indicated (abdominal route, if not sexually active)

Laboratory investigation:

1. Depending upon the findings from the above assessment

General advice:

1. Weight management
2. Dietician input as regards healthy and balanced diet
3. Lifestyle interventions
4. Input from relevant specialists if an endocrine or biochemical abnormality is suspected

Contraception:

1. Hormonal contraception for sexually active, although there are limited data on the effectiveness of contraception in obese adolescent.
2. Obesity is category 2/3 in the medical eligibility criteria for Combined hormonal contraceptives (CHC) use (increased risk of thrombosis).

Biological characteristics of adolescence

1. In adolescence the hypothalamo-pituitary-ovarian axis gradually develops but bleeding irregularities occurs
2. Hyperandrogenic conditions, manifesting with seborrhea and acne
3. Primary dysmenorrhea is a frequent symptom
4. Polycystic ovary syndrome manifesting during adolescence
5. The immune system is in a process of maturation and is still vulnerable to infection.

Psychosocial characteristics

1. Adolescents start their sexual life, may have several partners in a shorter duration
2. Adolescents have special behavioural patterns like mood swings, abrupt changing of behaviour
3. Adolescents often have body image concerns

Sexual health care
Frequent sexual health issues are:

1. Insecurity, lack of knowledge about one's own body and about sexuality
2. Painful sex
3. Experience of violence

Care includes:

1. Proactively asking about sexuality and sexual activities
2. Counselling, information, education (empowerment)
3. Correcting myths from social media and pornography
4. Ask about exposure to unwanted sexual contacts, drugs misuse, smoking, etc.

What is an ideal contraceptive for the adolescents?

1. Very effective and protect against STI
2. Independent of compliance
3. No negative impact on body image and appearance
4. Preserve and enhance fertility
5. Give additional benefits as regards menstrual irregularities, dysmenorrhea, and PCO-related symptoms.

Progestogen only pill
Category 1

1. *Advantage:*
   a. Effective, no cardiovascular risk and reduction of dysmenorrhea
2. *Disadvantage:*
   a. Large differences in PI between perfect use versus typical use
   b. Bleeding irregularities, mood effect, acne
   c. Offers no protection against STI

Long acting reversible contraceptive methods
(LARCs: Copper IUD, LNG-IUD, implant, and Depo Medroxy Progesterone—DPMA)
Category 1

1. *Advantages*
   a. They are very effective, independent of user adherence, and independent of partner
   b. Lead to significant reduction in dysmenorrhea (implant, LNGIUD, DMPA)
2. *Disadvantages*
   a. Needs specialist intervention for insertion and removal of these methods
   b. PID risk dependent on sexual behaviour (copper IUD, LNG-IUD)
   c. Irregular bleeding (implant),
   d. Heavy and painful period (copper IUD)
   e. Rarely mood effect and acne

Combined hormonal contraceptives
Category 2/3.
BMI $\geq$ 30–35 kg/m^2
BMI $\geq$ 35 kg/m^2

1. *Advantages:*
   a. Effective (user dependent)
   b. Independent of partner
   c. Preserve and enhance fertility
   d. Protection against ovarian, endometrial, and colon cancer
   e. Small increase in breast cancer, cervical cancer dependent on duration of use.
   f. Additional benefits
      i. Control of bleeding irregularities and dysmenorrhea
      ii. Control of PCO-related symptoms such as acne
2. Issues related to CHC use in obese individuals:
   a. Oestrogen containing oral contraceptives increase the plasma concentration of clotting factors II, VII, X, XII, factor VIII, and fibrinogen.
   b. Oestrogen increases plasma concentrations of these clotting factors by increasing gene transcription. Higher doses of oestrogen appear to confer a greater risk of venous thrombus formation.
   c. CHC increase plasma insulin and glucose levels and reduce insulin sensitivity.
   d. Obesity is associated with increased risk for deep venous thrombosis and risk increases with the presence of other comorbidities (smoking, immobility), and dose of oestrogen.
   e. Offers no protection against STI.
   f. CHC appears to be less efficacious in obese with an odds ratio of 1.65 (95% CI 1.09−2.50).
   g. The vaginal ring appears to have more stable hormones.
   h. Contraceptive efficacy of the patch in obese adolescent is unreliable.
   i. Barrier methods, fertility awareness-based method—may be unreliable because of cycle irregularities and lack of motivation

Postcoital contraception:

1. Ulipristal appears to a better choice than levonorgestrel
2. Insertion of an intrauterine device
3. Offer advice as regards use of condom and describe its overall benefits

Menstrual issues:

1. Menstrual abnormalities are particularly common in obese adolescents in the form of amenorrhea/oligomenorrhea, heavy periods, and sometimes irregular bleeding;
2. they are caused by anovulation (immaturity of the hypothalamic-pituitary −ovarian axis, and PCOS);
3. or because of excessive oestrogens production by the abundant adipose tissue, which stimulates the normal endometrium and disrupts the normal feedback mechanism.

Management:

1. Chronic anovulation can lead to endometrial hyperplasia, therefore it is important that there is regular withdrawal bleeding induced four times a year.
   a. The cyclical use of progestogens
2. Hormonal contraceptives
3. Insertion of levonorgestrel-releasing intrauterine system
4. Nonsteroidal anti-inflammatory drugs for dysmenorrhea,

If irregular bleeding persists despite treatment, then consider the followings:

1. Exclude local causes;
2. investigate the endometrium by ultrasound scanning; and
3. by endometrial sampling (depending upon the scan findings).

Polycystic ovarian syndrome in adolescence:

1. Although most adolescents and adults with PCOS are obese, only 20% of obese women have PCOS.
2. PCOS is associated with insulin resistance and the metabolic syndrome.
3. PCOS may present with any, or a combination of obesity, menstrual abnormalities, hirsutism, Acanthosis nigricans, acne, hair loss, or premature adrenarche.
4. Oligo anovulation and hyperandrogenism must be present, and ultrasound is not recommended for diagnosis now (Teede et al., 2018).

Management:
Life style interventions, including weight loss: a significant amount of weight loss of 6.5% can lead to the followings:

1. a significant decrease in Insulin resistance;
2. significantly lower levels of Testosterone, LH, LH/FSH ratio;
3. the prevalence of amenorrhea, and/or oligomenorrhea decreased significantly;
4. combined oral contraceptives (COCP): improve symptoms by increased production of SHBG, and lowers levels of androgens;
5. progestins such as drospirenone and cyproterone acetate have antiandrogenic effects;
6. COCP containing desogestrel or cyproterone are associated with decreased insulin sensitivity and a better lipid profile.

Insulin sensitisers:

1. These drugs tend to reduce insulin levels (Metformin) and increase insulin sensitivity (Metformin and Thiazolidine-diones).

Metformin leads to:

1. increased insulin sensitivity by the liver;
2. increased peripheral glucose uptake;
3. decreased fatty acid oxidation;
4. decreased glucose absorption from the gut;
5. lower testosterone levels; and
6. an improvement in menstrual cyclicity

1.3 Summary

Obesity in childhood and adolescence is a global epidemic with health implications. All efforts should be made to reduce the healthcare burden of chronic illness associated with it. Adolescent obesity is associated with early onset of puberty, and its associated gynaecological symptoms, especially menstrual issues. These young ladies should be managed sensitively and be supported.

Reference

1. Gungor NK. Overweight and obesity in children and adolescents. *J Clin Res Pediatr Endocrinol*. 2014;6(3):129−143.

Further reading

Bumbuliene Z, Tridenti G, Vatopoulou A. Obesity and the onset of adolescence. In: Mahmood T, Arulkumaran S, Chervenak F, eds. *Obesity and Gynaecology*. 2nd ed. London: Elsevier; 2020:3−13. Available from: https://doi.org/10.1016/B978-0-12-817919-2.00001-2.

Marshall WA, Tanner JM. Variations in pattern of pubertal changes in girls. *Arch Dis Child*. 1969;44(235):291−303. Available from: https://doi.org/10.1136/adc.44.235.291. PMC 2020314. PMID 5785179.

Seif MW, Busby G. Obesity in adolescence. In: Mahmood T, Arulkumaran S, Chervenak F, eds. *Obesity and Gynaecology*. 2nd ed. London: Elsevier; 2020:15−22. Available from: https://doi.org/10.1016/B978-0-12-817919-2.00002-4.

Teede HJ, Misso ML, Costello MF, et al. Recommendations from the international evidence-based guideline for the assessment and management of polycystic ovary syndrome. *Fertil Steril*. 2018;110:364−379.

Witchel SF, Plant TM. Puberty: Gonadarche and Adrenarche (In). In: Strauss JF, Barbieri RL, eds. *Yen and Jaffe's Reproductive Endocrinology*. 6th ed. Boston: Elsevier; 2009:395−431. (ISBN 978-1-4160-4907-4).

Polycystic ovary syndrome

2

Tahir Mahmood
Department of Obstetrics & Gynaecology, Victoria Hospital, Kirkcaldy, United Kingdom

2.1 Introduction

Polycystic ovary syndrome (PCOS) is one of the most common endocrine disorders in women of reproductive age.

According to the National Institute of Health (NIH) criteria, the prevalence of PCOS ranges between 6% and 10% and with utilisation of the ESHRE/ASRM consensus criteria, it is as high as 15%.

2.1.1 Clinical presentation

PCOS may appear for the first time during adolescence. The symptoms usually start in adolescence and progress gradually over time.

Some women may have some signs of hyperandrogenism during the early stage of puberty, but a clear diagnosis of the syndrome during puberty is difficult, since all the symptoms have not been fully established.

1. *Irregular cycles/anovulation:*
    a. Most recent guidance suggests that irregular menstrual cycles are defined as <21 or >35 days or <8 cycles in a year from the point of 3 years postmenarche to perimenopause.
    b. Within the first 3 years postmenarche, the cycles should be >45 days and within 1 year postmenarche >90 days.
    c. Primary amenorrhea by age 15 or >3 years postthelarche.
    d. Secondary amenorrhea may also occur occasionally and is defined as a lack of menstrual periods for at least 6 months in a history of oligomenorrhea.
    e. It is recognised that irregular menstrual cycles and other features of PCOS can overlap with those observed in the normal pubertal transition.
    f. 79% women with PCOS have either history of oligomenorrhea or secondary amenorrhoea.

2. *Signs of hyperandrogenism:*
    a. Acne, hirsutism, and male pattern baldness, are suggestive of PCO.
    b. Hirsutism may be one of the symptoms, but not in all cases according to the Rotterdam definition.
    c. When present, hirsutism is scored according to modified Ferriman−Gallway system.
    d. A score equal to or greater than 4−6 is considered clinical hyperandrogenism depending on ethnicity.

Handbook of Obesity in Obstetrics and Gynecology. DOI: https://doi.org/10.1016/B978-0-323-89904-8.00038-X

e. Acanthosis nigricans is a profound sign of IR and is usually associated with PCO and obesity.

3. *Obesity:*
   a. About 40%−50% of women with PCOS are overweight.
   b. Increased abdominal obesity and waist-to-hip ratio is correlated with reduced menstrual frequency, subfertility and insulin resistance (IR).
   c. Women in the United States with PCOS have a higher body mass index (BMI) than their European counterparts. Anovulation and hirsutism can be exacerbated by weight gain.
   d. Acne may be also one of the manifestations of hyperandrogenism if it persists after adolescence, while alopecia may develop in the case of severe hyperandrogenism.

2.1.2 Diagnostic criteria

1. Currently there are three criteria used for the diagnosis of PCOS.
2. These criteria give different weights to hyperandrogenism (biochemical and/or clinical), oligoanovulation and polycystic morphology of the ovaries assessed by ultrasound.
3. The aetiology of the syndrome is unknown, and it is also associated with insulin resistance and hyperinsulinaemia, elevated Luteinising Hormone (LH) levels, raised LH/follicle stimulating hormone (FSH) ratio, and elevated anti-Mullerian hormone (AMH).
4. The NIH definition relies on two criteria and does not include the ovarian morphology.
5. The Rotterdam definition considers all three criteria but only two are required for the diagnosis.
6. The Androgen Excess and PCOS Society definition considers hyperandrogenism as the main criterion associated with either oligoanovulation or polycystic morphology of the ovaries or both.
7. However it is important to note that none of the criteria includes either AMH levels, LH/FSH ratio, or insulin resistance.

Characteristic features of three criteria have been summarised in Table 2.1.

1. Based on the many combinations of the above criteria, several phenotypes of PCOS have been recognised.
2. There are four phenotypes in the Rotterdam criteria, as shown in Table 2.2.

Table 2.1 Diagnostic criteria described for the diagnosis of polycystic ovary syndrome.

	National Institute of Health (1990)	Rotterdam (2003)	Androgen Excess and PCOS Society (2009)
Criteria	1. Hyperandrogenism 2. Oligo/anovulation DX: Both required for diagnosis	1. Hyperandrogenism 2. Oligo/ovulation 3. Polycystic ovarian morphology (ultrasound) DX: Two out of three required	1. Hyperandrogenism 2. Oligo/anovulation 3. Polycystic ovarian morphology by (ultrasound) DX: Number 1 plus 2 or 3

Table 2.2 Four phenotypes of PCOS as described in the Rotterdam criteria.

Phenotype 1	Oligo/anovulation
	Hyperandrogenism: clinical hirsutism or biochemical
	Polycystic ovaries
Phenotype 2	Oligo/anovulation
	Hyperandrogenism: clinical hirsutism or biochemical
Phenotype 3	Hyperandrogenism: clinical hirsutism or biochemical
	Polycystic ovaries
Phenotype 4	Oligo/anovulation
	Polycystic ovaries

2.2 Diagnosis

The diagnosis of PCOS is made when two out of three of the following criteria are met:

1. *Clinical and/or biochemical evidence of excess androgen* after the exclusion of other related disorders as listed in the differential diagnosis section.
   a. Measurement of free testosterone, total testosterone measurement, and free androgen index.
   b. Measurement of 17α-hydroxyprogesterone [to differentiate between PCOS and late onset congenital adrenal hyperplasia (CAH)].
   c. Adrenal androgens, such as dehydroepiandrosterone sulfate (DHEAS) may be increased in PCOS; however, its importance in the diagnosis is lower, as exclusive overproduction of this is not common.
   d. Measurement of DHEAS or Δ4-androstenedione could be considered when total or free testosterone levels are normal.
   e. Prolactin is also mildly elevated in PCOS women.

2. *Oligoovulation or anovulation:*
   Very challenging to look for the evidence of ovulation with irregular cycles, serum progesterone >35 nmol/L is diagnostic of ovulation.

3. *Ultrasound appearance of the ovaries:*
   a. The Amsterdam (2003) criteria mandated the presence of >12 follicles in each ovary measuring 2−9 mm and/or increased ovarian volume (> 10 mL).
   b. In 2018, the cut-off for follicle number was raised to 20 or more in either ovary, to reflect the improvement in ultrasound technology (transvaginal scanning with a frequency band width including 8 MHz).
   ESHRE evidence-based PCO guideline states the following:
   "For adolescents, who have features of PCOS but do not meet diagnostic criteria, an 'increased risk' could be considered and reassessment advised at or before full reproductive maturity, 8 years post menarche. This includes those with PCOS features before combined oral contraceptive pill (COCP) commencement, those with persisting features and those with significant weight gain in adolescence."
   "Ultrasound should not be used for the diagnosis of PCOS in those with a gynaecological age of < 8 years (< 8 years after menarche), due to the high incidence of multifollicular ovaries in this life stage."

4. *Ethnic variation:*
a. Health professionals should also consider *ethnic variation* in the presentation and manifestations of PCOS, including:
  i. a relatively mild phenotype in Caucasians;
  ii. higher BMI in Caucasian women, especially in North America and Australia;
  iii. more severe hirsutism in Middle Eastern, Hispanic, and Mediterranean women;
  iv. increased central adiposity, insulin resistance, diabetes, metabolic risks, and Acanthosis nigricans in Southeast Asians and Indigenous Australians;
  v. lower BMI and milder hirsutism in East Asians; and
  vi. higher BMI and metabolic features in Africans
  vii. (https://www.monash.edu/__data/assets/pdf_file/0004/1412644/PCOS_Evidence-Based-Guidelines_20181009.pdf).

2.2.1 Differential diagnosis

Other causes of amenorrhoea and anovulation and hyperandrogenism must be considered in the differential diagnosis. Table 2.3 shows summary of these clinical conditions.

1. Thyroid dysfunction, particularly hyperthyroidism
2. Nonclassic CAH-raised 17-hydroxyprogesterone
3. Androgen-secreting ovarian tumours.

Table 2.3 Differential diagnosis of clinical conditions which should be considered.

Condition	Symptoms	Blood tests
Congenital adrenal hyperplasia	Irregular periods/ amenorrhea, hirsutism	High 17α-hydroxyprogesterone
Cushing's syndrome	Central obesity, irregular periods, hirsutism	High androgens
Androgen producing tumour (ovarian/Adrenal)	Amenorrhea, hirsutism, virilisation	High testosterone, high androgens
Thyroid disease	Menstrual irregularities Infertility	Abnormal thyroid and TSH values
Hyperprolactinaemia	Galactorrhoea	High prolactin levels,
Weight-related/exercise-related amenorrhoea	Amenorrhoea	Suppressed levels of FSH, LH, and oestradiol
Hypogonadotrophic hypogonadism	Amenorrhoea	Suppressed levels of FSH, LH, and oestradiol
Hypothalamic amenorrhea	Low body weight, eating disorders	Low FSH, LH, and oestradiol
Primary ovarian insufficiency	Hot flashes	High FSH, LH, and low oestradiol
Acromegaly	Hirsutism, weight gain	High GH, high serum IGF-1

Key: *DHEA*, Dehydroepiandrosterone; *FSH*, follicle stimulating hormone; *GH*, growth hormone; *IGF-1*, insulin-like growth factor-1; *LH*, luteinising hormone; *TSH*, thyroid stimulating hormone.

4. Androgen-secreting adrenal tumours
5. Ovarian hyperthecosis
6. Cushing's syndrome (especially in women with severe hyperandrogenism)
7. Premature ovarian failure (very raised LH, FSH, but very low oestradiol and AMH)
8. Hyperprolactinaemia (high prolactin, suppressed FSH/LH/oestradiol)
9. Weight-related and exercise-related amenorrhoea (very low FSH and oestradiol levels)
10. Hypogonadotrophic hypogonadism (very low LH/FSH/oestradiol)
11. Growth hormone disorders (acromegaly)—hirsutism and weight gain.

Pathophysiology
Diet

1. Recent experimental data in rodents have shown that a high-fat, high-sugar diet during the intrauterine life may be the trigger for the onset of ovarian disturbances and induce PCO.
2. Obesity seems to have a negative effect on spontaneous and induced ovulation in PCOS.

Inheritance
Although a single gene autosomal dominant transmission has been postulated, it is widely accepted that PCOS is a multiple gene condition.

Hyperandrogenism

1. Increased androgen secretion comes from enhanced intrinsic steroidogenetic capacity of theca cells.
2. LH hypersecretion stimulates theca androgen secretion and obesity amplifies the stimulating action of LH on theca cells.
3. Lower FSH result in reduced aromatase activity.
4. Hyperinsulinaemia synergises with LH for thecal androgen production.
5. Intraovarian mechanisms such as AMH inhibition of FSH and of aromatase activity may further deteriorate hyperandrogenemia.
6. Obesity also amplifies the stimulating action of LH on theca cells.
7. Obese women with PCOS have higher total and free Testosterone levels and reduced levels of sex hormone binding globulin (SHBG) as compared to nonobese PCOS women.
8. There is an increased steroidogenic activity in follicle theca cells [17α-hydroxylase/ 17,20-lyase and 3β-hydroxysteroid-dehydrogenase (3β-HSD)].
9. There is decreased activity of 17β-hydroxysteroid-dehydrogenase (17β-HSD).
10. These changes in enzymatic activity lead to increased production of Δ4-Androstenedione in women with PCO.

2.2.2 Hypersecretion of luteinising hormone

1. A majority of women with PCO and anovulation have high LH levels and an elevated LH/FSH ratio.
2. It has also been demonstrated that the changes in LH secretion occur only in women with anovulation. Women with PCOS and normal ovulatory cycles have normal levels of basal LH and the LH to FSH ratio.
3. There is accelerated LH pulse frequency and amplitude and elevated LH response to gonadotrophin-releasing hormone (GnRH). However LH secretion is blunted in obese women with PCO, acting at the pituitary and not at the hypothalamus level.

4. The negative effect of BMI on LH secretion may be mediated by hyperinsulinaemia, since insulin infusion decreases basal and Gn-RH-induced LH secretion.
5. There is an inverse relationship between Leptin levels and LH levels and LH pulse frequency.
6. Following a short-term caloric restriction, a decrease in Leptin levels and an increase in LH pulse amplitude.

2.2.3   Hyperinsulinaemia

1. Insulin resistance occurs in 50%−70% of both obese and lean women with PCOS.
2. Insulin resistance leads to increased levels of fasting insulin in blood as compared to controls.
3. Insulin acts directly on the ovary and stimulates increased androgen production, while also decreasing the production of SHBG from the liver.
4. Hyperinsulinaemia augments LH-stimulated androgen production.
5. Insulin also reduces IGFBP-1 production, which leads to increased free IGF-I which, by acting on the ovary, further stimulates the increased production of androgens.
6. Insulin also acts on muscles and fat tissues, where it induces the production of TNFα and free fatty acids.

2.2.3.1   Anovulation-follicular development in polycystic ovary syndrome

1. The ovaries of women with PCOS contain a large pool of antral follicles 2−8 mm in diameter.
2. Although the mechanism for this is not clear, it is likely that multiple factors (hyperinsulinaemia, hypersecretion of LH, hyperandrogenism, elevated AMH, and levels of FSH below the threshold for follicle recruitment selection) are involved, which lead to the failure of dominant follicle selection leading to follicle growth arrest.
3. Hyperinsulinaemia is possibly responsible for a premature response of the granulosa cells to elevated LH, which occurs at an early stage in contrast to >10 mm in ovulatory cycles.
4. Therefore the granulosa cells from small antral follicles cease to divide and undergo premature terminal differentiation, hence ovulation does not occur.
5. In women with PCOS, high production of androgens and reduced aromatase leads to a raised intrafollicular androgenic environment which leads to follicular atresia.
6. AMH levels can be up to three times higher in PCOS women than in normal controls.
7. The mechanism of the increased production of AMH in PCOS is not clear, but it has been associated with overproduction of many substances, such as androgens, LH, insulin, and advanced-glycation end products (AGEs) or with a decrease in AGEs' receptor.
8. In PCOS, AMH suppresses aromatase expression and reduces the sensitivity of granulosa cells to FSH. This facilitates the establishment of hyperandrogenism and anovulation.
9. Recruitment of small antral follicles normally takes place under the intercycle rise of FSH.
10. There are significantly lower concentrations of FSH in PCOS than in normal women.
11. In patients with PCOS, an intercycle type of FSH rise does not take place which explains an inability of follicles to grow to the preovulatory stage.

2.3 Obesity, polycystic ovary syndrome, and related indices

Obesity aggravates phenotypic features of PCOS. Many genes are dysregulated in adipocytes of obese women compared with nonobese individuals. In omental fat of obese women with PCOS, there were different expression patterns in genes compared to nonobese PCOS women. Many effects are exerted through the secretion of various adipokines from adipose tissue upon fertility in women with PCOS.

2.3.1 Adipokines

2.3.1.1 Leptin

1. Serum leptin levels are significantly higher in women with PCOS than in normal controls.
2. Leptin affects the function of hypothalamo−pituitary−ovarian axes and the secretion of hormones produced by it.
3. At the level of ovary, leptin was found to modulate basal and FSH-stimulated steroidogenesis in cultured human lutein granulosa cells, with high concentrations suppressing the secretion of oestradiol and progesterone, and inhibiting folliculogenesis.

2.3.1.1.1 Adiponectin

1. Adiponectin levels are reduced in obesity and are even lower in women with PCOS than in normal controls.
2. The main role of adiponectin is to increase insulin sensitivity and therefore its diminished production in PCOS intensifies insulin resistance and hyperinsulinaemia.
3. In vitro experiments have demonstrated that adiponectin may adversely affect steroidogenesis.

2.3.1.1.2 Resistin

Resistin increases insulin resistance, but in PCOS circulating levels show diversity between different studies.

2.3.1.1.3 Interleukin-6

Circulating levels of IL-6 increase in obesity and they are associated with increased insulin resistance.

In the rate model, IL-6 has been observed to prevent LH-triggered ovulation, and inhibit LH-/FSH-induced oestradiol production.

2.3.1.1.4 Plasminogen activator inhibitor type-1

Circulating plasminogen activator inhibitor type-1 (PAI-1) levels are increased in overweight/obese patients with PCOS compared to BMI-matched controls.

PAI-1 has been associated with miscarriage in women with PCOS.

2.3.1.1.5 Other adipokines

The role of visfatin, vaspin, apelin, kisspeptin, and retinol-binding protein 4 in the pathogenesis of PCOS is unspecified.

2.4   Comorbidities

Infertility

1. Women with chronic anovulation may have difficulties conceiving without ovulation induction treatment.
2. PCOS is the cause of almost 90% of cases with anovulatory infertility.
3. PCOS also poses problems prior and during pregnancy.
4. It has been reported that the time for conception is prolonged,
5. During assisted production treatments (ovarian stimulation for IVF), the risk of ovarian hyperstimulation syndrome is markedly increased.

During pregnancy:

1. There are increased risks for miscarriage, stillbirth, and preterm birth.
2. There is also an increased risk of gestational diabetes and preeclampsia.
3. All these morbidities are exacerbated by obesity.

Cardiovascular disease:

1. Cardiovascular disease (CVD) remains one of the leading causes of death in obese women.
2. Metabolic syndrome and CVD risk factors are clearly increased in PCOS.
3. However, there are limited current data on clinical events, overall CVD risk, and optimal screening for additional risk factors.

Review of literature:

1. There was no statistical difference between PCOS and non-PCOS groups in terms of myocardial infarction, stroke, CVD-related death, and coronary artery/heart disease.
2. One study suggests that when a group of women with PCOS are compared to a UK-wide population, the risk of myocardial infarction (but not angina) was increased in women with PCOS over 45 years.
3. When they compared the same women with PCOS to a local community population, the risk of myocardial infarction and angina was increased in women with PCOS.

Gestational diabetes, impaired glucose tolerance, and type 2 diabetes:

1. Women with PCOS had increased prevalence of IGT (OR 2.48, 95% CI 1.63, 3.77; BMI-matched studies OR 2.54, 95% CI 1.44, 4.47), DM2 (OR 4.43, 95% CI 4.06, 4.82; BMI-matched studies OR 4.00, 95% CI 1.97, 8.10).
2. Consistently, DM2 was four times higher in a recent Danish registry study and was diagnosed 4 years earlier in PCOS.
3. Oligomenorrhea, along with clinical or biochemical hyperandrogenism, obesity, or a family history of risk of DM2 may be indicators of risk of DM2.
4. In a low-risk northern European ethnic group, lean women did not develop DM2 by 46 years, with risk increased in the majority who were overweight or obese.
5. 47% of Asian women with PCOS had IGT or DM2 by 41 years, despite limited obesity.
6. The prevalence of metabolic syndrome is in all PCOS phenotypes higher than in controls and even higher in the hyperandrogenic than in the normoandrogenic phenotype, as has been found in a cohort of Chinese population

7. The concept of absolute versus relative risk is also important as in low-risk populations (Caucasian, healthy weight), a fourfold increased risk from PCOS equates to a low incidence of DM2, yet in high-risk Southeast Asians or obese women, PCOS significantly impacts on DM2 incidence.
8. The ESHRE guideline states that screening is warranted in all adults with PCOS and in adolescents with additional risk factors at baseline.
9. Optimal tests remain unclear and fasting glucose, HbA1c, or OGTT can be used.
10. Frequency of testing should be a minimum of 3-yearly, informed by additional risk factors.

Obstructive sleep apnoea:

1. Obstructive sleep apnoea (OSA) is characterised by repetitive occlusions of the upper airway during sleep with futile ventilatory efforts, oxygen desaturations, sleep arousal, and the resumption of ventilation, fragmenting sleep and causing daytime sleepiness.
2. OSA appears more common in PCOS and in obesity.
3. OSA screening is currently warranted in those with symptoms, where treatment benefit has been demonstrated in the general population.

Endometrial cancer

1. The risk of endometrial cancer has been shown to be between two to six times higher in women with PCOS, with most adenocarcinoma (>95%) including type I and type II cancers, with type I increased in PCOS.
2. The increased prevalence of endometrial cancer in PCOS is related to prolonged endometrial exposure to unopposed oestrogen in anovulation.
3. Endometrium in PCOS may exhibit progesterone resistance.
4. Associations between PCOS and endometrial cancer are complex and comorbid conditions such as obesity, infertility, DM2, and metabolic syndrome are relevant, whilst PCOS treatment options may influence cancer risk.
5. Metformin has no association or a protective association with endometrial cancer.
6. Clomiphene studies are limited by power, but a small nonsignificant increased risk of endometrial cancer has been shown.
7. Letrozole, yet to be explored in relation to endometrial cancer, is used as an adjuvant treatment for hormone receptor positive postmenopausal breast cancer and may decrease hormonal related cancer risk.
8. Oral contraceptives reduce risk for endometrial cancer in general populations and effects may be enduring.
9. Routine screening for endometrial hyperplasia or cancer in PCOS is not warranted.
10. Endometrial surveillance by transvaginal ultrasound or endometrial biopsy is indicated for those women with PCOS who have thickened endometrium, prolonged amenorrhoea, unopposed oestrogen exposure, or abnormal vaginal bleeding, based upon clinical suspicion.

Quality of life in women with PCOS

1. The polycystic ovary syndrome questionnaire (PCOSQ) has 26 items across emotions, body hair, weight, infertility, and menstrual abnormalities and the modified polycystic ovary syndrome questionnaire (MPCOSQ) adds acne.

2. A meta-analysis of five studies using SF-36 and three studies using the WHO tool in adult women suggests that women with PCOS have lower quality of life compared to women without PCOS.
3. In PCOS, HRQoL occurs in the context of the multitude of clinical features and is affected by anxiety, poor body image and low self-esteem, depressive symptoms, delayed diagnosis, and inadequate education and information provision by health professionals.

Depressive and anxiety symptoms

1. The prevalence and severity of depressive and anxiety symptoms are increased in PCOS.
2. Psychological conditions impact on QoL and are likely to influence engagement in lifestyle interventions and self-management in PCOS.
3. A large international survey has shown that most women report psychological issues are under recognised and less than 5% are satisfied with emotional support and counselling.
4. Depressive symptoms and depression are more common in PCOS, with daily fatigue, sleep disturbances, and diminished interest prominent.
5. A meta-analysis of 10 studies reported increased depressive symptom scores in 44% with PCOS versus 17% in controls.

Anxiety

1. Anxiety symptoms are increased in PCOS.
2. A meta-analysis of 10 studies showed increased moderate/severe anxiety symptoms in PCOS (OR: 5.38; 95% CI: 2.28, 12.67), with a prevalence of 41.9% (IQR: 13.6, 52.0%) in PCOS and 8.5% (IQR: 3.3, 12.0%) in controls.
3. A large population-based study of 24385 women with PCOS, matched for sex, age, and country of birth to 10 controls, showed increased anxiety disorder (OR 1.37, CI: 1.32, 1.43).

Psychosexual function in women with PCOS

1. Psychosexual dysfunction refers to sexual problems or difficulties that have a psychological origin based in cognitions and/or emotions such as depression, low self-esteem, and negative body image and both risk factors for and prevalence of psychosexual dysfunction appear increased in PCOS.
2. The prevalence of psychosexual dysfunction varies from 13.3% to 62.5% in PCOS. It appears that women with PCOS suffer from greater psychosexual dysfunction than women in the general population in most studies.

Body image in women with PCOS

1. Women with PCOS feel less physically attractive, healthy, or physically fit and are less satisfied with their body size and appearance, and this negative body image predicts both depression and anxiety.
2. Infertile women with PCOS have lower body satisfaction than noninfertile women with PCOS.
3. Hirsute women experienced lower self-esteem than nonhirsute women.
4. Overall, PCOS features, in particular hirsutism and increased weight, impact negatively on body image and QOL, and negative body image is strongly associated with depression in women with PCOS, even after controlling for weight.
5. Approaches for screening and assessment that are easy to use and widely applicable are needed.

Eating disorders and disordered eating in women with PCOS

1. Available data suggests an increased prevalence of eating disorders and disordered eating among women with PCOS.
2. Women with PCOS also have more identified risk factors for eating disorders across obesity, depression, anxiety, self-esteem, and poor body image, and higher prevalence than in the general community of any eating disorder (21% vs 4%) but not bulimia nervosa.

Management

1. *Clinical management*
a. The management of PCOS depends on their symptoms and wishes:
   i. Control of menstrual disorders, hirsutism, fertility control, or infertility.
2. As the first approach, obese women with PCOS should lose weight and this is independent of whether they wish to conceive or not.
3. It has been shown that energy-restricted diet alone or in combination with exercise can restore reproductive function, probably by enhancing insulin sensitivity.
4. The safety of any pharmacological agent used to achieve weight loss should be taken into consideration, especially in infertility patients.
5. Even a 5% reduction in body weight can improve the clinical and biochemical abnormalities in PCOS women.
6. If all the above measures do not resolve morbid obesity, bariatric surgery needs to be considered.

2.4.1 Women not wishing a pregnancy

1. Treatment aims to cope with the symptoms of the syndrome, since a definite cure does not exist.
2. The use of COCPs is considered first-line treatment, especially in adolescents, as they will regulate menstruation and will also decrease hyperandrogenism.
3. The lowest effective dose of ethinyloestradiol and natural oestrogen are preferred along with, preferably, nonandrogenic progestogens.
4. Cyproterone acetate should not be considered as first line due to the increased risk of thromboembolism.
5. Antiandrogens could be considered for the treatment of hirsutism when COCPs are contraindicated, or for androgen-related alopecia.
6. Metformin could be also recommended in addition to lifestyle for the control of weight excess and hormonal and metabolic outcomes both in adult women and adolescents with a clear diagnosis of PCOS or with symptoms of PCOS even without a diagnosis.
7. A combination of COCPs with metformin or with antiandrogens can be considered in selective cases for the management of metabolic features or to treat hirsutism, respectively.
8. Cosmetic interventions including laser hair removal may also be of help for the management of hirsutism.
9. Metformin may provide greater benefit for women belonging to the high metabolic risk group, such as diabetes risk factors or impaired glucose tolerance.
10. Finally, for morbid obesity, bariatric surgery is an option.

2.5   Other contraception consideration

2.5.1   Contraceptive methods

2.5.1.1   LNG IUS

1. Highly effective
   Category 1
2. No additional cardiovascular risk
   a. No beneficial effect on menstrual irregularities
   b. No beneficial effect on hyperandrogenic skin symptoms
   c. Possible beneficial effect on risk of endometrium carcinoma.

2.5.1.2   Progestogen-only oral contraceptives: (ovulation inhibitors desogestrel, drospirenone)

1. Highly effective user dependent
   Category 1
2. No additional cardiovascular risk
   a. Effect on menstrual irregularities: potentially improvement by leading to amenorrhoea
   b. No beneficial effect on hyperandrogenic skin symptoms.

2.5.1.3   Copper IUD

1. Highly effective
   Category 1
2. No additional cardiovascular risk
   a. No beneficial effect on menstrual irregularities
   b. No beneficial effect on hyperandrogenic skin symptoms
   c. Possible protective effect on endometrium.

Treatment of Hirsutism: has been discussed at length in Chapter [xx].

Women wishing a pregnancy: management approach has been discussed at length in chapter [xx].

Options are as follows:

More detailed description is given in the chapters on infertility and assisted conception.

1. Diet and lifestyle changes,
2. *Ovulation induction,*
   a. *Clomiphene citrate:* antioestrogen effect, acts by blocking the oestrogen receptors in the hypothalamus, increases GnRh pulsatility, leading to increased pituitary secretion of gonodotrophins which promote ovarian follicular development.
   b. Aromatase inhibitors: *letrozole* acts by inhibiting the aromatisation of androgen to oestrogen and thus releases the hypothalamic−pituitary axis from negative feedback.
   c. *Low dose FSH:* for those women who fail to respond to first-line treatment with oral agents.
   d. *Laparoscopic ovarian drilling:* used for clomiphene-resistant PCOS, especially those with high LH levels.

    e. *Metformin*: acts as an insulin sensitiser and is superior to placebo in inducing ovulation, but is inferior to clomiphene.

    f. *Metformin in combination with clomiphene*: metformin in combination with clomiphene might be useful in cases of clomiphene resistance cases, and be considered before initiating treatment with FSH.

    g. *In vitro fertilisation (IVF):* when all the abovementioned treatment modalities fail to induce pregnancy, then IVF is an option. In women with PCOS, outcome is worse in those with high BMI compared with women with normal or low BMI.

Further reading

Bitzer J. Long term contraceptive care in obese and super obese women. In: Mahmood T, Arulkumaran S, Chervenak FA, eds. *Obesity and Gynaecology*. 2nd ed. Elsevier; 2020.

International evidence-based guideline for the assessment and management of polycystic ovary syndrome <PCOS_Evidence-Based-Guidelines_20181009.pdf>; 2018.

Messinis IE, Messinis CI, Dafopoulos K. Obesity in polycystic ovary syndrome and infertility. In: Mahmood T, Arulkumaran S, Chervenak FA, eds. *Obesity and Gynaecology*. 2nd ed. Elsevier; 2020.

The American College of obstetricians and Gynaecologists. Polycystic ovary syndrome (PCOS): frequently asked questions. <https://www.acog.org/womens-health/faqs/polycystic-ovary-syndrome-pcos>.

The Royal College of Obstetricians and Gynaecologists. Polycystic ovary syndrome, long-term consequences. Green-Top Guideline No. 33, London <https://www.rcog.org.uk/en/guidelines-research-services/guidelines/gtg33/>, 2014.

The Royal College of Obstetricians and Gynaecologists. Metformin therapy for the management of infertility in women with polycystic ovary syndrome. Scientific Impact Paper No. 13, London <https://www.rcog.org.uk/en/guidelines-research-services/guidelines/sip13/>, 2017.

Hirsutism

3

Tahir Mahmood
Department of Obstetrics & Gynaecology, Victoria Hospital, Kirkcaldy, United Kingdom

Hirsutism affects a significant proportion of women of reproductive age. The incidence of hirsutism is highly variable depending on location and ethnicity, being higher in those of African or Mediterranean origin. It affects self-confidence, image and the quality of life of the woman.

Hirsutism is defined when there is an excessive amount of terminal hair growth in androgen-sensitive areas in women. Excessive hair growth occurs as a result of increased androgen production, increased skin sensitivity to androgens, or both. It must be differentiated from nonsexual pattern of excessive hair growth (hypertrichosis), which can be generalised and is independent of androgens.

Growth cycle of hair:

There are *three phases* of hair growth:

1. *anagen*: the active cellular mitosis phase occurs in the basal matrix of the follicle and leads to rapid hair growth and this *is the target stage of pharmacological agents used to treat hirsutism*;
2. *catagen*: the involuting or regressing phase; and
3. *telogen*: the resting or quiescent phase.

There are three types of hair:

1. *Lanugo*: soft, nonpigmented hair;
2. *Vellus*: short, fine, light-coloured hair, barely noticeable but covers most parts of the body;
3. *Terminal*: thicker, longer, pigmented hair, can be found on the scalp, axillae, the genital region, eye brows, and eye lashes;
4. The morphological basis of hirsutism is the excess transformation of vellus to terminal hair in a male pattern distribution.

Terminal hair is androgen dependent and their life cycle is influenced by the genetic and racial factors as well.

Facial hair has a longer growth phase, therefore therapy effect is seen after about 6−9 months of treatment.

The role of androgens:

1. The androgens stimulate the conversion of vellus into terminal hair and prolong the anagen phase.

Handbook of Obesity in Obstetrics and Gynecology. DOI: https://doi.org/10.1016/B978-0-323-89904-8.00009-3

2. Testosterone is the most important circulating androgen secreted from the ovaries and adrenal glands.
3. Androgens are also produced through peripheral conversion of androgen precursors which is more marked in overweight/obese women.
4. Free testosterone is the main bioactive portion as most circulating testosterone is bound by serum albumin and sex hormone-binding globulin (SHBG) which has a high affinity for testosterone and can modulate the bioavailability of free testosterone.
5. Dihydrotestosterone (DHT) is more potent and is generated from testosterone by 5α reductase in the hair follicle and stimulates the dermal papilla to produce terminal hair instead of vellus hair.
6. Hair follicles on the face, chest, lower abdomen, lower back and upper thighs are particularly androgen-sensitive; pubic and axillary hair is less so.
7. Increased androgen levels transform vellus follicles in sex specific areas resulting in an increase in follicle size, diameter and the amount of time in the anagen phase.

Causes of hirsutism:

1. *Polycystic ovary syndrome (PCOS)*: this is the most common cause of hirsutism, accounting for 70% to 80% of cases of hirsutism.
2. *Idiopathic hyperandrogenism*: this accounts for 6%−15% of the causes of hirsutism. These women have elevated androgen levels, normal menses and normal ovaries at ultrasound.
3. *Idiopathic hirsutism*: this is often due to an ethnic or familial trait and accounts for 4%−7% of cases of hirsutism. These individuals have regular menses and normal circulating androgen levels. An increased 5α reductase activity in the hair follicle or alteration in androgen receptor function may be the underlying cause.
4. *Late onset congenital adrenal hyperplasia (CAH)*: this is an uncommon cause of androgen excess in 2%−4% of cases.
5. *Androgen-secreting ovarian and adrenal tumours* are rare causes of hirsutism (0.2%).
   Ovarian tumours include: granulosa cell tumours, Leydig cell tumours, Sertoli leydig cell tumours, luteomas of pregnancy and thecoma.
6. Other *endocrine causes include* Cushing's syndrome, acromegaly, thyroid dysfunction, and hyperprolactinaemia.
7. *Drugs*: anabolic steroids, testosterone therapy, some oral contraceptives, danazol, cyclosporine, sodium valproate, phenytoin, streptomycin, diazoxide and minoxidil can lead to hirsutism.
8. *Obesity* through increased peripheral conversion of androgen precursors which is more marked in overweight/obese women.
9. *Genetic cause*: XY female.
10. Menopausal women:
    a. Relatively stable levels of testosterone production from ovaries can lead to an increase in hair growth.
    b. Very little is known about acne and alopecia in these women.
    c. Postmenopausal women with PCOS have higher 17-hydroxyprogesterone, androstenedione, DHEAS, total testosterone and FAI than women without PCOS.
    d. In some women, raised levels of luteinising hormone (LH) can lead to stromal hyperplasia, high testosterone levels and severe hirsutism.

3.1 Clinical features and symptoms

Women with hirsutism may consult general practitioners, gynecologists, endocrinologists and dermatologists. A multidisciplinary approach might be necessary. Clinicians must undertake a focused history to differentiate hirsutism from virilism. Table 3.1 describes Rotterdam criteria for making the diagnosis of hirsutism.

1. Hirsutism adversely impacts quality of life and most women readily treat hirsutism complicating assessment, hence health professionals should be prepared to assess any woman who complains of excess hair,
2. Acne is associated with biochemical hyperandrogenism, yet the predictive value of acne alone is unclear and there is no accepted assessment tool.
3. Most studies of women with alopecia reveal a relatively low prevalence of hyperandrogenemia and the predictive value of alopecia alone remains unclear, in part as there are many causes that can contribute to alopecia aside from hyperandrogenism. The signs and symptoms could be nonspecific or specific to the pathological causes (Table 3.2).

3.2 Examination

Mapping for the excessive terminal hair in androgen-dependent areas, including the face, chest, linea alba, lower back, buttocks, and anterior thighs by using Ferriman—Gallwey score to assess the degree of hirsutism.

It is important to differentiate between terminal hair and vellus hair, as vellus hair does not indicate hirsutism.

Examination of these patients should also include:

1. measurement of body mass index (BMI);
2. examination for acne, seborrhoea, presence or absence of virilisation;
3. abdominal and pelvic masses;
4. features found in endocrine disorders such as Cushing's syndrome (full moon face and buffalo hump), galactorrhoea, thyroid enlargement, and hair loss;
5. changes in quality of life; and
6. assessment for symptoms of depression.

Ferriman—Gallwey scoring system (FG):

1. In the original paper (1961), 11 body sites were recommended for assessment:
    a. (Upper lip, chin, chest, upper and lower back, upper and lower abdomen, upper arm, forearm, thigh, and lower leg).

Table 3.1 The Rotterdam 2003 PCOS consensus diagnostic criteria.

This criteria requires two out of three of the following to diagnose PCOS: 1. Polycystic ovaries [either 12 or more follicles of 2—9 mm size or increased ovarian volume (> 10 cm^3)]. 2. Oligo-ovulation or anovulation. 3. Clinical and/or biochemical signs of hyperandrogenism.

Table 3.2 Focused history taking for women presenting with hirsutism.

Focused History	Data collected
History elicited	Age of onset (gradual or rapid), sites, current management, racial background, family history of (hirsutism, PCOS, androgenetic alopecia, type II diabetes, cardiovascular disease, late onset CAH, male-pattern balding before 30 years of age)
	Other known current medical conditions and treatments
Menstrual history	History of menarche and menstrual symptoms(oligomenorrhea/amenorrhea), galactorrhea
	History of Acne, androgenetic alopecia, seborrhoea
History of treatment with medications	Oral contraceptives with androgenic progestins (norgestrel, levonorgestrel, norethindrone)
	Anabolic steroids, danazol, glucocorticoids, androgen therapy (testosterone)
	Valproic acid
PCOS and HAIR-AN (hyperandrogenism, insulin resistance, and Acanthosis nigricans)	Weight gain, *Acanthosis nigricans*
	Polydipsia/polyuria related to glucose intolerance
	History of hypertension or hyperlipidaemia
Late-onset CAH	Onset prepuberty, premature pubarche, menstrual irregularities, primary amenorrhea
Hyperprolactinemia	History of galactorrhoea (spontaneous or expressible)
Pituitary tumor	Visual disturbance, headache
Adrenal or ovarian tumour	Virilisation (increased libido, clitoromegaly, voice—male like)
Thyroid dysfunction	Hot or cold intolerance, tremors, menstrual irregularities, diffuse scalp hair loss

2. And each site is scored on a 4-point scale (0 = "no hair" to 4 = "frankly virile").
3. *The modified FG score* excludes the forearm and lower leg as these areas are less or not sensitive to androgens.
   a. A score of less than 8 is regarded as normal;
   b. 8 to 15 indicates mild hirsutism;
   c. 16 to 25 moderate hirsutism; and
   d. > 25 indicates severe hirsutism.
   e. Approximately 5%−10% of women of reproductive age are hirsute as assessed by this scoring system.

Limitations of FG scoring:

1. Subjective in nature with wide interobserver variation;
2. does not include areas such as sideburns and buttocks; and
3. does not consider racial and ethnic variation.
4. Middle Eastern and South Asian women have higher cut-off scores for hirsutism than those of Eastern Asian origin.
5. Acanthosis is more common in women of Southeast Asian background, reflecting increased insulin resistance.

Investigations:
NICE (2014) recommends the following:

1. No investigations for women with mild hirsutism and no other signs of PCOS or other underlying conditions.
2. Recommends checking early morning testosterone levels in women with moderate-to-severe hirsutism how have no other signs of PCOS or other underlying conditions.
3. The reference ranges for different laboratories vary widely.
4. Clinical decisions should be guided by the reference ranges of the local laboratory.
1. *Testosterone concentration:*
   a. There is poor correlation between Testosterone concentrations and the severity of hirsutism, because of individual variations in hair follicle response to free testosterone.
   b. Oral contraceptives users will have a falsely lower testosterone level.
   c. A testosterone concentration of >5 nmol/L may suggest an androgen producing tumour.
   d. A testosterone level >7 nmol/L should be seen at endocrinology clinic for further assessment.
2. *Free androgen index (FAI):* This reflects the active form of circulating testosterone:
   a. FAI is calculated as: FAI = [total testosterone ÷ SHBG] × 100.
   b. Obese and PCOS patients may have an elevated FAI, even with normal testosterone concentrations.
   c. High androgens and low SHBG have also been linked to increased risk of cardiovascular disease.
3. *17 hydroxyprogesterone (17-OH P4):*
   a. Assessed to screen for late onset CAH.
   b. Raised levels require confirmation by using a Short Synacthen test.
   c. *Short Synacthen test* involves baseline assessment of 17OHP4, followed by an intramuscular injection of 250 micrograms of Synacthen. The levels of 17OHP4 are rechecked an hour later.
   d. A significant rise in 17OHP4 levels is diagnostic of CAH.
4. *Dehydroepiandrosterone sulfate (DHEAS):*
   a. Where an adrenal cause is suspected.
   5. *Dexamethasone suppression test or 24 hour urinary free cortisol*:
   a. For suspected cases of Cushing's syndrome.
   6. *Computed tomography (CT)/magnetic resonance imaging (MRI):*
   a. Pelvic CT and MRI to exclude the possibility of ovarian or adrenal androgen-producing tumour when testosterone concentrations are >5 nmol/L.
7. *Selective venous sampling of ovarian and adrenal veins*
   a. In cases where androgen-secreting tumours are suspected but pelvic imaging is negative.

8. *Prolactin (PRL)/thyroid function test (TSH):*
  a. To rule out hyperprolactinaemia and hypothyroidism for those presenting with both amenorrhea and hirsutism.
9. *Pregnancy test:*
  a. Pregnancy should always be ruled out before initiating any treatment in women with absent or irregular menstruation.
10. *Clinical management:*
  a. Treatment should be guided by the degree of severity of hirsutism, the woman's preferences, her reproductive status, the underlying cause and any potential adverse effects.
  b. Management strategies should aim to improve the quality of life, reduce the free androgen level, block peripheral androgens and improve cosmetic appearance by the removal of existing hair.
  c. For pregnant or lactating women, no safe pharmacological treatments are available.
  d. Condition specific clinical features and investigations have been summarised in Table 3.3.
11. *Lifestyle modification:*
  a. Women with PCO and who are obese should be advised to lose weight through a combination of exercise and diet.
12. *Cosmetic measures:*
  a. Cosmetic methods are required for the management of existing hair as medical treatment do not treat existing hair:
13. *Physical methods:*
  a. Shaving, chemical depilatories, bleaching, plucking, tweezing or threading, and waxing are all rapid hair removal methods.
  b. These methods are effective temporarily.
  c. The most important side-effects include: irritation or dermatitis.
  d. Waxing can also be painful and lead to folliculitis.
14. *Mechanical methods:*
  a. Electrolysis is delivered either in the form of galvanic electrolysis or thermolysis, or a combination of both.
  b. *Galvanic electrolysis* involves passing an electric current down through a needle inserted into the follicle to destroy the hair bulb. It is effective, but painful, slow and expensive.
  c. *Thermolysis* is faster, but it is somewhat less effective and uses a high-frequency alternating current, which produces heat in the hair follicle and leads to destruction.
  d. The combination or blend method combines both of the above methods.
15. *Laser and photoepilation:*
  a. *Laser* (alexandrite, diode, neodymium, YAG, and ruby lasers):
    i. Photoepilation (intense pulsed light):
    ii. Most effective in people with lighter skin and dark-coloured hairs.
    iii. The long-term effect is uncertain.
    iv. Blonde, red, and white hairs are not suitable for laser.
    v. There is a short-term effect of approximately 50% hair reduction with alexandrite and diode lasers up to 6 months after treatment.
    vi. Limited evidence for the effectiveness of intense pulsed light, neodymium: YAG or ruby lasers.
    vii. Adverse side effects include pain, skin redness, swelling, burned hairs and pigmentary changes.

Table 3.3 Condition specific clinical features and investigations.

Clinical condition	Specific features	Investigations/diagnostic markers
Polycystic ovarian syndrome(PCO)	Menstrual irregularities, obesity, infertility, acne, acanthosis nigricans	Rotterdam criteria, normal or elevated testosterone, free androgen index
PCO and HAIR-AN (hyperandrogenism, insulin resistance and acanthosis nigricans)	Weight gain, pigmented skin patches in the groin, neck or axillae, polyuria/polydipsia related to glucose intolerance, history of hypertension or hyperlipidaemia	Glucose tolerance test, raised insulin levels, abnormal lipid profile, renal and liver function test
Idiopathic hyperandrogenenemia	Regular cycles, normal ultrasound scan of ovaries	High androgen levels
Idiopathic hirsutism	Regular menstrual cycles (excluding medication such as anabolic steroids, testosterone, danazol, cyclosporine, sodium valproate, Phenytoin, minoxidil, diazoxide)	Normal androgen levels
Adrenal hyperplasia	Family history of congenital adrenal hyperplasia(CAH), classic form: ambiguous genitalia. Late onset: menstrual issues, oligoanovulation, sub-infertility	Elevated 17-hydroxy progesterone Elevated cortisol levels
Androgen secreting tumours	Raid onset and progression to hirsutism and virilisation, palpable pelvic or abdominal mass	Early morning testosterone level >6.94 nmol/L (>200ng/dL), androstenedione, DHEAS, CT/MRI
Cushing syndrome	Central obesity, facial plethora, acne, hypertension	Raised 24-hour urine free cortisol, impaired glucose tolerance, dexamethasone suppression test
Thyroid dysfunction	Hot or cold intolerance, diffuse scalp hair loss, weight change, textural skin changes	Abnormal thyroid function test
Hyperprolactinaemia	Galactorrhea, amenorrhea, subfertility	Elevated prolactin levels, MRI pituitary
Genetic	XY female	Chromosomes
Acromegaly	Clinical features of pituitary tumour	Detailed endocrinology including GH measurements and imaging

16. *Pharmacological treatments:*
   a. Treatment should be used for at least 6–9 months before an effect can be observed.
      i. *Topical therapy:*
17. *Eflornithine cream (Vaniqa) 13.9%:*
   Applied twice daily for the treatment of facial hirsutism in women.
   a. Inhibits an enzyme, ornithine decarboxylase which is involved in controlling hair growth and proliferation irreversibly.
   b. Hair regrows to pretreatment levels within 8 weeks once treatment is stopped.
   c. The most frequently reported adverse reaction include mild rash and acne.
   d. Contraindicated for use during pregnancy and breast feeding.
18. *Combined oral contraceptive pill (COCP):*
   Oral contraceptives used as first line of treatment of hirsutism.
   The COCP inhibits androgen secretion by the ovaries and increases SHBG production by the liver, both leading to lower circulating levels of free androgens.
   Dianette *(Co-cyprindiol)* is a combination preparation of ethiny-estradiol (EE) and Cyproterone acetate (CA): *(EE35 μg + CA 2 mg);*
   a. CA acts by inhibiting gonadotrophin induced androgen production; and
   b. by increasing androgen clearance by the liver.
   c. Can be prescribed to women with hirsutism provided there are no contraindications (such as uncontrolled hypertension and current breast cancer).
   d. Treatment with Dianette is stopped 3–4 months after the woman's hirsutism has completely resolved because of an increased risk of venous thromboembolism (VTE).
19. *Relapse of hirsutism after stopping Dianette:*
   a. Dianette can be used intermittently;
   b. or an alternative combined oral contraceptive containing drospirenone, (Yasmin) can be considered.
20. *Drospirenone in Yasmin (EE30μg + drospirenone 3 mg):*
   a. This progestogen is an analogue of Spironolactone.
   b. It has an antiandrogen effect by inhibiting androgen production and blocking androgen receptors.
21. *Be aware:*
   Not to use second generation COCPs containing the most androgenic progestins: levonorgestrel or norethisterone.
   a. *Alternatives to COCP*
   If COCP are contraindicated or have not shown any improvement in hirsutism after 6 months or more of treatment, then alternative specialist treatments, such as antiandrogens, insulin- sensitising drugs, or gonadotrophin-releasing analogues should be considered.
22. *Antiandrogens:*
   a. All antiandrogens carry the risk in pregnant women of feminisation of the male foetus,
   b. An effective method of contraception should be prescribed for women of child bearing age as antiandrogen treatment can lead to feminisation of the male foetus.
23. *Cyproterone acetate (CA):*
   a. CA is a 17-OHP4 acetate derivative, competes with dihydrotestosterone (DHT) for the androgen receptor and also inhibits 5α reductase to a lesser extent.
   b. It can be used alone in higher doses (50–200 mg daily) or along with COCP in a reverse sequential regimen as follows:
      i. CA 50–100 mg/day can be combined with EE in a reverse sequential regimen (e.g., add CA 50–100 mg on days 5–15 of COCP).

    c. Side-effects include liver toxicity, irregular menstrual bleeding, depression, fatigue, breast symptoms and decreased libido.

24. *Spironolactone:*
   a. An antiandrogen and aldosterone antagonist.
   b. Can be taken with (50−150 mg/day) or without the OCP.
   c. Competes with DHT to bind to the androgen receptor, and inhibits enzymes involved in androgen biosynthesis.
   d. Usually well tolerated.
   e. Side effects include irregular menstrual bleeding, headache, hypotension, nausea and decreased libido.
   f. Should not be used in women with renal insufficiency or hyperkalaemia.
   g. Considered to be more effective than other alternatives, such as finasteride, metformin and low-dose cyproterone acetate.

25. *Flutamide or bicalutamide:*
   a. Nonsteroidal, competitive inhibitors of androgen receptor binding.
   b. 125−250 mg twice daily.
   c. The most important rare side-effect is hepatotoxicity including fulminant liver failure.

26. *5α Reductase inhibitor (Finasteride):*
   a. An inhibitor of the 5α reductase enzyme.
   b. Reduces the conversion of testosterone into DHT.
   c. 1−5 mg/day.
   d. Can cause liver dysfunction.

27. *Insulin-sensitising agents:*
   a. Insulin sensitising drugs decrease hyperinsulinaemia by increasing insulin sensitivity.
   b. lead to in an increase in levels of SHBG.
   c. Reduce the levels of free circulating androgens.
   d. Side effects include gastrointestinal symptoms, increased risk of cardiovascular events, liver dysfunction, and lactic acidosis.

28. *Metformin:*
   a. Improve hirsutism through an improvement in insulin resistance.
   b. Metformin may reduce free androgen levels by around 11% in women with PCOS compared to placebo.
   c. 500−1000 mg twice daily.
   d. It is less effective when compared with OCPs and antiandrogens.

29. *Thiazolidinediones*
   Rosiglitazone and Pioglitazone:
   a. The European Medicines Agency withdrew rosiglitazone in 2010 after concerns about an increased risk of cardiovascular events.
   b. Pioglitazone has been associated with bladder tumours and has been withdrawn in some countries.

30. *Gonadotropin-releasing hormone analogs (GnRH)*
   a. This regimen includes intramuscular administration of leuprolide acetate (7.5 mg monthly intramuscularly, combined with 25−50 μg transdermal estradiol).
   b. GnRH analog agonists suppress the hypothalamic−pituitary−ovarian axis.
   c. Inhibits LH and follicle-stimulating hormone (FSH).
   d. decrease the secretion of androgens by the ovaries.
   e. No specific advantages over other therapies.

31. *Glucocorticoids*
  a. *Prednisolone*:
    i. Sometimes used in cases of nonclassic CAH.
    ii. Suppress adrenocorticotropic hormone-dependent adrenal androgen synthesis.
    iii. 5−10 mg/day.
    iv. Side-effects are weight gain, osteoporosis and adrenal suppression.
      bA pragmatic approach to management:
      i. Lifestyle changes, including diet, exercise and weight loss, should be initiated as the first line of treatment for hirsute women with PCOS.
      ii. There is a correlation between use of weight reduction drugs (e.g., orlistat) and reduction of hyperandrogenenemia.

32. *Treatment approaches for virilism:*
  a. *According to the underlying cause:*
    i. Adrenal suppression for CAH.
    ii. Surgical removal of ovarian and adrenal tumours.
    iii. Hypophysectomy for Cushing's syndrome.

33. *Monitoring:*
  a. *Subjectively*, based upon the patient feedback whether there is a decrease in the growth of hair, or less need to use other methods to remove the hair.
  b. *Objectively*, by using the Ferriman−Gallway score to rate hirsutism and compare it with previous values.

34. *Long term prognosis:*
  a. Most medications must be taken for 6 months before a noticeable improvement occurs.
  b. The existing hair can be mechanically removed or bleached, and some women continue to use these methods in combination with medication.
  c. After 6 months, if the patient perceives that the response has been suboptimal then alternative options be considered, which include a change in dose or drug, or the addition of a second agent.
  d. Treatment will never be curative, and therefore chronic treatment will likely be necessary.
  e. Androgen hormone concentrations decrease with age.
  f. Women in their 20s may need multiyear treatment to control the hirsutism.
  g. Women in their 30 and 40s have decreasing androgen concentrations and some women may decide that hormonal therapy is no longer necessary.
  h. When pregnancy is desired, all pharmacological treatments for hirsutism must be discontinued.
    i. Antiandrogens, in particular, are contraindicated in women trying to conceive because of potential adverse effects on male sexual development.

35. *Virilism and hirsutism:*
  Hirsutism may occur together with symptoms of defeminisation (virilism).

36. *Signs and symptoms of virilism include:*
  a. Acne
  b. Deepening of voice
  c. Infrequent menstruation
  d. Breast atrophy
  e. Clitoromegaly
  f. Increased libido
  g. Increased muscle mass in shoulder girdle
  h. Malodorous perspiration.

3.2.1 Causes of virilism

1. *Androgen producing ovarian and adrenal tumours:*
   a. A serum *testosterone* level >5nmol/L warrants further investigation for adrenal or ovarian tumours.
   b. If serum testosterone is elevated despite a normal DHEAS level, an ovarian source is more likely.
   c. If a DHEAS level >700 µg/dL is present despite a normal serum testosterone level, an adrenal source should be suspected.
   d. Pelvic ultrasonography can be done to detect an ovarian neoplasm or a polycystic ovary.
   e. CT scan or MRI of the adrenal region is useful for diagnosis as shown in Flow diagram 1.
2. *Adult onset CAH:*
   a. CAH is more common in Ashkenazi Jews, Hispanic and Slavic people.
   b. Mildly elevated serum testosterone and DHEAS are often present.
   c. An elevated 17-hydroxyprogesterone (50−300 nmol/L) is seen in women with late-onset CAH (best measured in the early morning and during the follicular phase of the cycle).
3. Cushing's syndrome and acromegaly
4. *XY females with functioning testicles:*
   Usually present around puberty with primary amenorrhea and signs of virilism.
   The diagnosis can be confirmed with karyotyping.
5. *Iatrogenic: androgen therapy.*

Treatment flowchart for female virilism (Flow diagram 1)

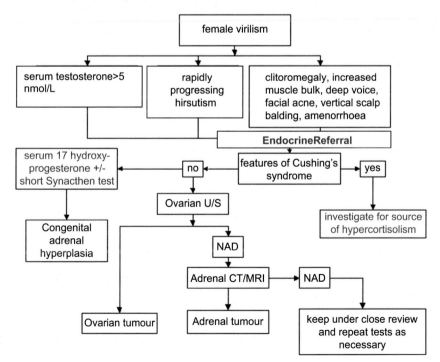

3.3 Summary

1. Hirsutism affects 5%−10% of women of reproductive age, and hyperandrogenemia is the key trigger for excess hair growth.
2. In the female, androgens originate from the ovaries and the adrenal glands.
3. The most common cause of hirsutism is PCOS accounting for 70%−80% of the cases.
4. Treatment is important to improve the self-esteem of the patients.
5. The combined oral contraceptive pill is the commonest used treatment for hirsutism.
6. Second generations COCP containing levonorgestrel or norethisterone can potentially increase androgenic symptoms.
7. In cases of rapidly progressing hirsutism (1 year from onset to medical referral), virilisation or when testosterone >5, CAH and androgen secreting tumours must be excluded.
8. High androgen and low SHBG are independent risk factors for CVD.
9. Patients should be warned that long-term treatment may be necessary before a noticeable effect is seen.

Further reading

National Institute for Health and Care Excellence (NICE) − Hirsutism, Clinical Knowledge summaries, last revised in December 2014. http://cks.nice.org.uk/hirsutism

Meek CL, Bravis V, Don A, Kaplan F. Polycystic ovary syndrome and the differential diagnosis of hyperandrogenism. *Obstetric Gynaecol*. 2013;15:171−176.

RCOG. Long-term Consequences of Polycystic Ovary Syndrome, Royal College of Obstetricians &Gynecologists', Green-top Guideline No. 33, November 2014. https://www.rcog.org.uk/globalassets/documents/guidelines/gtg_33.pdf

VanZuuren EJ, Fedorowicz Z. Interventions for hirsutism excluding laser and photoepilation therapy alone: a bridged Cochrane systematic review including GRADE assessments. *Br J Dermatol*. 2016;175(1):45−61.

ESHRE: International evidence-based guideline for the assessment and management of polycystic ovary syndrome 2018 (PCOS_Evidence-Based-Guidelines_20181009.pdf) (https://www.eshre.eu/Guidelines-and-Legal/Guidelines/Polycystic-Ovary-Syndrome)

Contraception for obese and super obese women

Omar Thanoon[1], Asma Gharaibeh[2] and Tahir Mahmood[1]
[1]Department of Obstetrics & Gynaecology, Victoria Hospital, Kirkcaldy, United Kingdom,
[2]Royal infirmary of Edinburgh, Edinburgh, Scotland

1. Obesity continues to be a major public health concern across the globe.
2. The prevalence of obesity has doubled over the past 30 years with 15% of women worldwide classified as obese as of 2014.
3. It is reported that obese women have less contraceptive usage, more contraceptive failure, and lower intake of preconceptional folic acid, which can greatly compromise prepregnancy and pregnancy care.
4. Prevention of untimed pregnancy in obese women is a major priority for health care professionals.

4.1 Risks associated with obesity during pregnancy

Maternal obesity is linked with a range of serious maternal and foetal outcomes

1. Miscarriage
2. Preterm delivery
3. Gestational diabetes and metabolic syndrome of pregnancy
4. Pregnancy-induced hypertension
5. Preeclampsia
6. Venous thromboembolism
7. Induction of labour
8. Prolonged labour
9. Caesarean section
10. Postpartum haemorrhage
11. Wound infection
12. Macrosomia
13. Birth injury (shoulder dystocia)
14. Stillbirth
15. Neonatal death

Handbook of Obesity in Obstetrics and Gynecology. DOI: https://doi.org/10.1016/B978-0-323-89904-8.00023-8

4.2   Classification of obesity

BMI	Classification
<18.5	Underweight
18.5−24.9	Normal weight
25.0−29.9	Overweight
30.0−34.9	Class I obesity
35.0−39.9	Class II obesity
≥ 40.0	Class III obesity

4.3   Metabolic disorders associated with obesity in nonpregnant obese

1. Metabolic syndrome
2. Diabetes mellitus
3. Essential hypertension
4. Cardiovascular disease including myocardial infarction,
5. Venous thromboembolism
6. Breast cancer
7. Endometrial cancer
8. Changes in the metabolism of sex steroids used in hormonal contraception
9. May influence half-life, clearance (area under the curve), and time to reach steady state

4.4   Mechanisms by which obesity could potentially affect contraceptive efficacy

1. Obesity can have profound effects on different physiologic processes, including absorption, distribution, metabolism, and excretion of contraceptive drugs.
2. Obesity is also associated with altered body composition with an increase in fat mass, which can affect the distribution of hydrophilic and lipophilic drugs.
3. Other physiological alterations in obesity that can have a potential impact in contraceptive drug metabolism and excretion include increased splanchnic and renal flow.
4. In spite of all the potential mechanisms by which obesity could affect contraceptive efficacy, there have been few studies to date that have investigated the pharmacokinetics of contraceptive steroids in obese women.

4.4.1   How obesity may affect contraceptive efficacy?

1. In one study, the half-life of levonorgestrel (LNG) in obese subjects was twice that of normal BMI subjects.
2. In one study of oral contraceptives, obese women had a lower area under curve and lower maximum value for ethinylestradiol than normal weight women.

3. In a longitudinal study of depot Medroxy progesterone (DMPA) in different classes of obese women, median MPA was consistently lowest among class 111 obese women, but above the levels needed to inhibit ovulation.
4. European Society of Contraception has concluded that there is no robust evidence for decreased efficacy of different contraceptive methods in overweight or obese women.

4.5   Potential concerns with obesity and contraception

1. Historically, overweight and obese women have been excluded from trials in contraception, leading to a lack of robust evidence.
2. As a generalisation, women tend to blame contraception for weight gain. This perceived weight gain is a leading cause of discontinuation of contraception at least in some parts of the world.
3. Obesity doubles the risk of venous thromboembolism as compared with someone with a normal BMI.
4. In principle, choice for contraception should take account of possible adverse metabolic effects associated with various hormonal methods of contraception,
5. Procedure-dependent contraceptive methods [intrauterine devices (IUDs) and sterilisation] are technically more challenging to perform in an obese woman than their normal BMI counterparts.

4.5.1   Contraceptives and weight gain

1. Many women and clinicians worldwide believe that an association exists between weight gain and oral contraceptives.
2. Perception about weight gain can also lead to early discontinuation among users of contraception.
3. More importantly, most of those who discontinued, failed to adopt another method of contraception, exposing themselves to an increased risk of pregnancy,
4. Weight gain is due to one of the following factors: fluid retention, fat deposition, or muscle mass.
5. A causal relationship between combined oral contraceptives and weight gain has not been clearly established.
6. Progestogen-only contraception is ideally suited for women who have contraindications to or who are unable to tolerate oestrogens.
7. There is limited evidence for weight gain when using progestogen-only contraception.

4.5.2   Potential mechanisms by which contraceptives can cause weight gain

1. Treatment with hormonal contraceptives may lead to activation of renin−angiotension−aldosterone system.
2. Fluid retention may be induced by the mineralocorticoid activity of contraceptive steroids.
3. Oestrogens increase the size and number of subcutaneous adipocytes, which can be associated with increased subcutaneous fat in breast, hips, and thighs.

4. The anabolic properties of COCs can have an effect on satiety and appetite that could result in increased food intake and weight gain.
5. A Cochrane review of 49 studies, however, did not find any large association between oral COC and weight gain.
6. Some longitudinal studies have suggested that the perceived weight gain with COC may be related to the natural changes in weight from a lifetime perspective.
7. In a Cochrane review of oral progestogen contraceptives, mean weight gain at 6 and 12 months was less than 2 kg for most studies.
8. There is some evidence that obese adolescent users of DMPA may gain more weight compared to those with normal weight.

4.5.3 Safety of hormonal contraceptives in obese women

4.5.3.1 Cardiovascular disease

1. Obesity is associated with an increased risk of different health conditions, including diabetes mellitus, dyslipidaemia, heart disease, stroke, venous thrombo-embolism, hepatobiliary disease, and cancer.
2. Current users of oral contraceptives had a moderately increased risk of hypertension.
3. Some studies report that oral contraceptives may be associated with an increased risk of myocardial infarction and stroke is more marked in obese women, especially if they smoke heavily.
4. A systemic review evaluating the effects of CHC use in women with BMI > 30 on MI and stroke reported inconclusive and conflicting evidence.

4.5.3.2 Venous thromboembolism

1. Exogenous oestrogens and obesity may increase blood coagulability with an increase in procoagulant factors (factors VII, VIII, XII, fibrinogen).
2. Oral contraceptives among smokers may be associated with increased levels of fibrinogen and intravascular fibrin deposition and may also increase risk of arterial thrombosis.
3. The risk of VTE increases in obese women (especially with BMI >35) and the use of combined oral contraceptives (3.46 times increased risk).
4. It appears that the risk of VTE is associated with higher dose ethinylestradiol (> 50 μg).
5. There is no evidence for increased risk of VTE with progestogen-only pill.

4.5.3.3 Cancer

1. Oral contraceptive use has been associated with increase in risk of cervical cancer.
2. There are conflicting results with breast cancer and the use of COC.
3. This increased risk of cervical and breast cancer among current users appears to be lost within 5 years of stopping oral contraceptive.
4. There is a decrease in risk of ovarian, uterine, and colorectal cancers.
5. On balance, many women benefit from significant reductions in risks of certain types of cancer that seems to persist many years after stopping treatment (lymphatic and haematopoietic cancers).
6. Obesity, per se, has been associated with increased risk of cancer, especially breast, endometrial, ovarian, and colorectal.

7. It is, therefore, theoretically possible that oral contraceptive use in obese women may have significant effects on incident cancer.

The UKMEC criteria (2016) for contraceptive use clearly outlines the criteria for contraceptive use in obesity.

UKMEC criteria (2016) for contraceptive use—obesity and selected clinical conditions that are of particular relevance to obese women.

UKMEC	DEFINITION OF CATEGORY
Category 1	A condition for which there is no restriction for the use of the method
Category 2	A condition where the advantages of using the method generally outweigh the theoretical or proven risks
Category 3	A condition where the theoretical or proven risks usually outweigh the advantages of using the method. The provision of a method requires expert clinical judgement and/or referral to a specialist contraceptive provider, since use of the method is not usually recommended unless other more appropriate methods are not available or not acceptable
Category 4	A condition which represents an unacceptable health risk if the method is used

CONDITION	Cu-IUD	LNG-IUS	IMP	DMPA	POP	CHC
	I = Initiation, C = Continuation					
Obesity						
a) BMI ≥30–34 kg/m²	1	1	1	1	1	2
b) BMI ≥35 kg/m²	1	1	1	1	1	3
Smoking						
a) Age <35 years	1	1	1	1	1	2
b) Age ≥35 years						
(i) <15 cigarettes/day	1	1	1	1	1	3
(ii) ≥15 cigarettes/day	1	1	1	1	1	4
(iii) Stopped smoking <1 year	1	1	1	1	1	3
(iv) Stopped smoking ≥1 year	1	1	1	1	1	2
History of bariatric surgery						
a) With <30 kg/m² BMI	1	1	1	1	1	1
b) With ≥30–34 kg/m² BMI	1	1	1	1	1	2
c) With ≥35 kg/m² BMI	1	1	1	1	1	3

CARDIOVASCULAR DISEASE (CVD)						
Multiple risk factors for CVD (such as smoking, diabetes, hypertension, obesity and dyslipidaemias)	1	2	2	3	2	3
Hypertension						
a) Adequately controlled hypertension	1	1	1	2	1	3
b) Consistently elevated BP levels (properly taken measurements)						
(i) Systolic >140–159 mmHg or diastolic >90–99 mmHg	1	1	1	1	1	3
(ii) Systolic ≥160 mmHg or diastolic ≥100 mmHg	1	1	1	2	1	4
c) Vascular disease	1	2	2	3	2	4

Diabetes						
a) History of gestational disease	1	1	1	1	1	1
b) Non-vascular disease						
(i) Non-insulin dependent	1	2	2	2	2	2
(ii) Insulin dependent	1	2	2	2	2	2
c) Nephropathy/retinopathy/neuropathy	1	2	2	2	2	3
d) Other vascular disease	1	2	2	2	2	3

CONDITION	Cu-IUD		LNG-IUS		IMP	DMPA	POP	CHC	
	I = Initiation, C = Continuation								

Cervical cancer									
a) Awaiting treatment	I 4	C 2	I 4	C 2	2	2	1	2	
b) Radical trachelectomy	3		3		2	2	1	2	
Breast conditions									
a) Undiagnosed mass/breast symptoms	1		2		2	2	2	I 3	C 2
b) Benign breast conditions	1		1		1	1	1	1	
c) Family history of breast cancer	1		1		1	1	1	1	
d) Carriers of known gene mutations associated with breast cancer (e.g. BRCA1/BRCA2)	1		2		2	2	2	3	
e) Breast cancer									
(i) Current breast cancer	1		4		4	4	4	4	
(ii) Past breast cancer	1		3		3	3	3	3	
Endometrial cancer	I 4	C 2	I 4	C 2	1	1	1	1	
Ovarian cancer	1		1		1	1	1	1	

4.5.4   Contraceptive issues after bariatric surgery

1. Bariatric surgery is believed to be the most effective treatment method for the morbidly obese.
2. The general consensus from major associations is that pregnancy should be avoided for 12–24 months after bariatric surgery.

3. There are concerns regarding the efficacy and safety of hormonal contraception in women who have undergone bariatric surgery.
4. In one study, the author recommends that low-dose progestogen-only minipills should not be used after jejunoileal bypass.
5. There are concerns around the malabsorption of oral contraceptives, especially in women undergoing malabsorptive procedures such as Roux-en-Y gastric bypass and biliopancreatic diversion with duodenal switch.

4.5.5 *Intrauterine contraceptive devices in obese women*

4.5.5.1 *Types*

4.5.5.1.1 Copper intrauterine device

1. There are many devices available with different frames, structures, different delivery rates of copper ions, and different duration of action (5−12 years).
2. There are no studies indicating a reduced efficacy of copper IUDs in obese women compared to normal weight women.
3. The usual side effects are prolonged bleeding, dysmenorrhea, lower abdominal pain, or discomfort.
4. Its use should be cautiously considered for women with previous history of Wilson's disease, endometriosis, and sexually transmitted infection.
5. Copper IUDs have a protective effect on the endometrium.

4.5.6 *Levonorgestrel-containing intrauterine systems*

There are currently three systems available (LNG52, LNG 14, LNG 20), with different periods of effectiveness.

The progestin within the IUS thickens cervical mucus.

The progestins also increase expression of glycodelin A in endometrial glands, which inhibits binding of sperm to the egg.

There may be partial inhibition of ovarian follicular development and ovulation, but most women continue to have ovulatory cycles.

There is no evidence of impaired contraceptive effectiveness of IUS in obese women.

There is no increased risk of VTE in obese IUS users.

The association between IUS use and breast cancer remains controversial. At the best, there may be a minor degree increased absolute risk for breast cancer (up to 1.3 times, reported by a Finnish cohort study), but there are no data specifically looking into the potentially different risk for obese women.

4.5.6.1 *Practical issues with intrauterine systems in obese women*

1. Intrauterine device insertion presents some challenges in the obese woman. It may be difficult to ascertain the size and direction of the uterus. In addition, visualisation of the cervix may be difficult.
2. Severe uterine anatomic distortion creates additional difficulty at insertion with increased risk of expulsion,

3. Simple measures such as use of a larger speculum, placing a condom with the tip removed over the blades of the speculum, comfortable positioning of the patient and use of ultrasound during the device insertion can be helpful.
4. There is evidence of some benefit with the LNG intrauterine system in selected obese women with abnormal uterine bleeding with no huge dropout rates or complications compared to other BMI groups.
5. The European Society of Contraception Statement (2015) for obese women suggests that IUD contraception is highly recommended in obese women with LNG-intrauterine system a safe and effective contraceptive method for obese women.
6. There are no specific contraindications to hormone-containing devices or copper IUDs in the obese woman.
7. Hormone-containing devices apart from being effective contraceptives may produce additional benefit in obese women with abnormal uterine bleeding or endometrial hyperplasia.
8. The risk of complications during placement is minimal and includes uterine perforation, and pelvic inflammatory disease. The risk of expulsion is 10% over 3 years of use.

4.5.7   Sterilisation procedures in obese women

1. Obesity can complicate tubal sterilisation procedures.
2. Obese women have a higher risk to general anaesthetic compared to women with normal BMI.
3. Obesity can be associated with higher incidence of difficulties during surgery, technical failure rate, and longer operating times compared to nonobese controls.
4. In obese women vasectomy for the woman's partner may provide a safer alternative and should be explored.
5. Patients with morbid obesity may have challenges during laparoscopic approach because of anaesthetic issues related to adequate ventilation. Conversely, an open approach will be even more challenging as described in Chapter 20.

4.6   Methods of sterilisation

4.6.1   Laparotomy

1. Tubal ligation through an open abdominal incision can be carried out during either elective caesarean section or at the time of other elective abdominal surgery.
2. Mini-Laparotomy (2−5 cm length suprapubic transverse incision) is usually used when equipment for laparoscopic or transcervical tubal occlusion is not available or too expensive.
3. The most widely used technique of abdominal sterilisation is the Pomeroy technique; it involves grasping the tubal isthmic part with an atraumatic clamp, ligation of the loop of the fallopian tube with absorbable suture and excision.
4. Both tubal ends separate after resorption of the suture and fibrotic tissue is formed between distal and proximal parts of the fallopian tube.
5. The alternative is to use occlusive mechanical devices such as the Filshie clip, Hulka Clements clips, Fallope rings, etc.

6. Patients with extensive history of abdominal surgery should be counselled on the possibility that their fallopian tubes may not be accessible at the time of abdominal or laparoscopic surgery.

4.6.2 Laparoscopy

1. Laparoscopic sterilisation is usually performed under general anaesthesia, with carbon dioxide being used to provide abdominal distension.
2. It requires a two-puncture technique, but refinement of the instruments has led to procedures being carried out through a single, subumbilical puncture.
3. Tubal occlusion can be achieved in different ways: unipolar electrocoagulation or the use of different occlusion,
4. Laparoscopic bilateral total salpingectomy is another method of sterilisation and it could potentially decrease ovarian cancer risk.
5. Theoretically, bilateral salpingectomy is 100% effective.
6. Salpingectomy is the preferred method for women who have become pregnant after a tubal ligation.

4.6.3 Hysteroscopy

1. The two most widely used hysteroscopic sterilisation devices are Essure hysteroscopic sterilisation (Bayer) and Adiana hysteroscopic sterilisation (Hologic).
2. The Essure hysteroscopic sterilisation device was approved by the FDA in 2002, it involves inserting a nickel-steel alloy implant in the fallopian tubes causing tubal occlusion by inducing fibrosis.
3. The Adiana hysteroscopic sterilisation system was approved by the FDA in 2009. It involves delivering less than 3 W of bipolar radiofrequency energy within the lumen of fallopian tubes, thus creating a superficial lesion. After the radiofrequency energy is delivered, a silicone matrix is placed into the tubal lumen in the region where the lesion was formed.
4. In 2012 the Adiana system was withdrawn from the market due to high failure rates and high legal costs.
5. Complication rates following Essure were high (device breakage, uterine perforation, distal placement of the device, allergic reactions, and chronic pain); therefore, the Essure hysteroscopic sterilisation device was withdrawn from the European market in 2017 and the US market in 2018.
6. Currently there is no licensed hysteroscopic sterilisation technique in the UK.

Further reading

Committee on Gynecologic Practice. ACOG Committee Opinion Number 540: risk of venous thromboembolism among users of drospirenone-containing oral contraceptive pills. *Obstet Gynecol*. 2012;120(5):1239−1242.

Damadoran S, Swaminathan K. Obesity and contraception. In: Mahmood T, Arulkumaran S, Chervenak FA, eds. *Obesity and Gynaecology*. 2nd ed. Elsevier; 2020:43−56. Available from: https://doi.org/10.1016/B978-0-12-817919-2.00005-X.

Mamun AA, Callaway LK, O'Callaghan MJ, et al. Associations of maternal pre-pregnancy obesity and excess pregnancy weight gains with adverse pregnancy outcomes and length of hospital stay. *BMC Pregnancy Childbirth.* 2011;11:62.

McKeating A, O'Higgins A, Turner C, McMahon L, Sheehan SR, Turner MJ. The relationship between unplanned pregnancy and maternal body mass index 2009−2012. *Eur J Contracept Reprod Health Care.* 2015;20(6):409−418.

Merki-Feld GS, Skouby S, Serfaty D, et al. European society of contraception statement on contraception in obese women. *Eur J Contracept Reprod Health Care.* 2015;20 (1):19−28.

Suchon P, Al Frouh F, Henneuse A, et al. Risk factors for venous thromboembolism in women under combined oral contraceptive. The PIL1 Genetic RIsk Monitoring (PILGRIM) study. *Thromb Haemost.* 2016;115(1):13542.

Contraception after bariatric surgery

Tahir Mahmood
Department of Obstetrics & Gynaecology, Victoria Hospital, Kirkcaldy,
United Kingdom

5.1   Introduction

Bariatric surgery is the most effective treatment for morbid obesity. National Institute of Health has laid out criteria for a group of patients who should be considered suitable for bariatric surgery:

1. Body mass index (BMI) > 40.
2. BMI of 30–40 plus one of the following obesity-related comorbidities: severe diabetes mellitus, Pickwickian syndrome, obesity-related cardiomyopathy, severe sleep apnoea, or osteoarthritis interfering with lifestyle.

Bariatric surgeries lead to weight loss and comorbidity improvement by the following mechanisms:

1. gastric restriction: reduction of amount of food that can be consumed;
2. malabsorption: impaired food-digestion by smaller stomach and nutrients absorption by shortened intestine; or
3. a combination of both.

These procedures result in changes in gut hormones that promote satiety and suppress hunger.

Most weight-loss surgeries today are performed using minimally invasive techniques (laparoscopic surgery).

The most common bariatric procedures are:

1. Adjustable gastric band: it is a commonly performed restrictive procedure, by placing an inflatable band around the upper portion of stomach, thus creating a small stomach pouch above the band.
   a. Vertical band gastroplasty: sectioning off the cardia of the stomach by a longitudinal staple line and placing a band or mesh around the outlet.
   b. Laparoscopic sleeve gastrectomy: removing approximately 80% of stomach.
   c. Intragastric balloon.
   d. Endoliminal gastroplasty.
2. Biliopancreatic diversion (BPD): this involves dividing the jejunum and connecting it near the ileocecal valve, thus bypassing a long segment of small bowel (this operation is no longer performed).

Handbook of Obesity in Obstetrics and Gynecology. DOI: https://doi.org/10.1016/B978-0-323-89904-8.00039-1

3. Roux-en-Y gastric bypass: this operation involves creating a small stomach pouch and then connecting to the small intestine further down.
4. BPD with duodenal switch: this procedure involves creating a smaller, tubular stomach pouch and connecting further down, bypassing a large portion of the small intestine.

5.1.1 Reproductive and general health consequences of bariatric surgery

1. Increased fertility in a short period of time postoperatively. Weight-reduction has a positive effect on sex hormone profiles and ovulation.
2. Conception during rapid weight loss (12−24months) is associated with higher rate of nutritional deficiencies and obstetrics complications including higher incidence of still birth.
3. Effective and safe contraception is important to prevent an untimely pregnancy while the women is trying to achieve an optimal and a stable weight.

5.2 Obesity and contraception issues

Historically, overweight and obese women have been excluded from trials in hormonal contraception. Therefore data regarding the safety and efficacy of contraceptive methods is lacking.

The effect of obesity on pharmacokinetics of steroidal contraceptives especially the risks of failure are poorly understood,

Obesity can have profound effects on absorption, enterohepatic metabolism, distribution, and excretion of hormones. Even our knowledge of pharmacokinetics of oral steroid contraceptives following bariatric surgery is poor.

Women need contraceptives with high efficacy in the context of physiological alterations within GI system, new pharmacokinetic adjustment, and weight stabilisation following bariatric surgery.

Research evidence:

There have been few studies to date that have investigated the pharmacokinetics of contraceptive steroids in obese women with or without bariatric surgery.

A Cochrane review of 17 studies involving different methods of contraception did not generally find any robust evidence for decreased efficacy of different contraceptive methods in overweight or obese women. However, it is recognised that combined hormonal contraceptives are associated with significantly increased risks such as venous thromboembolism (VTE).

5.2.1 Immediate postsurgery contraceptive advice

1. Women should be warned about increased fertility within a mean of 3 months postoperatively by resumption of normal ovulatory cycles after weight reduction surgery.

2. It has been reported that conceiving during the period of rapid weight loss seen in the first 12−24 months following surgery is associated with higher rates of nutritional deficiencies and obstetric complications.
3. Women should be prescribed an effective and reliable method of contraception for the following 12−18 months.

Combined Hormonal contraception (UKMEC Category 2/3)

5.2.2 Oral preparations

1. Use of Combined Oral Contraceptive as well as other combined hormonal contraceptives (CHC) leads to an increase in the cardiovascular and metabolic risk proportionally to the BMI.
2. During postoperative recovery period, long-term diarrhoea and/or vomiting could probably decrease OC effectiveness.
3. COCs are metabolised by bacterial enzymes and the enzymes in the intestinal mucosa, before they enter the portal circulation. Very little is known about the metabolism of oral contraceptives following malabsorptive and restrictive malabsorptive bariatric procedures.
4. Oral CHC leads to an increase in the cardiovascular and metabolic risk proportionally to the BMI.
5. Due to unknown and potentially reduced efficacy of oral CHC in postbariatric patients, their use is not advised in patients after bariatric surgery.

Nonoral combined hormonal contraceptive (Patch, Ring)

1. Nonoral hormonal contraceptives have been studied in only small numbers of obese patients, but appear to be effective.
2. The American College of Obstetricians and Gynecologists suggests that women undergoing bariatric surgery with a significant malabsorption component who want to use hormonal contraception should preferentially consider nonoral delivery of the contraceptive hormones after weight stabilization and exclusion of other risk factors for cardiovascular disease.

5.3 Progestogen-only pill

The POP contains only A progestin.
Efficacy:

1. The Pearl Index (PI) is 9 for typical use, and 0.3 by perfect use.
2. The efficacy of POP is not effected by body weight or BMI.
3. There are theoretical concerns about reduced absorption after bariatric procedures.
4. The mean plasma levels of NET and LNG were significantly lower in the operated patients who had jejunoileal bypass at 1−8 hours after oral intake.
5. In another study, the plasma levels of ENG, using 75 µg DSG before and after Roux-en-Y bypass surgery remained unchanged.
6. POP are not considered suitable for women after bariatric surgery.

Safety:

1. According to UKMEC, obesity alone does not restrict use of POP (category 1).
2. According to UKMEC, in obesity associated with other risk factors of CVD, the use of POP is qualified as category 2.
3. There are no data about the risk of VTE in this group of women.

5.3.1 Copper intrauterine device

Category 1 by the UK Medical Eligibility Criteria for Contraceptive Use (UKMEC Category).

1. Their principle of action is by local inflammatory and spermicidal effect of copper ions. As this is a local action, it is expected not to be affected by BMI or changes in weight.
2. There are no studies comparing specifically the efficacy of Cu-IUD in obese women and women with normal BMI.
3. Although there are no research data available on use of the Cu-IUD after bariatric surgery, theoretically the effectiveness of the device should be the same due to its local mechanism of action.
4. There are no theoretical reasons for Cu-IUD to cause any specific health risks in postbariatric surgery women.

Types of devices:
There are different types of Cu-IUDs with variable amount of copper content, frames, and duration of action between 5 and 12 years.
The most commonly used are:

1. ParaGard
     Flexi-T
     Multiload-T
2. GyneFix.

Pearl Index: for ideal use is 0.6; and for typical use 0.8.

1. The real-life user failure is less than 1%.
2. The cumulative pregnancy rate is 1.6% at 7 years, and 2.2% at 8 and 12 years.

Health benefits:

1. Cu-IUD is not associated with any increased risk for cardiovascular diseases (VTE or cardiac event), diabetes, metabolic syndrome, or weight increase.
2. Cu-IUD has a protective effect on the endometrium against endometrial carcinoma.

Risks:

1. Uterine perforation at placement (0.1%)
2. Pelvic inflammatory disease (0%−2%), caused by a preexisting vaginal and cervical infection,
3. The overall cumulative risk of IUD expulsion is 10% over 3 years of use
4. No data available about perforation or expulsion rates in postbariatric surgery women

Side effects:

1. Prolonged/heavy menstrual bleeding
2. dysmenorrhea, lower abdominal pain/discomfort

Contra-indications:

1. Severe uterine cavity anatomic distortion
2. Cervical stenosis
3. Uterine fibroids
4. Active pelvic infection
5. Suspected pregnancy
6. Wilson's disease or copper allergy
7. Unexplained abnormal uterine bleeding

Levonorgestrel containing intrauterine device (LNG-IUS)
Category 1 UKMEC
Mode of action:

1. The progestin secreted from the IUS causes thickening of cervical mucus.
2. The progestin in the uterine cavity also cause increase in expression of glycodelin A in endometrial glands which inhibits binding of sperm to the egg cell.
3. Serum concentrations of progestin can partially inhibit the ovarian follicular development and lead to anovulation (a majority of women tend to have ovulatory cycles).
4. UK Faculty for Sexual and Reproductive Health recommended the use of long-acting reversible contraceptive in the first 2 years after surgery.

Types of IUS currently available:

1. LNG 52 with an average daily release of 20 μg LNG and is effective for up to 5 years.
2. LNG 14 with an average daily release of 6 μg LNG and is effective for up to 3 years.
3. LNG 20 with an average daily release of 9 μg LNG and is effective for up to 5 years.

Efficacy:

1. PI for all the three IUS is around 0.2.
2. The cumulative pregnancy rate is 0.5%−1.1% after 5 years of continuous use with the LNG 20 IUD, 0.9 with LNG 14.
3. Available evidence suggests that LNG-IUS effectiveness is not reduced by higher BMI or body weight. It seems likely that malabsorptive surgery do not reduce its efficacy.
4. There are no statistically significant differences in contraceptive failure rates during the first 2−3 years of use among IUC users (Cu-IUD and LNG52) even among obese users. The overall failure rate was <1 per 100 women-years.
5. The presence of coexisting risk factors (smoking, diabetes, hypertension, history of VTE) in addition to obesity puts LNG-IUS into UKMEC category 2.

Health benefits:

1. There is no increased risk of cardiovascular diseases.
2. The risk of confirmed VTE was not increased in LNG-IUS users.

3. There are contradictory data about slightly increased risk of lobular and ductal cell cancer of Breast among women where LNG-IUS has been used for heavy menstrual bleeding (up to 1.3 times).
4. No impact on weight gain.
5. Reduction in heavy menstrual bleeding and dysmennorhea.
6. Protection from PID due to cervical mucus thickening.
7. Treatment of endometriosis.
8. Can be used for treatment of endometrial hyperplasia and cancer.
9. Risk reduction of endometrial cancer.
10. Fewer painful crises in women with sickle cell disease.

Side effects:

1. Irregular bleeding during the first 3−6 months.
2. By 24 months, 50% women have amenorrhea, 30% have oligomenorrhea, and 11% have spotting.
3. Mood changes: increased risk of being prescribed antidepressants which is of importance in those women who suffer from preclinical or undiagnosed perimenopausal depression.

Contraindications:

1. Current or active breast cancer or personal history of breast cancer in the past 5 years with no active disease
2. Active liver disease
3. Severe anatomical deformity of the uterine cavity
4. PID
5. Suspected pregnancy
6. Unexplained vaginal bleeding

Progestogen-only implant
UKMEC Category 1/Category 2 (coexisting risk factors)
The matrices of implants consist of small flexible biologically degradable inserted under the skin of the upper arm.
Preparations:

1. Norplant II: two flexible silicone rods, each containing 75 mg of LNG, lasting for 5 years.
2. Implanon: a single rod, containing 68 mg ENG, lasting for 3 years.

Efficacy:

1. The PI is 0.05.
2. In the US contraceptive choice project, the pregnancy rate at 4 years follow-up of ENG implant use was extremely low in overweight and obese women.
3. Pharmacokinetics studies suggest that the serum levels of ENG depend on body weight and decrease at a higher rate in women with obesity due to increased volume of distribution, effects on plasma protein binding, and altered clearance.
4. There is evidence suggesting lower serum ENG concentration after malabsorptive surgery when the weight has decreased after surgery, but the implant is still protective for at least 6 months.

Safety:

1. Implant can be used in cases of ore-existing CVD risk factors.
2. There is no evidence of increased risk of VTE or MI.
3. Studies on procoagulatory values or blood pressure did not show any difference.
4. No negative effects of carbohydrates or lipid metabolism.
5. No concern as regards loss of bone mass.
6. There are no data available regarding the risk of breast cancer.
7. In one study, implants users had an additional weight increase as compared with Cu-IUD users.

Health benefits

1. Alleviates dysmenorrhea and ovulatory pain.

Challenges:

1. With significant weight loss after surgery, loose skin may make it challenging to insert implant correctly.

Contraindications:

1. Active or recent breast cancer
2. Active and chronic liver disease except nodular hyperplasia
3. If on treatment with antiviral drugs

5.3.2 *Progestogen-only injection*

UKMEC Category 2/3

Depot preparations of progestin-only contraceptive provide reversible contraception

Preparations: currently three are available

1. Medroxy-progesterone acetate (DMPA) as intramuscular, every 13 weeks
2. Medroxy-progesterone acetate (SC), administered subcutaneously every 13 weeks
3. Norethisterone enanthate as intramuscular, every 8 weeks

Efficacy:
The PI range for perfect use is 0.2 and 6 for typical use.

1. There are limited data relating to DMPA use in women with obesity or after bariatric surgery, but the available evidence suggests that its effectiveness is not reduced.
2. Prolonged DMPA use and bariatric surgery both appear to be associated with the loss of bone mineral density but there is no data available for an additive effect of those two factors relating specifically to risk for osteoporosis or fracture.
3. There are no data on the casual association between DMPA use and the risk of VTE.

Side effects:

1. DMPA use is associated with some weight gain

Health benefits:

1. Treatment of heavy menstrual bleeding, dysmenorrhea, and for pain associated with endometriosis.
2. Offers some protection against ovarian and endometrial cancers.

Practical tips:

1. DPMA—IM injection is given in deltoid muscle, or use a longer needle to reach the muscle layer in gluteal region.
2. Alternatively DMPA-Sc should be used.

Relative contraindication:

1. Young adolescent, <18 years with BMI >30 are at increased risk of weight gain, so alternative preparations should be considered.

5.3.3   Barrier methods

UKMEC Category 1
Methods:

1. Male and female condoms
2. Cervical caps
3. Diaphragms

Efficacy:

1. PI of condoms is 18 by typical use and 2 by perfect use.
2. PI of female diaphragm is 12 by typical use and 6 by perfect use.
3. The reliability of barrier methods is not affected by bariatric procedure.

Health benefit:

1. Offer additional protection against sexually transmitted infections.

Practical tip

1. Weight loss or gain can alter the fit of a cap or diaphragm.
2. Check regularly to see if it still fits during the period of rapid weight loss phase.

Emergency contraception (EC):
Women should be offered Cu-IUD if there are no contraindications.
There is no data available on the effectiveness of oral EC

Further reading

Birgisson NE, Zhao Q, Secura GM, Madden T, Peipert JF. Preventing unintended pregnancy: the contraceptive CHOICE project in review. *J Women's Health*. 2015;24(5):349–353.
Faculty of Sexual and Reproductive Healthcare Clinical Guidance. Intrauterine Contraception; 2015.

FSRH guideline. overweight, obesity and contraceptionApril 2019 *BMJ Sex Reprod Health.*
    2019;45:1−69.
Hillman JB, Miller RJ, Inge TH. Menstrual concerns and intrauterine contraception among
    adolescent bariatric surgery patients. *J Women's Health.* 2011;20:533−538.
Jurga-Karwacka A, Bitzer J. Contraceptive choices for women before and after bariatric sur-
    gery. In: Mahmood T, Arulkumaran S, Chervenak F, eds. *(In) Obesity and
    Gynaecology.* 2nd ed. Cambridge: Elsevier Inc; 2020:57−65. Available from: https://
    doi.org/10.1016/B978-0-12-817919-2.00006-1.
Merhi ZO. Impact of bariatric surgery on female reproduction. *Fertil Steril.*
    2009;92:1501−1508.
Ochner CN, Gibson C, Shanik M, Goel V, Geliebter A. Changes in neurohormonal gut pep-
    tides following bariatric surgery. *Int J Obes.* 2011;35:153−166.

Obesity and sexual health

6

Tahir Mahmood
Department of Obstetrics & Gynaecology, Victoria Hospital, Kirkcaldy, United Kingdom

1. Sexual health is an important part of overall health, well-being, and quality of life.
2. The association between obesity and physical illness is well established.
3. There is a growing recognition of the negative impact that obesity can have on sexual health.
4. This may be mediated directly through the physical and psychosocial effects or indirectly through concurrent comorbidities.
5. It can affect sexual behaviour, social function, and sexual health outcomes.

6.1 Sexual behaviour

National Longitudinal Survey of Youth (2011) reported that:

1. Obese white adolescent girls:
    a. were more likely than nonobese girls to have a partner at least three years older;
    b. more likely to have more than three sexual partners in 1 year; and
    c. and less likely to use condoms during their most recent sexual encounter.
2. Obese girls were also at higher odds of having coital debut before the age of 13.
3. These differences were not present among black adolescent obese females.

The French National Survey (Contexte de la Sexualite en France 2005/06) reported that:

1. Obese women were 30% less likely to report having a sexual partner in the past 12 months.
2. Obese women were less likely to have an occasional sex partner.
3. However, among women with a sexual partner, there was no difference in the frequency of sexual intercourse by BMI.
4. Obese women were more likely to have met a sexual partner through the internet than women of normal weight.
5. The authors of the paper suggested that women with obesity might find it more difficult to attract a sexual partner and/or that they can establish a rapport with a potential partner while at the same time concealing their weight.

Data from the National Health and Nutrition Examination Survey from the United States:

1. This survey was conducted using computer-assisted self-interview.
2. In this survey, individuals who were overweight or with obesity reported fewer sex partners than individuals of normal weight.

Handbook of Obesity in Obstetrics and Gynecology. DOI: https://doi.org/10.1016/B978-0-323-89904-8.00022-6

3. 18% of the male and 28% of the female respondents who reported no lifetime sex partners ever tested positive for antibodies to Herpes simplex type 2 that was used as a serological marker of sexual exposure.

6.2   Obesity and sexual function

1. In general, for women in the western countries, there is often a sociocultural association between slender physique and physical attractiveness.
2. Studies have shown that individuals perceive their obesity as a serious psychosocial handicap.
3. This can lead to some of the psychological manifestations of the negative body image such as low self-esteem, which could lead to difficulty in initiating a healthy sexual relationship.
4. In the French survey, men who were overweight or obese were more than twice as likely as normal weight men to have experienced sexual dysfunction.
5. With regards to sexual function among women, multivariate analysis of results found an association between obesity and lower sexual function scores.
6. Women with obesity may also suffer from lack of libido and reduced satisfaction with sexual life.
7. A self-reported Quality of Sexual Life study in the United States reported that obesity was associated with higher incidence of risky behaviour compared to normal weight.

6.3   Obesity and sexual health outcomes

1. Girls who are heavier will attain secondary sexual characteristics and menarche earlier than normal weight counterparts.
2. However, surveys from the United States showed that there was no association between age at first intercourse and BMI.
3. However, data from the French national survey of sexual behaviour showed that obese women who were under 30 years of age were four times more likely than women of normal weight to report an unintended pregnancy or an abortion.
4. The French study also showed that women with obesity were less likely to use oral contraceptive and relied on less effective methods such as "withdrawal" method.
5. It is possible that reliance on less effective methods may reflect difficulty in negotiating condom use with a partner, greater sexual risk-taking, or misconceptions about one's fertility status.
6. It is equally possible that women with obesity are less likely to access contraceptive services (contraception hesitancy).
7. In another cross-sectional study of sexually active female adolescents, increased BMI was linked to higher number of sexual partners and participation in riskier sexual practices.
8. One US study of postpartum women reported that among women who had been using contraception at the time of conception, women with raised BMI had almost twice the rate of unintended pregnancy compared to women with normal weight.

9. There is limited evidence regarding the effect of obesity on the incidence of sexually transmitted infections (STI).

10. The French national survey of sexual behaviour among men in their late teens and 20s found that the odds of contracting an STI in the previous 5 years were more than 10 times greater for men with obesity than for men of normal weight.

11. However, in that survey, there was no difference between women of different BMI groups with the history of STI.

12. Data from the Centers for Disease Control (CDC2010) indicate that 1 in 4 young women between the ages of 14 and 19 in the United States is infected with at least one of the most common sexually transmitted diseases —human papillomavirus (HPV), Chlamydia, herpes simplex virus, and trichomoniasis.

13. Averett et al.[1] hypothesised that girls who are perceived (or perceive themselves) as less attractive will be willing to incur greater risks in order to attract a partner. Boys, according to this line of thinking, would move to less physically attractive matches in order to find willing partners for risky sex.

    a. Their findings confirm previous research indicating that overweight or obese girls are less likely to be sexually active than other girls. They also found that as a result of being less sexually active, overweight or obese girls are less likely to have vaginal intercourse without a condom.

    b. However, overweight or obese girls are not less likely to have sex under the influence of alcohol, and once they have had vaginal intercourse, their consistency of condom use is no different from that of their recommended-weight peers.

    c. However the most striking finding was that overweight or obese girls were at least 15% more likely than their recommended-weight peers to have had anal intercourse.

14. Kershaw et al. examined the effect of BMI groups (normal weight, overweight, and obese) at 6 months postpartum on STI incidence and risky sex (e.g., unprotected sex, multiple partners, risky and casual partner) at 12 months postpartum.

    a. At 6 months postpartum, 31% of participants were overweight and 40% were obese.

    b. Overweight women were more likely to have an STI (OR = 1.79, 95% CI = 1.11−2.89, $P < .05$) and a risky partner (OR = 1.64, 95% CI = 1.01−2.08, $P < .05$) at 12 months postpartum compared to normal weight women.

    c. However, obese women were less likely to have an STI than normal weight women (OR = 0.57, 95% CI = 0.34−0.96, $P < .01$).

6.4 Managing sexuality issues of young obese women

- Provide a more holistic approach with greater focus on developing positive attitudes and values towards sex, sexuality and relationships.
- This in turn can help to develop the knowledge and behaviour necessary to reduce harm at both personal and societal level, as part of wider health and wellbeing education.
- Some field work in this area suggests that young people regard school as their primary and preferred source of information about sex.
- The school delivery system provides the opportunity to develop a tailored program, delivered at age-appropriate intervals, and with scope to build on previous learning over a number of years.
- Ideally, this should commence in early years and have clear parameters, developed in conjunction with parents and other stakeholders.

Table 6.1 Common sexually transmitted infections.

Infection	Signs/symptoms	Examination	Investigation	Treatment	Long-term effects
Chlamydia, (caused by bacterium *Chlamydia trachomatis*)	Mostly asymptomatic, vaginal discharge (30%), postcoital/intermenstrual spotting, burning feeling on passing urine (Urethritis or cystitis), pelvic inflammatory (PID) disease (10%–30%) as pelvic pain/lower abdominal pain, anal discharge, discomfort, pruritus (anal coitus)	Tenderness in both adnexa at vaginal examination, cervical excitation/ contact bleeding, Cervix shows hyperaemia and oedema	– Endocervical swab for ELISA, – MSSU, Self Vaginal swab, – Pharyngeal swabs (oral sex risk), – Rectal swab (risk of anal intercourse)	– Doxycycline 100 mg BD for 1 week, – Azithromycin 1 g single dose, – During pregnancy, Erythromycin 500 mg QID for 1 week	– increased risk of infertility (tubal), – Ectopic pregnancy, – Pelvic inflammatory disease, Chronic pelvic pain, – transmission to partner, During pregnancy-transmission to newborn
Gonorrhoea, Caused by Gram-negative diplococci *Neisseria gonorrhoeae*	May present as asymptomatic, acute PID (lower abdominal or pelvic pain), dysuria without increased frequency, abscess of Bartholin gland, mucopurulent discharge, intermenstrual spotting/bleeding after intercourse, anal discharge/anal or perianal pain	vaginal examination, cervical excitation/ contact bleeding from the Cervix, muco-purulent cervical discharge	Endocervical swab, – MSSU, – Self vaginal swab	– Ceftriaxone 1 g I/M as a single dose. – ciprofloxacin 500 mg or ofloxacin 400 mg (not to be used during pregnancy), Azithromycin 2 g as a single dose, – Acute PID: local protocol (add Co-amoxiclav/ erythromycin)	PID (follow local protocols), increased risk of infertility (tubal), – Ectopic pregnancy, – transmission during pregnancy-transmission to newborn

| Genital Herpes – Herpes simplex-type1. 2-Herpes simplex type2 | Type 1: Oro-genital lesion Type 2: genital herpes, recurrence quite common | – Could be asymptomatic, or present with painful blisters, leading to ulceration associated with dysuria, painful anogenital discharge, systemic symptoms in primary infection | – HSV detection by PCR, – NAAT Test, – HSV specific antibodies | – Simple analgesics, – Saline bathing, – Topical anaesthetic such as 5% lidocaine gel, – Anti-viral drugs within 5 days of starting of the episode, (Aciclovir 400 mgX3 daily, Valaciclovir 500 mgX 2 daily) – Managing urinary symptoms which may require catheterisation, – local protocols for recurrent infections | – Regular and correct use of condom, – Suppressive therapy, – Super infections, usually with candida and streptococcus, – Autonomic neuropathy causing urinary retention, – Autoinoculation to fingers and adjacent skin, – asymptomatic meningitis |
| Bacterial Vaginosis (BV) caused by anaerobic bacteria: *Gardnerella vaginalis*, *Atopobium vaginalis*, and some other bacteria | – Not an STI but associated with sexual activity, – Vaginal pH >4.5, – Common cause of vaginal discharge – Pain, itch or burn in the vagina, –burning sensation when urinating and -itching around introitus | – A thin watery, clear, or sometimes grey/ greenish coloured discharge, – It coats vaginal wall, Not usually associated with signs of inflammation | Two Criteria: *Amsel's Criteria*: at least 3 out of 4 are present, – Thin white homogenous vaginal discharge, – Clue cells on wet microscopy, – Vaginal pH >4.5, – Whiff of fishy odour. *Hay/lson Criteria: Grade 3 (BV)*- predominantly | – Metronidazole 400 mg orally X twice for 7 days, – Metronidazole 2 g oral single dose, – Intravaginal MetronidazoleGel (0.75%) daily for 5 days, – Intravaginal Clindamycin cream 2% daily for 7 days, | – Females with BV are at increased risk of acquiring STI, – probably implicated in aetiology of STI, – HIV positive women with BV have an increased risk of transmitting HIV, – recurrent infection at 12 months following treatment, – associated with new partner, |

(*Continued*)

Table 6.1 (Continued)

Infection	Signs/symptoms	Examination	Investigation	Treatment	Long-term effects
	– Associated with using vaginal deodorants, smoking and vaginal douches, – A recent change in sexual partner, – And in the presence of an STI.		Gardnerella and/or Mobiluncus morphotypes. Few or absent Lactobacilli, – *DNA-Probe based lab tests*	– Tinidazole 2 g single oral dose, – Clindamycin 300 mg oral X twice for 7 days	– Recurrent BV treatment is challenging and requires longer courses of treatment following consultation with Genitourinary Medicine specialist
Vulvovaginal candidiasis (caused by yeasts of genus *Candida*)—VVC	– Normal flora of female genital tract, – Over 20 candida species that can infect humans, – *C. albicans* is commonest (80%–90%), – around 75% will have at least one lifetime episode of VVC, – 40%–45% will have at least two or more episodes, – Risk of recurrent VVC is 10%–25%	– Vulval itch, – Nonoffensive vaginal discharge, – soreness or burning feeling, – Superficial dyspareunia, – Cyclical symptoms, – *On examination of skin:* – Erythema, fissuring, swelling/oedema, Satellite lesions and excoriation marks. – Nonoffensive and cottage cheese like vaginal discharge, – discharge may be thin or even absent,	– clinical picture, – High vaginal swab of the discharge, – Culture recommended only in recurrent VVC, Reassure that it is not sexually transmitted.	– Avoid local irritants, – soap substitute (emollients), – rule out iron deficiency, – Fluconazole capsule 150 mg-single dose, – Clotrimazole pessary 500 mg vaginally, – Itraconazole 200 mg oral X twice for one day, – recurrent or severe VVC will require long courses of treatment	Increased risks with: – Diabetes, – Steroid treatment, – Immunosuppression, – Pregnancy, – HRT, – Combined oral contraceptive, – Recurrent antibiotics treatment, – Persistence of Candida, – Significant effect on quality of life

Trichomonas Vaginalis (TV): TV is a protozoon	– Found in Urethra in 90% of women infected, and also in vagina and para urethral glands, – Transmitted through sexual intercourse	– Up to 50% may be asymptomatic, – frothy vaginal discharge, – Vulval itching, – Dysuria, – offensive odour, – Occasionally lower abdominal pain or Vulval ulceration. – Vulvitis and vaginitis, strawberry cervix,	– Vaginal swab from posterior fornix, – Self –obtained Vaginal swab, – Urine	– Metronidazole 2 g oral single treatment, – Metronidazole 400 mg BDX 7 days, Tinidazole 2 g oral single dose – Avoid intercourse for one week after completion of treatment,	– May be associate with negative impact on pregnancy, – may enhance HIV transmission,
Anogenital Warts, caused by Human papilloma Virus (HPV)- DNA Virus	– More than 100 genotypes, – 90% caused by HPV types 6 or 11, – Most infections resolve spontaneously, – Median incubation period is 3 months, – High risk HPV found in cervical squamous cell carcinoma are type 16, – Usually transmitted by sexual contact,	– Benign epithelial skin lesions, – Singe/multiple, – "Cauliflower"-like on moist nonhair bearing areas, – "smooth papules" on dry hair skin, – May occur at any genital or perigenital site, – Extra genital lesions can occur in oral cavity, larynx, conjunctiva and nasal cavity. – mostly asymptomatic, – can also cause local irritation, bleeding, discomfort, secondary infection and local tissue infiltration	– Classical appearance, rarely biopsy is required, – colposcopy, – suspicious lesions require biopsy to rule out anogenital intra-epithelial neoplasia	*Nonkeratinised lesions:* – Podophyllotoxin, – Trichloroacetic acid, – Imiquimod. *Keratinised Lesions:* – Cryotherapy, – Excision, – Elecrotcautery, – Imiquimod, *Other options:* – 5 Fluorouracil 5% Cream, – Interferons, – Cidofovir 1% cream, – Laser therapy	Prevention of Warts: HPV Vaccination. – Bivalent (HPV 16, 18), Quadrivalent (HPV 16, 18, 1, 16)

(Continued)

Table 6.1 (Continued)

Infection	Signs/symptoms	Examination	Investigation	Treatment	Long-term effects
Syphilis, caused by Treponema Pallidum	– Transmitted by direct contact with an infectious lesion, mainly via genital route, – vertical transmission during pregnancy, – route of transmission is extra-genital in homosexual sex,	– Congenital (Early in first 2 y, and late- after 2 y), – Early syphilis (Primary, Secondary, Early latent), – Late syphilis (Late latent, and Tertiary). *Primary syphilis:* – Chancre (ulcer)—superficial, single, painless and clean base, – Can be atypical as being multiple, painful and diffuse, – Untreated chancres resolve spontaneously 3–8 weeks later,	*Primary:* Serum from chancre, – Aspiration of regional lymph nodes if chancre is infected – *Secondary and early congenital syphilis:* – Serum from mucosal patches and ulcers, – PCR, – Positive serology around 4 weeks of infection, – VDRL Carbon antigen test/RPR test, – Specific treponemal tests for IgG and IgM, – Anti- treponemal IgM EIA and immunoblot, – Treponemal EIA/CLIA or TPPA, – Rapid treponemal tests, – All these tests have false positive and negatives, – All positive tests should be re-tested	– Offer full sexual screen, – No sexual contact till lesions of early syphilis are treated, – Longer duration of treatment in late syphilis as the bacteria divide slowly, – Parenteral rather than oral treatment, – *Macrolide antibiotics:* *Choices:* – Benzathine penicillin, – Doxycycline, – Azithromycin, – Procaine Penicillin, – Ceftriaxone, – Amoxicillin, – Erythromycin	*Secondary Syphilis:* – Multi system bacteraemia, – widespread rash, – mucus patches, – Condylomata lata, – Periostitis, – Liver, spleen, renal eye, CNS involvement, – *Latent Disease (Tertiary)* – Gummatous disease, – cardiovascular, – Late neurological disease, *Useful websites are:* https://iusti.org/regions/ guidelines https://www.bashh.org/ guidelines https://www.bhiva.org/ guidelines

				Differential diagnosis: #
Molluscum contangiosum (caused by this DNA virus)	– Benign epidermal eruption of skin, – Four subtypes, with MCV 1 is the commonest, – Routine physical contact or occasionally fomites, – Lesions usually affect face and neck, trunks or limbs, – *As STI:* – Affects young adults, – Usually affects genitals, pubic region, lower abdomen, upper thighs, buttocks, – *Severe infection in:* – Immunocompromised, – Late stage HIV,	– Incubation (2–12 weeks), – Smooth surfaced, firm, doom shaped papules, – Usually asymptomatic, – White colour, but can be pink to yellow, – Can get infected, – Appear as clusters, – Rarely affect oral cavity or sole of foot, – Can be widespread in immunocompromised.	– Clinical diagnosis, – Rarely biopsy may be required with atypical presentation	– General advice, – Risk of autoinoculation, – Not to shave/wax genital area, – Not to squeeze the lesion as the central plug is full of infectious material, – Expectant treatment, – Podophyllotoxin 0.5% for genital lesions, – Liquid nitrogen, – Curettage for nonfacial nongenital lesions	– basal cell carcinoma, – cysts, – abscesses, – genital warts, – Keratoacanthoma, – fungal infection
Mycoplasma genitalium	– Self-replicating bacterium, – Lacks a cell wall, not visible by Gram stains, – Infection may persist for months or longer	Majority do not develop a disease, – Transmission is genital-genital/penile-anal contact, – Responsible for nongonococcal urethritis (dysuria), – Can cause post coital bleeding, uterine infection and PID, – painful intermenstrual bleeding	– Vulvo-vaginal swabs, – Endocervical swab, – First void urine sample, – NAATs detects M. genitalium specific DNA/RNA in clinical specimen	– Macrolide antibiotics, – Doxycycline, – Azithromycin, – Moxifloxacin, Patient advice: – Abstain from sexual intercourse for 14 days after start of treatment, – Attend for test of cure, – Partner treatment	Nongonococcal/nonchlamydial urethritis, – Pelvic inflammatory disease, – Muco-purulent cervicitis, – Sexually acquired proctitis, – Effect on fertility (lack of good evidence)

- School-based delivery provides scope for regulation and consistency, and has been shown to be a cost-effective approach.
- Structured sex education programs should be focused on increasing knowledge about sexual and reproductive health-related issues, prevention of sexually transmitted infection, and condom use.
- Sexuality education should focus on the key components:
  - Correct use of different methods of contraception especially around use of condoms, both male and female.
  - The availability of postcoital contraception.
  - The use of various methods of contraception and from where to access them.
  - The effectiveness of long acting reversible hormonal methods of contraception.
  - The risks of acquiring STI following unprotected intercourse with a new partner.
  - Should be advised that using a condom during intercourse helps to prevent STI.
  - Contact tracing and treatment is warranted for the male partner as well.
  - And be aware of signs and symptoms of common STIs for which they should be seeking investigations and treatment (Table 6.1).

Table 6.1 do not include information as regards HIV and Hepatitis B and C.

Reference

1. Averett S., Corman H., Reichman N. Effects of Overweight on Risky Sexual Behaviour of Adolescent Girls. NBER Working Paper 16172, 2010.

Further reading

Bitzer J. Long term contraceptive care in obese and super obese women. In: Obesity and Gynaecology (eds) Mahmood T., Arulkumaran S., Chervenak F.A., Elsevier, 2020, 2nd ed., pp 67−76 <https://doi.org/10.1016/B978-0-12-817919-2.00007-3>.

Cameron S., Cooper M. Obesity and sexual health, (in)Obesity and Gynaecology (eds) Mahmood T., Arulkumaran S., Chervenak F.A., Elsevier, 2020, 2nd ed., pp 37−42. <https://doi.org/10.1016/B978-0-12-817919-2.00004-8>.

Cameron S, Cooper M, Kerr Y, Mahmood T. EBCOG position statement—public health role of sexual health and relationships education. *Eur J Obstet Gynaecol Reprod Biol.* 2019;234:223−234.

Cawley J, Joyner K, Sobal J. Size matters: the influence of adolescents' weight and height on dating and sex. *Ration. Soc.* 2006;18:67−94.

<https://iusti.org/regions/guidelines>.

<https://www.bashh.org/guidelines>.

<https://www.bhiva.org/guidelines>.

Kirshaw TS, Arnold A, Lewis JB, Magriples U, Ickovics JR. The skinny on sexual risk: the effects of BMI on STI incidence and risk. *AIDS Behav.* 2011;15(7):1527−1538.

Uk Medical Eligibility Criteria for Contraceptive Use(UKMEC). Faculty of sexual and reproductive healthcare <http://www.fsrh.org/standards-and-guidance/documents/ukmec-2016/>; 2016.

Obesity: male and female infertility

7

Suresh Kini
Assisted Conception Unit, Department of Obsterics and Gynaecology, Ninewells Hospital and University of Dundee, Dundee, Scotland

7.1 Introduction

- Obesity has become a global epidemic, affecting more than 650 million adults worldwide.
- The prevalence of obesity has consistently increased to the point where more than 35% of adults are now considered to be obese.
- Women are generally more prone to obesity than men possibly because of their lower basal metabolic rate.
- The effect of obesity on reproductive function reflects through complex endocrinological changes resulting from an interaction between the fat compartment and hypothalamic pituitary gonadal axis with an ultimate effect on sex steroids mediated through the effect of circulating adipokines.
- Overweight is defined by the World Health Organization (WHO) as a body mass index (BMI) ≥ 25 kg/m^2 and obesity as ≥ 30 kg/m^2.
- Obesity brings out many problems such as social, psychological, demographic, and long-term health issues.

7.2 Epidemiology

- The prevalence of obesity has increased in developed countries because of a change in lifestyle, including reduced physical activity, changes in nutrition style, and an increased calorie intake.
- Other factors such as endocrine disorders, hormonal disorders, psychological disorders, and use of some drugs such as steroids and antidepressants may lead to obesity.
- Rates of obesity in the United States are significantly higher than in other developed nations.
- The number of obese Americans has doubled since 1960.
- The WHO reported that in the United States and most European countries, 60% of women are overweight (≥ 25 kg/m^2), of these, 30% are obese (≥ 30 kg/m^2) and 6% are morbidly obese (≥ 35 kg/m2).

7.3 Female infertility

- Obesity exerts a negative influence on female fertility.

Handbook of Obesity in Obstetrics and Gynecology. DOI: https://doi.org/10.1016/B978-0-323-89904-8.00006-8

- Obese women are more likely to have ovulatory dysfunction due to dysregulation of the hypothalamic pituitary ovarian (HPO) axis.
- Obese women have reduced fecundity even when eumenorrheic.
- Obesity plays a significant role in reproductive disorders in women.
- It is associated with anovulation, menstrual disorders, infertility, difficulties in assisted reproduction, miscarriage, and adverse pregnancy outcomes.
- *In obese women, gonadotrophin secretion is affected for the following reasons:*
- Increased peripheral aromatisation of androgens to oestrogens,
- The insulin resistance and hyperinsulinaemia in obese women lead to hyperandrogenaemia.
- The sex hormone binding globulin, growth hormone, and insulin-like growth factor binding proteins are decreased, and leptin levels are increased.
- The neuroregulation of the HPO axis deteriorates.
- These alterations may explain impaired ovulatory function and so reproductive health.
- Because of lower implantation and pregnancy rates, higher miscarriage rates, and increased maternal and foetal complications during pregnancy, obese women have a lower chance to give birth to a healthy new-born.

7.3.1   Transgenerational inheritance

- There is a mounting body of evidence suggesting that maternal obesity may confer a risk of metabolic dysfunction through multiple generations.
- Children of obese mothers are more likely to develop obesity, type 2 diabetes, and cardiovascular disease as adults.

7.3.2   Pathophysiological basis of infertility in obese women

- It is complex and multifactorial. Several mechanisms are involved in the relationship between fertility and obesity.
- Adipokines are the signalling molecules (hormones) secreted by adipose tissue, and abnormalities in adipokines can cause inflammation and abnormal cell signalling, and thus can lead to deterioration in cell metabolism and function.
- Some of these adipokines are leptin, adiponectin, interleukin-6 (IL-6), plasminogen activator inhibitor-1 (PAI-1), tumour necrosis factor-α (TNF-α), resistin, visfatin, chemerin, omentin, and ghrelin.
- High levels of leptin seen in obesity may inhibit folliculogenesis. High levels of leptin interfere with endometrial receptivity and embryo implantation. Leptin levels have been found to be positively correlated with insulin resistance in women with polycystic ovarian syndrome (PCOS).
- Adiponectin levels increase with weight loss. In the absence of adiponectin in obese women, plasma insulin levels increase. Consequently, high levels of insulin lead to hyperandrogenaemia.
- IL-6, in the high levels seen in obese women, may contribute to impaired fertility in women with PCOS.
- PAI-1 has been associated with miscarriage in women with PCOS.
- TNF-α may affect several levels of the reproductive axis: inhibition of gonadotrophin secretion, ovulation, steroidogenesis, corpus luteum regression, and endometrial development.
- The mechanism of other adipokines on reproductive functions such as resistin and ghrelin has not been fully understood.

- Increased resistin levels seen in obesity leads to insulin resistance and leads to decreased insulin sensitivity.
- Visfatin, shows insulin-mimetic effects, that increases glucose uptake in adipocytes and muscle cells, and decreases glucose release from hepatocytes.
- Chemerin can impair follicle-stimulating hormone (FSH)-induced follicular steroidogenesis and thus can play a role in the pathogenesis of PCOS.
- Almost all of the adipokines seem to have their effects on reproduction by causing insulin resistance (Table 7.1).

7.3.3   The clinical effects of obesity on female infertility

- Obesity has a negative effect on reproductive potential, primarily thought to be due to functional alteration of the HPO axis.
- Obese women often have higher circulating levels of insulin, which is a known stimulus for increased ovarian androgen production. These androgens are aromatised to oestrogen at high rates in the periphery owing to excess adipose tissue, leading to negative feedback on the HPO axis and affecting gonadotropin production.
- This manifests as menstrual abnormalities and ovulatory dysfunction.
- Obese women with a BMI > 27 kg/m^2 have a relative risk of anovulatory infertility of 3.1 (95% CI, 2.2−4.4) compared with their lean counterparts with a BMI 20.0−24.9 kg/m^2.
- Obese women have a lower chance of conception within 1 year of stopping contraception compared with normal-weight women (i.e., 66.4% of obese women conceive within 12 months, compared with 81.4% of those of normal weight).
- Multiple studies have demonstrated that obese women have increased time to pregnancy.
- Obese women remain subfertile even in the absence of ovulatory dysfunction. Study showed reduced fecundity in eumenorrheic obese women and the probability of spontaneous conception declined linearly with each BMI point > 29 kg/m^2.

Table 7.1 The effects of the adipokines on reproduction.

Adipokines	Effects on reproduction in obesity	Serum levels in obesity
Leptin	Inhibits insulin-induced ovarian steroidogenesis	Increases
	Inhibits LH stimulated oestradiol production by the granulose cells	
Adiponectin	Plasma insulin levels increase	Decreases
IL-6	Causes insulin resistance	Increases
PA1−1	Causes insulin resistance	Increases
TNF-α	Impairs insulin action-hyperinsulinaemia	Increases
	Inhibits insulin signalling	
Resistin	Causes insulin resistance	Increases
Visfatin	Increased insulin sensitivity	Increases
Omentin	Increased insulin sensitivity	Decreases
Chemerin	Negatively regulates FSH-induced follicular steroidogenesis	Increases

FSH, Follicle-stimulating hormone; *IL-6*, Interleukin-6; *LH*, Luteinising hormone; *PA1−1*, Plasminogen Activator Inhibitor Type-1; *TNF-α*, tumour necrosis factor-α.

- In normogonadotropic anovulatory women, increased BMI and abdominal obesity are associated with decreased odd ratios of ovulation in response to clomiphene citrate.

7.3.4   Effects on the oocyte

- In women with obesity, there is high rates of meiotic aneuploidy with fragmented disorganised meiotic spindles and improper alignment of chromosomes.
- Obesity appears to disrupt the mitochondrial function in the oocyte.
- There is also evidence of endoplasmic reticulum (ER) stress in the obese state.
- The continued dietary excess of fatty acids accumulates in the tissues and exerts toxic effects, which is termed lipotoxicity.
- Obese women have higher levels of reactive oxygen species (ROS) that induce mitochondrial and ER stress leading to apoptosis.
- Lipotoxicity plays a role in the development of insulin resistance and a heightened inflammatory state.
- Obesity is considered to be a chronic low-grade inflammatory state.
- Obese women have higher circulating levels of C-reactive protein, a marker of systemic inflammation.
- The developing blastocyst produces adiponectin, IL-1, and IL-6. The altered inflammatory milieu in obese women likely exerts an influence on follicle rupture at the time of ovulation and invasion of the trophoblast into the receptive endometrium.
- This effect of obesity at the level of the oocyte could have downstream effects on endometrial receptivity and embryo implantation.

7.3.5   Effects on the embryo

- Obese women are more likely to create poor-quality embryos.
- Embryos may also be susceptible to lipotoxicity.
- Elevated leptin levels in obese women may exert a direct negative effect on the developing embryo.

7.3.6   Effect on the endometrium

- Some studies suggest that decidualisation defects seen in obesity may contribute to compromised endometrial receptivity and poor implantation and may negatively affect the placentation process.
- Many of the pregnancy complications seen in obese women are linked to placental dysfunction.
- Studies found significantly higher spontaneous abortion rates in obese women.
- In women with a history of recurrent pregnancy loss, obesity is a known risk factor for miscarriage in a subsequent pregnancy.
- Chronic dysregulation of leptin pathways in obesity may negatively affect implantation.

7.3.7   Challenges of managing obese women

- Overweight and obese subfertile women have a reduced probability of successful fertility treatment.

- Examination: The utility of the clinical examination is often limited in the obese woman, which results in a greater reliance on imaging.
- Obese patients have difficult venous access.

7.3.8 Funding

- Around the world, fertility treatment is withheld from women above a certain BMI, with a threshold ranging from 25 to 40 kg/m^2.
- The proponents of this policy use three different arguments to justify their restrictions: risks for the woman, health and wellbeing of the future child, and importance of society.
- The opponents feel that the obese women should be informed about the consequences and encouraged to lose weight. If, however, they are unable to lose weight despite effort, there should not be any argument to withhold their treatment.
- Based on available evidence, it may be appropriate to consider morbid obesity as a contra-indication for public-funded treatment where the aim is to maximise the value for money.

7.3.9 Pelvic ultrasound

- Obesity can contribute to missed diagnoses, nondiagnostic results, imaging examination cancellation because of weight or girth restrictions.
- Recognition of equipment limitations, imaging artefacts, optimisation techniques, and appropriateness of modality choices is critical to providing good patient care.

7.3.10 Ovulation induction

- In ovulatory but subfertile women, the chance of spontaneous conception decreases by 5% for each unit increase in the BMI.
- National guidelines in the United Kingdom advise for a weight loss to a BMI of < 30 kg/m^2 prior to the start of any ovulation induction.
- Overweight and obese women respond poorly to clomiphene induction and require higher doses of gonadotropins for superovulation.

7.3.11 Tubal investigations

- Operators have encountered difficulty in completing the Hysterosalpingo Contrast Sonography procedures in obese participants, when the uterus was acutely retroverted or oblique, when loops of active bowel were present, or the adnexa were located beyond the penetration of the ultrasound signal.
- Significant technical difficulty and increased radiation exposure have been associated with hysterosalpingography.
- Laparoscopy and dye is not contraindicated in obese patients. Despite being associated with increased operating times, complication rates in obese patients are comparable to their nonobese counterparts. However, these procedures should be performed by a skilled surgeon in a special hospital setting.

7.3.12   Ovarian reserve

* The systematic review and metaanalysis suggest that antimullerian hormone and FSH are significantly lower in obese women and are inversely correlated with BMI.
* There is limited number of published studies for any evidence supporting an association between BMI and inhibin B.
* Antral follicle count does not appear to differ according to BMI.

7.4   Treatment options

* Obese women wishing to conceive should consider a weight management programme that focuses on preconception weight loss (to a BMI $< 35 \text{ kg/m}^2$), prevention of excess weight gain in pregnancy, and long-term weight reduction.
* Weight management in all individuals is best achieved through a lifestyle modification programme that combines dietary modification, physical activity, and behavioural interventions.

7.4.1   Weight loss

* With a goal of 10% weight loss, some studies had significantly higher conception rates and live birth rates (LBRs).
* Other studies in obese women with PCOS, demonstrating improved ovulation and LBRs following lifestyle intervention and weight loss.

7.4.2   Physical activity

* Current recommendations are to increase physical activity to at least 150 minutes weekly of moderate activity such as walking.
* Physical activity has been shown to decrease systemic inflammatory mediators that may contribute to the improvement in fertility.

7.4.3   Barriers to weight loss in infertile women

* The perceptions that exercise can cause fatigue and it is a hard work as well as depression seem to decrease with continuation of an exercise programme in overweight infertile women.
* Coached sessions of achievable frequency, for example, weekly, for up to 6 months should be considered to increase compliance.
* Motivational interviewing techniques might also be useful. Dual enrolment may result in better adherence as partners tend to motivate each other.

7.4.4   Dietary factors

* A 500−1000 kcal/day decrease from usual dietary intake should lead to a 1- to 2-lb weight loss per week.
* With a low-calorie diet of 1000−1200 kcal/day, achieving an average 10% decrease in total body weight over 6 months.

- Adherence to 'Mediterranean' diet, characterised by higher intake of unsaturated fats, lower intake of animal fats, and lower ratios of omega-6 to omega-3 fatty acids for 2 years in patients with metabolic syndrome significantly decreases insulin resistance and serum concentrations of inflammatory markers.
- Studies have been published extensively on the 'fertility diet', characterised by less consumption of trans fats and animal protein and more consumption of low-glycaemic carbohydrates, high-fat dairy, and multivitamins.

7.4.5 Bariatric surgery

- Bariatric surgery in women can restore menstrual regularity, correct ovulation, shorten folliculogenesis with ovulation, reduce serum testosterone levels, diminish percent body fat, and improve both sexual function and chance of pregnancy.
- Surgically induced weight loss only partially improves deficient luteal progesterone production with a rise in LH secretion, suggesting the persistence of corpus luteum dysfunction.
- Delaying pregnancy until $1-2$ years after bariatric surgery has been recommended to avoid foetal exposure to nutritional deficiencies from rapid maternal weight loss.
- A study examining pregnancy outcomes after bariatric surgery demonstrated lower risk of gestational diabetes and large-for-gestational-age infants. However, it also showed a concerning increased risk of small-for-gestational-age infants and a trend toward higher risks of stillbirth and neonatal death.
- Bariatric surgery can be recommended for the women with a BMI of ≥ 35 kg/m^2.
- Bariatric surgery does appear to improve the PCOS phenotype. Metabolic parameters, including insulin sensitivity and blood pressure, were also improved.

7.4.6 PolyCystic Ovarian Syndrome (PCOS) women

- Women with PCOS are commonly ($35\%-80\%$) overweight (BMI > 25 kg/m^2) or obese (BMI > 30 kg/m^2).
- Women in the United States with PCOS have a higher BMI than their European counterparts. Obesity may intensify the severity of the phenotypic characteristics of the PCOS, including disturbed menstrual cycle.
- The severity of hyperandrogenism seems to be amplified in obese women with PCOS.
- In obese women with PCOS, insulin resistance and hyperinsulinaemia are higher than in lean women with PCOS.
- In the case of noncompliance, various treatments or interventions, including clomiphene citrate, gonadotrophins, insulin sensitizers, and laparoscopic ovarian drilling, are applied.

7.4.7 Types of weight-loss medication

- Oral medications used included orlistat, a lipase inhibitor; sibutramine, a selective serotonin and norepinephrine reuptake inhibitor; and acarbose, an alpha-glucosidase inhibitor shown to induce modest weight loss.
- Metformin: the metaanalyses showed that weight-loss interventions have a nonsignificant advantage over metformin with respect to achievement of pregnancy or improvement of ovulation status.

- Herbal: many obese women may also self-medicate with herbal supplements, although their safety and effectiveness have not been demonstrated.
- Ephedra containing supplements have potentially life-threatening cardiovascular side effects and have been banned by the FDA.

7.5   Male infertility

- With the ever-increasing incidence of obesity, it is important to be aware of the adverse impact of obesity on male fertility.
- There are now several population-based studies showing that overweight and obese men have up to 50% higher rate of subfertility when compared with normal-weight men.
- Observational studies have shown associations between male obesity and a variety of sperm parameters, including concentration, motility, abnormal morphology, and DNA damage.
- However, the association between obesity and sperm parameters is not conclusive, with some showing this association only in a proportion of men with severe obesity.
- Endocrine abnormalities (including increased plasma levels of oestrogen, leptin, insulin resistance, and reduced androgens and inhibin B levels) are likely to be important in the aetiology of sperm dysfunction in obese men.
- Other factors may also contribute, including genetic abnormalities, sexual dysfunction, testicular hyperthermia, and ROS.
- There are two main mechanisms by which ROS can affect sperm function: DNA damage resulting in defective paternal DNA being passed onto children and sperm membrane damage resulting in decreased motility and ability to fuse with the oocyte.
- The primary management must be to achieve weight reduction, using a reduced calorie diet in combination with an exercise programme.
- Such regimes are difficult for patients to follow, and considerable psychological support and active monitoring is required.
- In extreme cases, bariatric surgery can be considered, although, at present, there is no long-term data on semen analysis or fertility outcomes following surgery.
- More specific treatments to correct endocrine abnormalities associated with obesity are being evaluated, but disappointingly, no effective treatment has yet been proven.
- It is perhaps not surprising that male BMI is not usually used to restrict access to fertility treatment.
- In those couples that require assisted conception treatment, it could be argued that ICSI should be offered when there is male obesity, to overcome the negative effects demonstrated on sperm function.

7.6   Conclusion

- Overweight and obese women seeking fertility should be educated on the effects of being overweight on the ability to achieve pregnancy and the benefits of weight reduction, including improvement in pregnancy rates, and a reduced need for ovulation induction and assisted conception.
- A combination of a reduced calorie diet, which is not overly restrictive, and aerobic exercise, intensified gradually, should form the basis of programmes designed for such individuals.

- Lifestyle interventions in women and men should still be considered the first line therapy, with drug use reserved for monitored trials.
- Clinicians while taking care of obese men with infertility should recognise that it is likely that the obesity contributes to the reproductive dysfunction.
- While weight loss can restore normal testicular endocrine function, the value of weight reduction to improve male fertility remains very uncertain.

Further reading

Belan M, Harnois- Leblanc S, Laferrere B, et al. Optimizing reproductive health in women with obesity and infertility. *CMAJ*. 2018;190(24):742−745.

Mahmood TA, Arulkumaran S, Chervenak FA. *Obesity and Gynecology*. 2nd ed. Oxford: Elsevier; 2020.

Practice Guideline. Obesity and reproduction: a committee opinion. *Fertil Steril*. 2015;104 (5):1116−1126.

Obesity—recurrent miscarriage 8

Mythili Ramalingam
Assisted Conception Unit, Department of Obstetrics and Gynaecology, Ninewells Hospital and University of Dundee, Dundee, Scotland

8.1 Introduction

Recurrent miscarriage (RM) is defined as loss of three or more consecutive pregnancies prior to 20 weeks of gestation, though some authors describe RM as two or more consecutive pregnancy losses.

The incidence of RM has been reported to be around 0.5%−2.3% and is rising in relation to the prevalence of obesity. It has been estimated that 1%−2% of second-trimester pregnancies miscarry before 24 weeks of gestation.

Approximately in 50% of couples with RM, the underlying cause remains unexplained after all investigations. Though this is reassuring it can be distressing for the couple.

In the United Kingdom, obesity affects one-fifth of the female population. Maternal obesity has been reported as a risk factor for adulthood obesity in offspring. Obesity may also lead to a poor pregnancy outcome like as sudden and unexplained intrauterine death.

Increased body mass index (BMI) has also been suggested as the second most significant factor predicting early pregnancy loss after advanced female age.

8.2 Obesity and miscarriage

Obesity is associated with low levels of adiponectin but raised levels of leptin in both serum and follicular fluid.

Higher levels of leptin can impair ovarian steroidogenesis. Low levels of adiponectin lead to higher levels of serum insulin levels which is one of the factors that increases levels of circulating androgens.

Therefore, obesity itself creates quite a hostile biochemical environment for the early stages of the developing gamete and pregnancy.

8.3 Obesity and recurrent miscarriage

A recent systematic review and meta-analysis has reported a significant association between excess weight and RM, independent of age as a risk factor.

Handbook of Obesity in Obstetrics and Gynecology. DOI: https://doi.org/10.1016/B978-0-323-89904-8.00016-0

The exact mechanism by which obesity increases the risk of miscarriage and RM is still not very clear. It is unlikely that the increased risk of miscarriage among obese patients can solely be attributed to Polycystic ovarian syndrome (PCOS) because of the low prevalence of PCOS in the general population with mostly spontaneous miscarriage.

The possible theories for the association between obesity and RM include the effect of obesity on endometrial development and the effect on oocyte quality.

8.4   Aetiologies of recurrent miscarriage

Genetic

1. Embryonic chromosomal abnormalities
2. Parental balanced reciprocal translocations
3. Sperm DNA fragmentation
4. Maternal age

Thrombotic

1. Hereditary thrombophilia
2. Antiphospholipid syndrome
3. Alloimmunity

Uterine factors

1. Congenital uterine abnormalities
2. Acquired uterine anomalies (uterine fibroids)
3. Cervical incompetence

Endometrial

1. Endometrial receptivity disorders
2. Luteal phase defect
3. Decidualisation defects

Hormonal

1. Hypothyroidism
2. Diabetes mellitus
3. Hyperprolactinaemia
4. PCO

Metabolic

1. Obesity
2. Metabolic syndrome of obesity

Environmental

1. Excessive smoking
2. Caffeine consumption
3. Cocaine use
4. Heavy alcohol consumption

8.5 Polycystic ovarian syndrome

PCOS is the most common endocrine disorder in women of reproductive age. It is associated with menstrual irregularity, ovarian dysfunction, and symptoms of hyperandrogenism. Approximately half of women with PCOS are obese.

Polycystic ovary syndrome has been linked to an increased risk of miscarriage but the exact mechanism remains unclear. Polycystic ovarian morphology, elevated serum luteinising hormone levels, or elevated serum testosterone levels, although used as markers of PCOS, do not predict an increased risk of future pregnancy loss among ovulatory women with a history of RM who conceive spontaneously. The increased risk of miscarriage in women with PCOS has been recently attributed to insulin resistance (IR), hyperinsulinaemia, and hyperandrogenaemia. The prevalence of IR is increased in women with RM compared with matched fertile controls. An elevated free androgen index (FAI) appears to be a prognostic factor for a subsequent miscarriage in women with RM.

A combination of high serum levels of androgens (testosterone, dehydroepiandrosterone sulfate, androstenedione) including a high peripheral IR and compensatory hyperinsulinaemia. The raised leptin levels have been reported in obese women, causing a detrimental effect on ovarian steroidogenesis.

There is a possible negative association between FAI and oocyte quality or fertilisation in women with PCOS.

Hyperinsulinaemia is associated with increased levels of plasminogen activator inhibitor 1 (PAI-1) that is strongly associated with an increased risk of miscarriage and RM. Also hyperinsulinaemia by itself is a significant independent risk factor for miscarriage and also is believed to play a key role in implantation failure by suppression of circulating glycodelin and insulin-like growth factor binding protein.

8.6 Ovarian dysfunction

Studies comparing the outcome of assisted reproduction in obese women with non-obese women found that obese women have a significantly lower number of mature oocytes and oocytes with reduced diameter.

Overweight women's embryos also had a lower potential for development and IVF and a lower rate of blastocyst formation.

Intrafollicular human chorionic gonadotrophin concentration appears to be inversely related to BMI and may be related to concurrent decrease in embryo quality and rates of pregnancy.

8.7 Endometrial changes in obesity

Implantation of the embryo and a successful pregnancy require a receptive endometrium and obesity may affect the endometrium or its environment causing implantation failure and pregnancy losses.

The precise impact of obesity on the molecular and histopathological aspects of the endometrium is not fully understood.

Progesterone induces secretory changes in the lining of the uterus, which is needed for implantation of the embryo. It has been suggested that a causative factor in many cases of miscarriage may be inadequate secretion of progesterone.

8.8 Immunological factors

The hypotheses of human leucocyte antigen incompatibility between couples, the absence of maternal leucocytotoxic antibodies, or the absence of maternal blocking antibodies do not have clear evidence and therefore should not be offered routinely in the investigation of couples with RM.

Natural killer (NK) cells are found in peripheral blood and the uterine mucosa. Peripheral blood NK cells are phenotypically and functionally different from uterine NK (uNK) cells. There is no clear evidence that altered peripheral blood NK cells are related to RM. The uNK cells which have been investigated as playing a role in later stages in implantation may in turn result in miscarriage. Studies of peripheral blood and endometrial uNK cells have suggested that they may play a role, but it is unclear at present.

Therefore, testing for peripheral blood NK cells as a surrogate marker of the events at the maternal–foetal interface is inappropriate and should not be offered routinely in the investigation of couples with RM.

There is a suggestion that uNK cells may play a role in trophoblastic invasion and angiogenesis in addition to the local maternal immune response to pathogens. The largest study examining the relationship between uNK cell numbers and future pregnancy outcome reported that raised uNK cell numbers in women with RM was not associated with an increased risk of miscarriage.

This remains a research field and testing for uNK cells should not be offered routinely in the investigation of RM.

Cytokines are immune molecules that control both immune and other cells. Cytokine responses are generally characterised either as T-helper-1 (Th-1) type, with production of the proinflammatory cytokines interleukin 2, interferon, and tumour necrosis factor (TNF) alpha, or as T-helper-2 (Th-2) type, with production of the anti-inflammatory cytokines interleukins 4, 6, and 10. It has been suggested that normal pregnancy might be the result of a predominantly Th-2 cytokine response, whereas women with RM have a bias towards mounting a Th-1 cytokine response. A meta-analysis concluded that the available data are not consistent with more than modest associations between cytokine polymorphisms and RM. Further research is required to assess the contribution that disordered cytokines make to RM before routine cytokine tests can be introduced to clinical practice.

Obesity is associated with chronic inflammation. Women with idiopathic RM and some other obstetric complications are known to have higher levels of inflammatory markers (e.g., IL-6 and CRP). Chronic inflammation, therefore, could be a mechanism contributing to higher risk of RM in obese women.

8.9   Management of recurrent miscarriage

Couples with RM should be informed that maternal obesity or being significantly underweight is associated with obstetric complications and could have a negative impact on their chances of a live birth and on their general health.

Gradual weight loss has been shown to improve fertility and the outcomes of fertility treatments. Striving for a healthy normal range BMI is recommended.

The management of obesity is important because exercise and diet modification are of low cost.

Alternatively, bariatric surgery is increasingly offered to obese patients with comorbidities.

The increased risk of miscarriage in women with PCOS has been attributed to IR and hyperinsulinaemia. Weight loss in obese women with PCOS, through protein-rich, very-low-calorie diet, has been shown to significantly reduce serum fasting glucose and insulin, improve insulin sensitivity, and decrease PAI-1 activity.

Insulin-sensitising agents such as metformin have been used in the treatment of obese PCOS women who have shown improvements in hyperinsulinaemia and hyperandrogenism. Metformin use appears to be safe in the first trimester. Metformin has been found to reduce miscarriage rate in women with PCOS but these are based on uncontrolled small studies.

A meta-analysis of 17 randomised controlled trials of metformin, an insulin-sensitising agent, in women with PCOS and infertility showed that metformin has no effect on the sporadic miscarriage risk when administered pre pregnancy. However, there are no randomised controlled trials to assess the role of metformin in women with RM.

Progesterone is necessary for successful implantation and the maintenance of pregnancy. This benefit of progesterone could be explained by its immunomodulatory actions in inducing a pregnancy-protective shift from proinflammatory Th-1 cytokine responses to a more favourable antiinflammatory Th-2 cytokine response. A meta-analysis to assess progesterone support for pregnancy showed that it did not reduce the sporadic miscarriage rate. However, in a subgroup analysis of trials involving women with RM, progesterone treatment offered a statistically significant decrease in miscarriage rate compared with placebo or no treatment. However, this meta-analysis was based on three small controlled studies, none of which detected a significant improvement in pregnancy outcome.

PROMISE studied 836 women with unexplained RMs at 45 hospitals in the United Kingdom and the Netherlands, and found a 3% higher live birth rate with progesterone. PRISM studied 4153 women with early pregnancy bleeding at 48 hospitals in the United Kingdom and found there was a 5% increase in the number of babies born to those who were given progesterone who had previously had one or more miscarriages, compared to those given a placebo. The benefit was even greater for the women who had previous "recurrent miscarriages" (i.e., three or more miscarriages)—with a 15% increase in the live birth rate in the progesterone group compared to the placebo group. This might suggest that we could potentially use additional progesterone support in a small selected group of patients.

A recent study concluded that for women with unexplained RM, supplementation with progestogen therapy may reduce the rate of miscarriage in subsequent pregnancies. The PRISM trial has shown that for women who had three or more miscarriages, there was a 15% increase in births for those who were given progesterone compared to those who were given a placebo.

Immunotherapy is expensive and has potentially serious adverse effects including transfusion reaction, anaphylactic shock, and hepatitis. A Cochrane systematic review has shown that the use of various forms of immunotherapy, including paternal cell immunisation, third-party donor leucocytes, trophoblast membranes, and intravenous immunoglobulin, in women with unexplained RM provides no significant beneficial effect over placebo in preventing further miscarriage. A 2010 meta-analysis confirmed this conclusion with respect to intravenous immunoglobulin. The use of immunotherapy should no longer be offered to women with unexplained RM.

Anti-TNF agents could potentially cause serious morbidity including lymphoma, granulomatous disease such as tuberculosis, demyelinating disease, congestive heart failure, and syndromes similar to systemic lupus erythematosus. There are no published data on the use of anti-TNF agents to improve pregnancy outcome in women with RM. Therefore, these should not be offered routinely to women with RM outside formal research studies.

8.10   Conclusion

There is a positive correlation between obesity and RM based on current evidence. Exact mechanisms are uncertain. Weight loss would seem prudent for obese women contemplating pregnancy after RM. There is limited other specific advice or treatment for women at present based on current evidence.

Further reading

Boots, Stephenson. Does Obesity Increase the Risk of Miscarriage in Spontaneous Conception: A Systematic Review, 2011.

Boots et al. Frequency of Euploid Miscarriage is Increased in Obese Women with Recurrent Early Pregnancy Loss, 2014.

European Society of Human Reproduction and Embryology. Recurrent Pregnancy Loss. Guideline, 2017.

Lashen et al. Obesity is Associated with Increased Risk of First Trimester and Recurrent Miscarriage: Matched Case-Control Study, 2004.

Lo et al. The Effect of Body Mass Index on the Outcome of Pregnancy in Women With Recurrent Miscarriage, 2012.

Mahmood TA, Arulkumaran S, Chervenak FA. *Obesity and Gynecology*. 2nd ed. Oxford: Elsevier; 2020.

Zhang et al. Risk Factors for Unexplained Recurrent Spontaneous Abortion in a Population From Southern China, 2010.

Obesity and assisted conceptionion

9

Mythili Ramalingam
Assisted Conception Unit, Department of Obstetrics and Gynaecology, Ninewells Hospital and University of Dundee, Dundee, Scotland

9.1 Introduction

The World Health Organization defines obesity as abnormal or excessive fat accumulation that impairs health. Obesity is most commonly defined using body mass index (BMI). The percentage of obese women (BMI ≥ 30) in the United Kingdom has increased from 16.4% in 1993 to 23.8% in 2004. In the reproductive age group, a third of the women are overweight (BMI 25−30) and one in five are obese.

Overweight women are known to be at a higher risk of menstrual dysfunction and ovulatory problems. This is due to altered secretion of pulsatile GnRH, resulting in altered SHBG, ovarian/adrenal androgens, and luteinising hormone (LH). Obese women are more likely to experience reproductive problems and therefore seek assisted conception treatment.

In women undergoing assisted reproduction techniques (ART), obesity has been associated with the need for higher doses of gonadotrophins, increased cycle cancellation rates, and fewer oocytes retrieved. Lower rates of embryo transfer (ET), pregnancy and live birth have also been reported, as have higher miscarriage rates. However, other studies have been unable to find any negative impact of obesity on ART outcome. A recent survey of assisted reproduction clinics in the United Kingdom demonstrates a wide variation in their approach towards obese infertile women. Obstetrics data suggest that maternal and foetal risks increase in the obese individuals.

The main cause of infertility in the obese relates to disturbances in ovulation. These, for the most part, can be resolved with a combined approach involving weight reduction strategies together with pharmacologically induced ovulation induction. Refractory dysovulation occurs with greater frequency in the obese, and for them, the use of ART has to be considered. ART, specifically IVF, will address the issues of egg and sperm availability, as well as tubal infertility for the obese, just as it does for the general infertile population. There is no definitive evidence that unexplained infertility occurs with greater frequency in the obese; though given the abovementioned remarks with respect to oocyte, embryo, and endometrial factors, one might have expected this to be the case. The practical issues that arise through the use of these techniques in overweight women need to be considered carefully.

Handbook of Obesity in Obstetrics and Gynecology. DOI: https://doi.org/10.1016/B978-0-323-89904-8.00032-9

9.1.1   Cycle effects

The impact of obesity on reproductive function is complex. The association between obesity and anovulation is well established. Obesity induces a series of hormonal changes of insulin resistance, hyperinsulinemia, low sex hormone- binding globulin, elevated androgens, increased peripheral conversion of androgens to oestrogens, increased free insulin- like growth factor 1, and high leptin. The combined effect of these changes causes hypothalamic dysfunction, aberrant gonadotropin secretion, reduced folliculogenesis, and lower luteal progesterone levels resulting in anovulation.

Increased levels of serum and follicular fluid leptin are described with increasing BMI. High levels of leptin impair follicular development and reduce ovarian steroidogenesis through direct effects on theca and granulose cells. There is also an inverse relationship of increasing BMI with reduced serum adiponectin levels. The low adiponectin levels are associated with elevated serum insulin levels, which increase circulating androgen levels in part linked to a reduction in the production of sex hormone binding globulin by the liver.

The trend to hyperandrogenism in the obese is also contributed by IGF-1-mediated effects on LH-induced steroidogenesis by theca cells. Enhanced androgen production causes granulosa cell apoptosis with direct consequences for follicle function. The increased availability of androgens for peripheral conversion to oestrogens in adipose tissue has pituitary effects with impaired FSH production affecting the ovarian follicular development. The clinical manifestations of the biochemical disturbances described include anovulatory cycles and subfertility. Ovarian dysregulation associated with hyperandrogenism, insulin resistance, menstrual irregularity, and infertility is commonly found in women with polycystic ovarian syndrome, many of whom are obese.

However, even in ovulatory women, obesity appears to inhibit natural fecundity and prolong the time to conception. A Dutch study showed women with a BMI of 35 had a 26% lower likelihood of spontaneous pregnancy, and women with a BMI of 40 had a 43% lower likelihood of spontaneous pregnancy than women with a BMI between 21 and 29.

9.1.2   Effects on the oocyte

A number of studies have suggested that oocyte yield after stimulation for IVF may be affected in the obese. Quantitative effects have been described where increased doses of gonadotrophins are required to elicit an ovarian response, and the ultimate yield of cumulus—oocyte complexes may be less than in normal weight controls. This may be linked to disturbances in leptin production or sensitivity as described earlier. Some studies have suggested that fertilisation rates of oocytes retrieved may be impaired in the obese, but this observation has not been consistent. Prospective studies are needed to clarify this issue. The observation of increased risks of miscarriage in the obese after IVF has been attributed by some to qualitative effects on oocytes leading to aberrant embryo development.

9.1.2.1 Effects on embryos

As with oocytes, the literature is not consistent with respect to the effects of obesity on embryonic development. Some studies have suggested that markers of embryo quality differ in the obese. Furthermore, there may be less available surplus embryos for cryostorage potentially having an impact on cumulative pregnancy rates per episode of ovarian stimulation. Some have suggested that these observed effects are unreliable since studies may not have taken into account potential confounders such as age, parity, and duration of infertility. Further work is required to inform this controversial debate.

9.1.2.2 Effects on the endometrium

Obesity may also alter the endometrium. There is evidence of altered endometrial gene expression during the implantation window of natural cycles in obese women. Similarly, there is evidence of lower implantation and clinical pregnancy rates in obese donor egg recipients.

There is an increase in miscarriage rate in the obese both in natural conception and that associated with infertility treatment. Specific to IVF, a 50% increased risk of miscarriage in women with a BMI of 30 kg/m^2 has been described. While embryo quality will be an important determinant of implantation potential, studies using an egg donation model suggest that endometrial factors are likely to be involved in this phenomenon as well. The precise mechanism is not understood but ovarian steroid regulation of endometrial development, perturbations in inflammatory and coagulation pathways, perhaps linked to insulin resistance, have been suggested to be involved.

9.1.3 Patient selection

The selection of which patients to treat, and in whom treatment should be deferred until weight loss is achieved, should ideally depend on age, tests of ovarian reserve, and the presence of comorbidities. If tests of ovarian reserve, which might include age, serum anti-Müllerian hormone (AMH), and/or antral follicle count, suggest that there is good ovarian reserve with no other comorbidities, then it is appropriate to defer treatment up until the desired BMI is obtained. However, if there is evidence of ovarian aging, there is a limited time for weight loss. In these circumstances, it might be wiser to proceed with the treatment. Despite this, healthcare professionals have a duty of care not just to the patient but also to the potential child, and treatment should arguably not be provided if there are significant obstetric and perinatal risks such as in cases of extreme morbid obesity. There are data showing that levels of AMH are reduced and hence the egg number in obese women.

9.1.4 Stimulation regime

Obesity has been associated with a reduced response to gonadotropins. In a large retrospective cohort study gonadotropin/intrauterine insemination (IUI) cycles, BMI over 30 was associated with significantly higher gonadotropin requirements, prolonged gonadotropin stimulation, lower peak oestradiol levels, and fewer large and medium size follicles. The reduced responsiveness of obese women to gonadotropins is likely due to the increased volume of distribution.

In IVF patients, female obesity is associated with increased gonadotropin requirements (both increases starting dose and duration of gonadotropins), higher cycle cancellation rates, decreased peak oestradiol levels, and decreased oocyte yield. However, there has been little consensus regarding the impact of female obesity on IVF success rates. Some studies have reported reductions of clinical pregnancy and live birth rates on the order of 15%–30% in obese women undergoing IVF compared to nonobese controls. Other studies have reported reductions in clinical pregnancy and live birth rates of more than 50%. In contrast, at least nine studies have reported no discernible impact of female obesity on IVF pregnancy.

As alluded to the abovementioned fact, there are observational data suggesting that the requirement of gonadotrophins is increased by at least 20% if BMI is $>30 \text{ kg/m}^2$. Chong et al. demonstrated that patients who have normal or $<10\%$ ideal body weight (IBW) are more likely to respond to lower doses of hMG than patients whose weight is 10% above IBW and, in particular, those who are 25% above their IBW. A high BMI was associated with a higher FSH threshold dose. This observation is supported by findings that the total dose of gonadotrophins needed to induce ovulation is increased in parallel with body weight. Why heavier women may need more hormones to induce ovulation or for controlled ovarian hyperstimulation is not clear. It may be related to the greater amount of body surface, inadequate oestradiol metabolism, and decreased sex hormone binding globulin. Also, the intramuscular absorption of the drug may be slower and incomplete in obese patients because of increased subcutaneous fat or fat infiltration of the muscle.

The effect of FSH at the ovarian level is dependent on plasma concentrations of the hormone. This, in turn, is influenced not just by the dose administered but also by endogenous FSH secretion, metabolic clearance rate, and the volume of distribution, which are individual and differ from woman to woman and are influenced by BMI. Elimination of FSH is carried out largely by the kidneys and the liver. The clearance rate is dependent on filtration, secretion, and reabsorption. The extent to which a drug is bound to plasma proteins also determines the fraction of drug extracted by the eliminating organs, which, in turn, is dependent on BMI and weight.

However, there is no randomised controlled trial in the literature testing the hypothesis that increasing the dose of gonadotrophins in obese women improves the live birth rates.

There is no evidence to suggest that one regime of pituitary suppression (agonist or antagonist) is better in obese women compared to those with normal BMI.

9.1.5 Monitoring of stimulation

While there are no data in the literature quantifying differences of monitoring in those with higher BMI, it is accepted generally that the performance and interpretation of ultrasound scans can be difficult in the obese. Theoretically, were oestradiol to be used in monitoring response to stimulation, the levels might be expected to differ in the obese from those with a normal BMI. However, there is no evidence to suggest that with overweight patients, it is advantageous to use both ultrasound and oestradiol in monitoring stimulated cycles

9.1.6 Egg collection

Clinical staff will be sensitive to the challenges which the care of women with high BMI undergoing surgical procedures present. While there is no evidence from the literature that there are more problems in caring for those who are obese, this is probably because most units are not treating morbidly obese women. That said, obese women will require a higher dose of sedation, due to increased surface area, which potentially may lead to a higher risk of exposure to the side effects of the drugs utilised, but in the absence of any published data in the literature, the perceived increase in risk remains theoretical.

Obesity complicates the delivery of assisted reproductive technologies. In obese women undergoing controlled ovarian hyperstimulation, the ovaries may shift to a higher position in the pelvis, making them more difficult to visualise with transvaginal scanning and increasing the risk of complications with oocyte retrieval such as bleeding, infection, and injury to surrounding tissue. In addition, the risks of providing anaesthesia to obese patients is well described, and makes management of these patients through nonhospital centres a challenge. In a recent survey of obesity policies at IVF facilities in the United States, 62% of respondents cited anaesthesia concerns as the primary reason for their BMI cut-off.

9.1.7 Embryo transfer

For the most part the procedure of ET is simple. However, with moves toward ultrasound-guided (USG) ET that may involve the use of abdominal ultrasound, USG-guided ET will be difficult in obese women due to poor views. Whether this would lead to lower pregnancy rate remains unknown as there are no data in the literature to explore either difficulties with the procedure or lower pregnancy rates.

9.1.8 hCG trigger

Theoretically, bioactive levels of hCG used for the ovulatory trigger will be less in obese women. However, as long as more than equivalent of 1000 IU of recombinant hCG is given as the ovulatory trigger, oocyte fertilisation rates and luteal function are unlikely to be influenced by differences in bioavailable gonadotrophin. Most ovulatory triggered preparations now contain at least 6500 IU of hCG.

9.1.9 Luteal support

Luteal support for obese women should be the same as that for women with normal BMI. This is because the vaginal pessaries are locally absorbed and bypass first-pass metabolism. There are no data comparing luteal support and outcomes in various BMI groups.

9.1.10   Pregnancy rate

All systematic reviews of observational studies have repeatedly demonstrated a detrimental impact of obesity on pregnancy rates. The largest single series comes from the Society of Assisted Reproduction (SART) in the United States. This analysis showed that failure to achieve a clinical intrauterine gestation was significantly more likely among obese women.

9.1.11   Clinical pregnancy rates

Clinical pregnancy rates in women with obesity undergoing gonadotropin IUI study findings are mixed. Some studies report no difference in the clinical pregnancy rates in obese patients compared to nonobese controls, while several others report a paradoxical increase. Possible reasons for an increased effectiveness of gonadotropin/ IUI in women with obesity include correction of anovulation, and compensation for erectile dysfunction and decreased frequency of intercourse.

The SART registry shows that there is a slight decline in the number of oocytes retrieved and the number of high-quality embryos as the BMI rises over 40. Implantation, clinical pregnancy, and live birth rates all decline gradually with increasing severity of obesity. However, the absolute decline in pregnancy rates is small. The overall likelihood of a live birth per cycle start declined from 31.4% in women with a normal BMI, to 28% in women with a BMI of 30−34.9, to 24.3% in women with a BMI of 40−44.5, and down to 21.2% in women with a BMI of >50.

The exact mechanism by which obesity lowers IVF success rates is unclear. Some studies have demonstrated alterations in embryo development and day-3 spent culture media metabolomics, while others have not detected any changes of indicators of embryo quality between obese women and nonobese controls. Alternatively, obesity may alter endometrial receptivity. Perhaps the best model to help elucidate the impact of obesity on reproduction is oocyte donation. Several studies have suggested that obesity does not impact donor egg recipient implantation or live birth rates, while other studies have found a negative association.

9.1.12   Miscarriage rate

As discussed earlier, it is uncertain whether the cause of increased miscarriages is linked to oocyte quality or other factors within the endometrium involved in implantation. The pregnancy loss rate (8.6% with normal BMI to 13.5% with BMI over 40) in oocyte recipients is comparable to the change in pregnancy loss reported by SART in women using their own eggs: 11.3% with normal BMI to 14.8% with BMI of 40−45, 17.6% with BMI of 45−50, and 20.3% with BMI over 50, suggesting that changes in embryo quality are probably not the primary driver for the BMI-related increase in pregnancy loss rates after IVF. Obesity clearly increases miscarriage risk. However, the absolute risk of pregnancy loss in women with obesity undergoing IVF is still lower than the reported risk of spontaneous pregnancy loss in women with two or more prior pregnancy losses (25%) or women aged over 40 ($\geq$ 35%).

9.1.13   Live birth rate

The SART data demonstrated that the live birth was 1.27 times lower in obese women as compared to those with normal BMI. However, the results also indicate that there are significant differences in pregnancy and live birth rates after ART when analysed by race and ethnicity, even within the same BMI categories. Analyses done by the subgroups demonstrated that prognosis was poorer when obesity was associated with polycystic ovary syndrome, while the oocyte origin (donor or nondonor) did not modify the overall interpretation.

Many adverse maternal, foetal, and neonatal outcomes are known to be associated with obesity. Management of the infertile thus poses complex questions linked to the welfare of potential mothers and their offspring. Many pregnancy-associated complications occur with greater frequency in the obese, for example, pregnancy-induced hypertension and gestational diabetes. Need for intervention carries with the specifics of the difficulty of surgery in those who are morbidly obese, together with the potential for complications such as infection, venous thromboembolism, and anaesthetic hazards. Maternal mortality, while a rare occurrence, has associations with obesity, and a recent report highlighted the fact that many maternal deaths occurred in women with preexisting medical conditions, including obesity, which seriously affected the outcome of their pregnancies. Foetal risks in pregnancy are a concern with observed increased occurrence of foetal abnormality, macrosomia, low birth weight, neonatal mortality, and stillbirth. An influential report suggested that obesity is the principal modifiable risk factor for stillbirth in the developed world, greater than increased maternal age and smoking. Recent evidence suggests that maternal BMI was a significant risk factor for preterm delivery, even in pregnancies as a result of frozen ETs, following freeze all cycles.

Beyond these short-term outcomes, the long-term health of individuals born to obese mothers is a public health issue of concern. Children of the obese will grow up with greater risks of coronary heart disease, hypertension, glucose intolerance, and diabetes as well as themselves being obese, thereby perpetuating the problem for the subsequent generation.

The management of the obese infertile raises economic issues of note given increased costs not only associated with treatment but also those associated with the management of complicated pregnancies, particularly the need for increased surveillance, higher rates of operative delivery, and the management of women with gestational diabetes and hypertension.

Debate within the last few years has taken place as to whether these morbidities and adverse outcomes, together with higher costs, should play a part in whether the obese should be permitted the same access to infertility services as those who are not overweight. It could be argued with the prevalence of obesity being at the level it is that in fact the boundaries of what can be considered normal in the population have changed. Adverse outcomes however would suggest this is not the case. It has been argued that a restrictive policy would lead to stigmatisation of the obese, but genuine health hazards are being increasingly identified, which carry significant implications for the individuals concerned. It has even been suggested that the autonomy of the individual to determine their own health would be being infringed by policies to deny access to care. On the other hand, the identification of long-term health risks could be considered as an opportunity for the empowerment of the individual to make lifestyle adjustments that may have real health benefits for themselves. Bearing in mind the issues described earlier, it is clear that patients have responsibilities beyond themselves, and healthcare professionals similarly have responsibilities to offspring and to society at large. Scarce resources, particularly at the present time, should be used to maximum effect. Interventions to assist individuals to achieve and sustain weight loss are not always effective. However, it would be anomalous for the reproductive health sector not to share with other areas of medical practice the public health responsibility for health promotion messages relevant to weight. Losing weight may of course delay the initiation of treatment and this is important particularly in those who seek assistance in later reproductive years. However, in the younger patient the amount of weight loss to make a difference may not be substantial and the time taken to achieve a target may not adversely affect the chance of treatment being successful. That said, there is no randomised trial evidence at present that weight loss programs prior to IVF treatment have an appreciable effect on outcomes or pregnancy-associated complications.

9.2 Conclusion

There is irrefutable evidence that fertility potential is adversely affected in the obese. The proportion of patients accessing infertility services who are obese is increasing. Natural fecundity, responses to treatment, and pregnancy outcomes are suboptimal in this group of patients. The mechanisms whereby these effects are manifested are not fully understood, but it is likely that the causes are multifactorial, including endocrine, inflammatory pathways, as well as effects on oocyte quality. Interventions to address subfertility while offering increased potential for conception raise important questions relevant to the safety of mothers and offspring. While adverse outcomes are increased in this group of patients, the absolute risk to the individual of complications remains relatively small. Most conceptions will result in healthy live-born, but offspring will have increased lifetime health risks. Ethical issues in this sphere of reproductive medicine challenge principles of helping the individual while taking account of consequences for others, not least the potential child but, bearing in mind the costs of treatment, pregnancy care, and, beyond, the views of society at large.

More research in this area is needed with clearly defined patient populations, using standardised BMI criteria and uniform outcome measures. Access to individual patient data may allow more refined methods of analysis including the ability to adjust for confounders in generating combined OR. Further research is needed in determining the best measure of obesity for reproductive outcome.

Medical and surgical interventions to improve fertility outcomes

Suresh Kini
Assisted Conception Unit, Department of Obstetrics and Gynaecology, Ninewells Hospital and University of Dundee, Dundee, Scotland

10.1 Introduction

1. The prevalence of obesity is increasing worldwide with more than 600 million obese adults, including 15% of women, in 2014.
2. Being overweight in early adulthood increases the risk of menstrual irregularities, ovulatory dysfunction, and consequently subfertility.
3. An increasing number of men and women with high body mass index (BMI) are being referred for the evaluation and treatment of subfertility across the country.
4. Identifying and developing effective long-term reproductive health strategies for overweight and obese men and women is of paramount importance.
5. Only one-third of obese patients are found to receive advice from healthcare providers regarding weight reduction.
6. Prospective studies have demonstrated that high levels of central and overall adiposity are associated with decreased fecundability, even when adjusting for confounders.
7. Obesity's established negative impact on reproductive potential is multifactorial, and increased adiposity can influence almost every stage of fertilisation from ovulation to successful implantation and development of the embryo.

10.2 Lifestyle interventions

1. Weight loss is recommended for men and women with high BMI before attempting natural conception or fertility treatments to improve fertility outcomes, assist with fertility treatment funding, and to reduce the risks of obstetric complications.
2. Reduced calorie intake and increased physical activity are the two essential pillars of any weight-loss program.
3. Most guidelines recommend a target of 5%−10% body weight loss in overweight/obese women with long-term goals of 10%−20% weight loss and waist circumference < 80−88 cm tailored to the ethnicity.
4. Maintaining long-term weight loss can prove challenging and attention needs to be given to other areas of lifestyle, such as alcohol intake, smoking cessation, and stress-reduction techniques.
5. Strategies to promote sustained weight loss include self-monitoring techniques such as food diaries, pedometers, time management advice, relapse prevention techniques, engagement of social support, and goal setting.

Handbook of Obesity in Obstetrics and Gynecology. DOI: https://doi.org/10.1016/B978-0-323-89904-8.00003-2

6. There is some evidence that intensive weight loss immediately prior to in vitro fertilisation (IVF) is associated with adverse outcomes, including increased cycle cancellation and decreased rates of fertilisation, implantation, ongoing pregnancy, and live births in women with polycystic ovary syndrome (PCOS).
7. Weight-loss strategies should be encouraged well in advance of pregnancy planning by the individual woman.

10.2.1   Dietary interventions

1. Dietary interventions in overweight or obese men and women should consider the degree of obesity, dietary preferences, and food availability.
2. If an eating disorder is suspected, referrals to the dietitian and clinical psychologist should be considered.
3. Strategies such as face-to-face education sessions and practical advice on approaches to healthy eating tailored to the patient should be incorporated.
4. It is currently recommended that women with BMI > 25 should aim for weight loss via caloric restriction through balanced dietary approaches irrespective of diet composition.
5. In the general adult population, a target energy deficit of 2500 kJ daily is recommended for weight loss.

10.2.2   Diet

1. An individualised approach works best.
2. The aim should be to lose weight at a safe and sustainable rate of 0.5−1 kg a week, and for most women, the initial advice should be to reduce their energy intake by 600 cal a day.
3. To consider swapping unhealthy and high-energy food choices (fast food, processed food, sugary drinks, and alcohol) for healthier choices.
4. Very low-calorie diet which involves consumption of less than 800 cal a day, can lead to rapid weight loss, but may not be suitable for everyone. Such diets should not be followed for longer than 12 weeks at a time. They should only be recommended under the supervision of a suitably qualified healthcare professional.

10.2.3   Role of exercise

1. Evidence shows that exercise benefits overweight women even in the absence of significant weight loss.
2. Thrice-weekly moderate exercise for at least 30 minutes has been demonstrated to reduce BMI, waist circumference, and insulin resistance (IR) in young PCOS women.
3. A recent meta-analysis found that weight-loss interventions, particularly diet and exercise, improved pregnancy rates and ovulatory status.
4. Miscarriage rates remained unchanged by weight-loss interventions.
5. The meta-analyses also showed that weight loss had a nonsignificant advantage over weight loss medications such as metformin with respect to achievement of pregnancy or improvement of ovulation rates.

10.3 Medical interventions

10.3.1 Weight-loss medications and fertility outcomes

1. Pharmacological agents are mainly indicated when patients fail to lose significant weight despite lifestyle changes and a low-calorie diet.
2. These agents have been shown to induce modest weight loss but are not suitable for long-term weight maintenance.
3. These have mainly included metformin (an insulin sensitiser), orlistat (a lipase inhibitor), sibutramine (a selective serotonin and norepinephrine reuptake inhibitor), and liraglutide [a glucagon-like peptide-1 (GLP-1) receptor agonist].
4. When prescribing the appropriate weight-losing drug, it is paramount to consider the safety of these drugs should a woman conceive while taking them.
5. The safety of acarbose in pregnancy is not established.
6. The use of weight-loss medications is contraindicated during pregnancy.
7. Out of all the drugs mentioned previously, pharmacokinetics of the orlistat places it in a favourable position due to its low absorption and first-pass metabolism resulting in a bio-availability of less than 1%.
8. Lifestyle interventions should still be considered the first-line therapy, with drug use largely reserved for monitored trials.

10.3.2 Metformin

1. Metformin is a synthetically derived biguanide that decreases hepatic glucose production and intestinal absorption of glucose, while increasing the peripheral uptake and utilisation of glucose.
2. It also stimulates fat oxidation and reduces fat synthesis and storage. Metformin decreases IR and enhances insulin sensitivity at the cellular level.
3. Metformin appears to promote weight loss and offers protection from the macrovascular complications of diabetes.
4. Metformin is administered orally in doses of 1500−2000 mg in divided daily doses.
5. The most common side effects of metformin are gastrointestinal in nature such as nausea, vomiting, and diarrhoea. Such side effects decrease with time and can be lessened by dose reduction and taking the metformin with food.
6. The most serious side effect, lactic acidosis, is rarely seen.
7. No obstetric complications or congenital anomalies were described.
8. It is also recommended periconceptionally to reduce the risk of ovarian hyperstimulation syndrome (OHSS) with IVF.
9. It has been observed that nonobese women with PCOS respond better to metformin than obese women.
10. A Cochrane review in 2017 showed that combined therapy with metformin and clomiphene citrate has improved ovulation rate and clinical pregnancy than using clomiphene citrate alone in obese women with PCOS.
11. Metformin is mainly used in women with PCOS as a second-line option for ovulation induction (either for clomiphene resistance or combined with clomiphene), especially in those with a BMI of 35 and over.
12. Metformin is also used along with weight management strategies as a first-line agent in women with high BMI and wishing to pursue fertility treatment in near future.

10.3.3 Sibutramine

1. Sibutramine blocks the reuptake of the neurotransmitters dopamine, norepinephrine, and serotonin. Sibutramine is no longer recommended in clinical practice because of the risk of serious cardiovascular problems in some patients who take it. Sibutramine has been withdrawn in Europe and the United States but is still available on the Internet.

10.3.4 Orlistat

1. Orlistat inhibits pancreatic lipase, resulting in a 30% reduction in the absorption of ingested fat leading to weight loss.
2. Orlistat is recommended once the woman has made a significant effort to lose weight through diet, exercise, and lifestyle changes. It is recommended that the diet should be rich in fruit and vegetables. It is used in conjunction with a balanced low fat diet for the treatment of obese women with a BMI $\geq 30 \text{ kg/m}^2$ or overweight patients (BMI $\geq 28 \text{ kg/m}^2$) with associated risk factors.
3. If a meal is missed or contains no fat, the dose of orlistat should be omitted.
4. A single orlistat capsule (120 mg) should be taken with water immediately before, during, or up to 1 hour after each main meal (up to a maximum of three capsules a day).
5. Treatment should continue beyond 3 months if there is loss of 5% of body weight. If weight loss is demonstrated, orlistat should be continued for 12 months or more.
6. Women with type 2 diabetes may take longer to lose weight using orlistat, so the target weight loss after 3 months should, therefore, be slightly lower.
7. The gastrointestinal effects of orlistat result in an increase in faecal fat as early as 24−48 hours after dosing. Upon discontinuation of therapy, faecal fat content usually returns to pretreatment levels within 48 72 hours.
8. Common side effects of orlistat include steatorrhoea, diarrhoea, flatulence, abdominal discomfort, headaches, and upper respiratory tract infections.
9. Women taking the combined oral contraceptive pills should use an additional method of contraception if they experience severe diarrhoea while taking orlistat.
10. The effect of orlistat in patients with hepatic and/or renal impairment, children, and elderly patients has not been studied.
11. Contraindications to the use of orlistat include hypersensitivity to the active drug substance or to any of the excipients, chronic malabsorption syndrome, and cholestasis.
12. The study shows no risk of birth defects from orlistat use. Orlistat is not recommended for pregnant or breastfeeding women.
13. Orlistat appears to be equally effective with metformin in reducing weight, IR, and testosterone levels.

10.3.5 Liraglutide

1. Liraglutide (GLP-1 receptor agonist) stimulates insulin secretion and lowers inappropriately high glucagon secretion in a glucose dependent manner.
2. Liraglutide reduces body weight and body fat mass through mechanisms involving reduced hunger and lowered energy intake as GLP-1 is a physiological regulator of appetite and food intake.
3. Liraglutide has been utilised in the management of patients with BMI $> 30 \text{ kg/m}^2$ or BMI $27-30 \text{ kg/m}^2$ and obesity-related comorbidities.

4. To improve the gastrointestinal tolerability, the recommended starting dose is 0.6 mg daily. After at least 1 week the dose should be increased to 1.2 mg. Some patients are expected to benefit from an increase in dose from 1.2 to 1.8 mg and based on clinical response; after at least 1 week, the dose can be increased to 1.8 mg.
5. Common side effects of liraglutide include nausea, vomiting, stomach upset, decreased appetite, diarrhoea, and constipation.
6. Liraglutide should not be used during pregnancy.
7. A recent double-blind trial by Wilding et al. showed 2.4 mg of semaglutide once weekly (another GLP-1 receptor agonist) plus lifestyle intervention was associated with sustained, clinically relevant reduction in body weight.

10.3.6 Inofolic acid

1. This is a nutritional supplement which contains folic acid and myo-inositol. This has an antioxidant role and increases insulin sensitivity in obese women with PCOS and helps in weight loss. Taken during treatment it has been shown to improve egg and embryo quality and reduce the risk of OHSS. Evidence is accumulating that myo-inositol is efficient enough to change sperm parameters to increase the chance of fertility. It can be taken throughout pregnancy and reduces the incidence of gestational diabetes.

10.4 Surgical Interventions to improve fertility potential in obese men and women

10.4.1 Bariatric surgery as a weight-loss measure

1. Bariatric surgery represents the most successful treatment that results in sustained long-term weight loss.
2. The percentage of body weight loss at 2 years after bariatric surgery can approach 60%.
3. Indications for this procedure are well established by the American Bariatric Society and suggest that candidates should have at least a BMI >40 without serious comorbidities (e.g., diabetes, hypertension) or at least a BMI >35 in the presence of one serious comorbidity.
4. Candidates with BMI <35 are considered if they have uncontrolled type 2 diabetes or metabolic syndrome.
5. Absolute contraindications include serious depression or psychosis, eating disorder, alcohol abuse, heart disease, and coagulopathy.
6. It is estimated that only 1% of the population that meets the criteria for these procedures undergoes surgery.
7. Patients should be carefully screened to optimise the success of a considered procedure.
8. Psychological assessment and appropriate setting of expectations before surgery is essential.
9. Sustainable weight loss after a bariatric procedure requires significant lifestyle changes.
10. Preoperative assessments include reviewing a patient's previous attempts at weight loss, eating and dietary styles, physical activity, and history of substance abuse.
11. A review of the patients' medical comorbidities and consequent suitability for anaesthesia are the key determinants of candidacy for surgery.

10.4.2 Types of bariatric surgery

1. Bariatric procedures are characterised as restrictive, malabsorptive, or a combination of the two.
2. Restrictive surgeries such as the sleeve gastrectomy aim to create a smaller gastric pouch that reaches capacity soon after food consumption to induce satiety.
3. The sleeve gastrectomy is a partial gastrectomy in which most of the greater curvature of the stomach is removed, thus producing a more "tubular" stomach.
4. Sleeve gastrectomy has been described as a less complex procedure than bypass and is associated with shorter operating time, a lower rate of blood loss, and a lower rate of reoperation among other morbidities.
5. Evidence has accumulated supporting sleeve gastrectomy as the most performed bariatric procedure in the United States since 2013.
6. Roux-en-Y gastric bypass is the second most practiced type of bariatric surgery.
7. This procedure is characterised by the creation of a small gastric pouch divided from the distal stomach and anastomosed to a "Roux" limb of small bowel. The proximal small intestine is then divided, and a limb is created that drains secretions from the gastric remnant, liver, and pancreas. The net effect is to have two stomach chambers that anastomose to the small intestine at different points. The stomach remnant drains proximally, and the newly created smaller functional stomach drains distally through the Roux limb. The decreased size of the stomach acts to restrict caloric intake, and the bypass of a segment of small bowel has a malabsorptive effect.
8. Roux-en-Y gastric bypass limit energy intake by effectively shortening the length of the gastrointestinal tract.
9. Malabsorptive procedures and consequent nutritional deficiencies are thought to drive the increase in intrauterine growth restriction and small for gestational age (SGA) infants seen in initial observational studies of pregnancy in bariatric surgery patients.
10. Bypass, with its impact on absorption, can produce a higher degree of weight loss over a shorter time frame.
11. The expected 2-year weight loss after a Roux-en-Y procedure is about 70% compared to 60% for sleeve gastrectomy.
12. Other bariatric procedures which may be considered include the much less commonly performed biliopancreatic diversion with duodenal switch, the intragastric balloon, and vagal blockade.
13. The biliopancreatic diversion involves manipulation of the pylorus, duodenum, and ileum and is considered a technically difficult operation with a significant complication rate.
14. Intragastric balloon has promise as a bridge to another procedure and has been approved for patients at a lower BMI threshold than traditional bariatric surgery (BMI 30−34.9).
15. It involves the placement of a soft saline-filled balloon into the stomach that promotes the sensation of satiety and gradually degrades after about 6 months.
16. Vagal blockade involves the placement of an electric pulse generator that is designed to lead to the decreased sensation of hunger.

10.4.3 The impact of bariatric surgery on fertility

1. Initial data on the effects of weight-loss surgery on fertility have been encouraging, however limited.

2. Various studies support the conclusion that surgical treatment of obesity tends to reverse the altered reproductive hormone profile seen in this population.
3. Further, bariatric surgery has been associated with profound changes extending to the hypothalamic pituitary adrenal axis.
4. American Congress of Obstetricians and Gynecologists (ACOG) have highlighted the potential for bariatric surgery to improve fertility outcomes through restoration of ovulation and reversing pathologic changes in PCOS and spontaneous conception, however it should "not be considered a treatment for infertility."
5. The American Society for Reproductive Medicine suggests that bariatric surgery appears safe in a population looking forward to becoming pregnant and has potential to lead to improvement in markers of reproductive health.

10.4.4 The potential of bariatric surgery for a negative impact on fertility

1. Some studies have suggested a negative impact on fertility in some patients.
2. Nutritional deficiencies after malabsorptive procedures have been proposed as a potential mechanism for subfertility.
3. Bariatric surgery is noted to cause a significant decrease in AMH levels in women under 35.
4. Importantly, this effect was not seen in those above 35. This change has been attributed to stress involved with operation and decreased absorption of precursors relevant in AMH production.

10.4.5 Pregnancy after bariatric surgery

1. ACOG recommend delaying pregnancy for 12−24 months.
2. Post bariatric surgery patients who became pregnant within 2 years had higher rates of prematurity, neonatal intensive care unit admission, and SGA infants.

10.4.6 Assisted reproduction after bariatric surgery

1. Systematic review of studies indicates IVF of women with BMI > 25 has a 10% lower success rate than those with less than this figure.
2. If a woman develops OHSS, a complication of IVF, this can in turn increase the risk of known bariatric surgery complications, such as intestinal obstruction and internal hernia.
3. Another issue of concern is adherence to the recommended postprocedural delay before conception.

10.4.7 Obesity in the male

1. Obese men achieving weight loss through medical and surgical means have been shown to increase quality of life, decrease rates of erectile dysfunction, and possibly improve derangements in reproductive hormone profile.
2. There is no specific data that has been able to characterise improvements in a couple's fertile potential through surgical weight loss in the male.

3. In those couples that require assisted conception treatment, it could be argued that ICSI should be offered when there is male obesity, to overcome the negative effects demonstrated on sperm function.

10.4.8   Barriers to weight loss

1. Overweight subfertile men and women appear most deterred from exercise by the perception that it causes tiredness and is hard work. Such perceptions seem to decrease with continuation of an exercise program.
2. Effective weight management programs should include behavioural changes to increase the person's physical activity level. These could include setting goals, stimulus control, and relapse prevention.
3. A multidisciplinary, holistic approach to weight loss, including primary care physician, gynaecologist, endocrinologist, exercise physiologist, dietitian, and psychologist is recommended in women with PCOS who have established metabolic complications.

10.5   Conclusion

1. In overweight and obese subfertile men and women, weight loss is associated with improved chances of becoming pregnant naturally or through fertility treatment.
2. Weight loss also improves ovulation frequency and aids menstrual regularity.
3. Many women can conceive without further assistance through weight loss alone.
4. Lifestyle interventions remain the first-line therapy for improvement in ovulation and menstruation.
5. A combination of a reduced-calorie diet, which is not overly restrictive, and aerobic exercise, intensified gradually, should be recommended.
6. The effects of antiobesity agents on weight and obesity-related characteristics of the PCOS remain unclear. More studies are needed to clarify the role of antiobesity agents as weight-loss intervention prior to fertility treatment.
7. Weight loss after bariatric surgery has been shown to improve markers for reproductive health and studies have demonstrated improved fertility.
8. At present, there is a lack of high-level clinical evidence to consider bariatric surgery primarily for fertility-based indications.

Further reading

ESHRE. Guideline on the management of recurrent pregnancy loss <https://www.eshre.eu/Guidelines-and-Legal/Guidelines/Recurrent-pregnancy-loss>; 2019

Legro R.S. Effects of obesity treatment on female reproduction: results do not match expectations. Fertil Steril. 2017;107(4): 860–867.

Mahmood TA, Arulkumaran S, Chervenak FA. *Obesity and Gynecology*. 2nd ed Oxford: Elsevier; 2020.

Practice Guideline. Obesity and reproduction: a committee opinion. *Fertil Steril*. 2015;104 (5):1116–1126.

The role of Bariatric Surgery in Improving Reproductive Health, (Scientific Impact Paper 17)
—Royal College of Obstetrics and Gynecologists, UK. https://www.rcog.org.uk/globa-lassets/documents/guidelines/scientific-impact-papers/sip_17.pdf.

Wilding JPH, Batterham R, Calanna S, et al. Once weekly semaglutide in adults with over-weight or obesity. *N Engl J Med.* 2021. Available from: https://doi.org/10.1056/NEJMoa2032183. Feb 10.

Heavy menstrual bleeding

11

Tahir Mahmood
Department of Obstetrics & Gynaecology, Victoria Hospital, Kirkcaldy, United Kingdom

11.1 Introduction

Heavy menstrual bleeding (HMB) is a common condition that affects 20%−30% of women during their reproductive age and has a major impact on women's quality of life.

The International Federation of Gynaecology and Obstetrics (FIGO) has defined HMB as "the women's perception of increased menstrual volume regardless of regularity, frequency or duration."

FIGO also define HMB as part of broader terminology of abnormal uterine bleeding (AUB) which includes intermenstrual bleeding (IMB) and postcoital bleeding (PCB).

FIGO definition of (AUB) includes:

1. Disturbance of menstrual frequency and cycles shorter than 21 days are classified abnormal frequency of menses.
2. Irregular menstrual bleeding—cycles when the onset of menses is unpredictable.
3. Menstrual periods that exceed 8 days duration on a regular basis are classified as prolonged menstrual bleeding.
4. HMB—describes increased menstrual volume regardless of regularity, frequency, or duration.
5. Intermenstrual bleeding—episodes of bleeding that occur between normally timed menstrual periods.

11.1.1 Physiology of menstruation

1. The average menstrual cycle length is 28 days and most women bleed for approximately 4−5 days associated with shedding of the superficial stratum functionalis of the endometrium.
2. The endometrium is under the regulation of ovarian steroid hormones, mainly oestrogen and progesterone and their involvement in the monthly endometrial cycle is well established.
3. Following menstrual shedding, the repair process is mainly under the influence of oestrogen and local haemostatic mechanisms.
4. Mechanisms that interfere with the normal endocrine, paracrine or haemostatic functions of the endometrium as well as possibly any interference with myometrial contractility may cause HMB.

Handbook of Obesity in Obstetrics and Gynecology. DOI: https://doi.org/10.1016/B978-0-323-89904-8.00042-1

5. Obesity influences the development and progression of menstrual problems,
6. Obese women are three times more likely to suffer from menstrual abnormalities than women of a normal weight.
7. Significant weight loss can restore normal menstruation pattern.
8. AUB may be classified using the FIGO classification, using the PALM-COEIN paradigm. This acronym describes the aetiological basis of menstrual problems.

11.1.2   FIGO classification of AUB (PALM-COEIN)

The FIGO classification of AUB includes nine categories which are divided into two distinct subgroups:

1. The PALM group: it consists of structural abnormalities that can be visualised using imaging techniques or diagnosed by histopathology.
2. The COEIN group: it describes nonstructural disorders that cannot be imaged or diagnosed with histopathology (Fig. 11.1).

11.2   Polyp

1. Polyps are epithelial proliferations comprising variable vascular, glandular, fibromuscular and connective tissue components.
2. Polyps may be asymptomatic and may be responsible for AUB.
3. A survey of premenopausal women with endometrial polyps found that 82% reported AUB. In these women, obesity and hypertension were two risk factors,
4. Another study also found that obese women had a significantly higher prevalence of polyps compared to normal BMI women,
5. Obesity appears to be an important risk factor for the developing endometrial polyps,
6. One potential mechanism for this is possibly higher levels of circulating oestrogens secondary to peripheral conversion of androgens by adipose tissue aromatase enzyme to oestrogens.
7. Hence higher levels of both oestradiol and longer duration of exposure to unopposed oestrogens in obese women may have an augmented effect on the proliferative phase of endometrium.

PALM Group		COEIN Group
Polyp		Coagulopathy
Adenomyosis		Ovulatory dysfunction
Leiomyoma	Submucosal/other	Endometrial
Malignancy and hyperplasia		Iatrogenic
		Not yet classified

Figure 11.1 FIGO-PALM-COEIN

11.3 Adenomyosis

1. Adenomyosis is the presence of ectopic endometrial-like tissue within the myometrium.
2. Unlike endometriosis, there appears to be a higher incidence of adenomyosis in obese women.
3. The prevalence of adenomyosis varies widely, ranging from 5% to 70%.
4. This is probably related to inconsistencies in the histopathologic criteria for diagnosis at hysterectomy specimens.
5. MRI has a greater specificity and positive predictive value compared to transvaginal ultrasound and a greater ability to distinguish between adenomyosis and leiomyomas.
6. This additional benefit of MRI over ultrasound scan is especially more relevant in obese population.

11.4 Leiomyoma

1. Uterine fibroids (myomas, leiomyomas) are the most common benign tumours in women of reproductive age.
2. Women who have incidentally diagnosed with small fibroids and are asymptomatic do not require treatment.
3. Fibroids tend to be twice or even three times more common in non-white women as compared to other racial or ethnic groups.
4. Heavy menstrual bleeding is the most common symptom of a fibroid uterus and multiple factors are thought to contribute:
5. Increased endometrial surface area, increased uterine vascularity, impaired uterine contractility, and endometrial ulceration caused by submucosal fibroids may be possible mechanisms for menstrual symptoms.
6. Location of uterine fibroids may contribute towards symptoms, with submucosal leiomyomas having a greater association with HMB, although objective evidence for this is limited.
7. There is no consistent relationship between the size and location of fibroids and HMB.

11.5 Malignancy and hyperplasia

Obesity increases the risk of malignancy developing within an endometrial polyp,

1. It has been estimated that 40% of all endometrial cancer is attributable to obesity, and that 86% of women with complex hyperplasia were obese.
2. BMI is predictive of endometrial thickness on ultrasound scan and this is predictive of hyperplasia.
3. The risk of endometrial cancer varied from an almost fourfold increase in women with a $BMI > 25 \text{ kg/m}^2$ to an almost 20-fold increase in women with a $BMI > 40 \text{ kg/m}^2$.
4. A recent prospective study showed that bariatric surgery in women with $BMI > 40$ can reverse atypical hyperplasia (Ref 24,25).

11.5.1   *Coagulopathy*

1. Obese women are at increased risk of venous and arterial thromboembolism.
2. There is increased procoagulant activity, impaired fibrinolysis, increased inflammation, endothelial dysfunction, and altered lipid and glucose metabolism in metabolic syndrome.
3. Adipose tissue is known to produce several cytokines (known as adipokines), including Leptin, and adiponectin, tumour necrosis factor-α, and plasminogen activator inhibitor-1.
4. Interleukin-6 has been implicated in mediating the link between abdominal obesity and venous thromboembolism.
5. Approximately 13% of women with HMB have biochemically detectable systemic disorders of haemostasis, most often von Willebrand disease.
6. Long-term anticoagulation may contribute to HMB/AUB.

11.6   Ovulatory dysfunction

1. There is a strong association between menstrual cycle irregularities and anovulation with overweight and obesity,
2. Compared to nonobese women, obese women had at least twofold greater odds of having an irregular cycle defined as >15 days between the longest and shortest cycle in the last 12 months,
3. Contemporary studies from the United States and Australia that women with BMI $>35\,\mathrm{kg/m^2}$ had risk of long cycles compared to women with BMI $22-23\,\mathrm{kg/m^2}$, and these findings of independent of racial ethnicity.
4. Ovulatory dysfunction can be often secondary to other disorders resulting in hormonal fluctuations such as:
    a. polycystic ovary syndrome
    b. hypothyroidism
    c. hyperprolactinaemia
    d. mental stress
    e. obesity
    f. anorexia
    g. weight loss
    h. or extreme exercise
    i. hormonal fluctuations may be iatrogenic, caused by sex steroids or drugs that impact dopamine metabolism.

11.7   Endometrial

Due to a primary disorder of mechanisms regulating local endometrial haemostasis.

1. Deficiencies in local production of vasoconstrictors (endothelin-1 and prostaglandin F2α).
2. and/or accelerated lysis of endometrial clot because of excessive production of plasminogen activator, in addition to increased local production of vasodilators (prostaglandin E2 and prostacyclin I2).

Primary endometrial disorders (may be a manifestation of deficient mechanisms of endometrial repair):

1. may be secondary to endometrial inflammation or infection (Chlamydia Trachomatis).
2. abnormalities in the local inflammatory response and/or aberrations in endometrial vasculogenesis:
   a. There are minimal data available in the literature on the influence of obesity on the volume of menstrual blood loss.
   b. However raised BMI is associated with poor efficacy of hormonal contraception suggesting an effect of obesity on bioavailability or action of steroids.

11.7.1 Iatrogenic

1. Unscheduled bleeding that occurs during the use of sex steroid therapy is termed "breakthrough bleeding (BTB)."
2. Many episodes of unscheduled bleeding/BTB are related to compliance issues or reduced circulating hormone levels because of enhanced hepatic metabolism:
   a. missed, delayed, or erratic use of anticonvulsants;
   b. certain antibiotics (e.g., rifampacin and griseofulvin); and
   c. cigarette smoking.
3. Vaginal spotting/bleeding in the first 3—6 months of the use of LNG-IUS.
4. Tricyclic antidepressants (amitriptyline and nortriptyline) and phenothiazines impact dopamine metabolism by reducing serotonin uptake, resulting in reduced inhibition of prolactin release, leading to anovulation and AUB.
5. The use of anticoagulant drugs (warfarin, heparin, and low molecular weight heparin) leads to impaired formation of an adequate "plug" or clot within the vascular lumen. Not yet classified (AUB-N).

Not yet classified group:
Clinical entities maybe associated with or contribute to AUB/HMB, such as:

1. chronic endometritis
2. arteriovenous malformations
3. bleeding from a caesarean section scar defect
4. isthmocoele
5. myometrial hypertrophy.

However, there is limited evidence to support this hypothesis.
Obesity in the absence of polycystic ovary:

1. Obesity independently increases hyperandrogenism, hirsutism, insulin resistance and infertility.
2. The presence of insulin resistance predicts a thicker endometrium on ultrasound scan.
3. A high BMI is positively associated with the thickness of the endometrium in the absence of PCO.
4. Obesity has been associated with increased uterine blood flow as measured by Doppler uterine artery pulsatility index.

11.8   Structured history taking in women with AUB

A structured history should be taken to establish a cause and diagnosis.

It should include the nature of bleeding, the impact of bleeding on quality of life and related symptoms such as pelvic pain, postcoital bleeding and intermenstrual bleeding.

Symptoms of anaemia such as tiredness, fatigue, lethargy and breathlessness should be elucidated. A coagulation disorder may be considered by history of excessive bleeding since menarche, and/or history of postpartum haemorrhage, and surgery related bleeding, such as dental extraction, as noted in Table 11.1 are identified.

Table 11.1 Structured history.

Menstrual	Personal	Associated symptoms/family	Sexual/reproductive
Menstrual diary (frequency, duration, volume, intercycle interval)	smoking	*Coagulopathy Screen only if following history is elucidated:*	Parity and mode of birth
Duration of HMB/AUB	Alcohol intake	• Excessive menstrual bleeding since menarche	Need for fertility preservation
Intermenstrual bleeding	Drug history (medical/recreational)	• Excessive bleeding at tooth extraction	Need for contraception
Post coital bleeding	occupation	• History of PPH	
Age at menarche	Weight gain/loss	• Excessive bruising >5 cm once or twice per month	Sexually transmitted infections (STIs)
Cervical cytology (Pap smear)	History of dysmenorrhea	• Frequent gum bleeding	Previous diagnosis of PCO, diabetes, hypertension
BMI (kg/m$^{2)}$	Chronic pelvic pain	• Family history of bleeding symptoms	Family history of breast, colon, endometrial, Lynch syndrome
	Pressure symptoms (constipation, urinary frequency)	Family history of thromboembolic events (VTE)	History of unopposed Oestrogen exposure/Tamoxifen use
	Abnormal vaginal discharge	Cancer	

11.9   Assessment of women presenting with HMB

The assessment of mean blood loss during menstruation remains largely a subjective estimation.

Recommended methods include:

1. the length of the cycle, flooding, clots and frequent change of sanitary towels;
2. objective tools like alkaline-haematin method;
3. menstrual cup (Mooncup), bleeding score; and
4. pictorial blood assessment chart (PBAC), which has been widely used in research studies.

11.10   Investigations

Table 11.2 provides a summary of investigations (adopted from NICE Guidelines 2018).

11.11   Management of women with HMB/AUB

General guidance
- Discuss the risks and benefits of each treatment option
- Consider patient preference
- Consider current and future fertility desires
- Consider comorbidities and coexisting symptoms, for example, pressure, pain
- Consider referral to specialist care if symptoms are severe, if treatment is unsuccessful/ declined, and/or if fibroids are 3 cm or more in diameter

11.11.1   Nonhormonal treatments

11.11.1.1   Tranexamic acid

1. Women with HMB, have high fibrinolytic activity which is likely to be due to high levels of plasmin and plasminogen activators in the endometrium.
2. Antifibrinolytics (tranexamic acid) inhibit fibrinolysis by blocking the lysine binding sites on plasminogen.
3. They inhibit both plasminogen activation and plasmin activity, thus preventing blood clot breakdown and reduces mean blood loss.
4. A Cochrane review of 12 randomised controlled trials (RCTs) reported the reduction in mean blood loss (MBL) in women with HMB ranging from 34% to 56% in those treated with >3 g Tranexamic acid for 5 days, (administered 3−4 times a day from day 1 of menstruation).
5. Side effects are mainly gastrointestinal, with caution in women with a personal history of thromboembolism.

Table 11.2 Suggested investigations for HMB.

Type	Specific test	Indication
Blood tests	Full blood count	Baseline testing to exclude anaemia/iron deficiency
	Liver function test/ renal function test	Based upon history and medical treatment decision
	Thyroid profile	In women with symptoms of thyroid disease
	Coagulation screen	Those identified from the family/personal history
	Female hormone profile	Based on history only
	Serum ferritin	Not required as routine
Histology	Endometrial sampling and hysteroscopy	To rule out endometrial hyperplasia and or cancer/or history is suggestive of submucosal fibroids, and polyps
Imaging	Ultrasound (preferably transvaginal route)	To rule out structural abnormalities like fibroids and adenomyosis, If unacceptable, transabdominal ultrasound should be considered.
Imaging	MRI	If transvaginal scanning is not acceptable/ and or needs confirmation of ultrasound scan findings (uterine structural abnormality/ adenomyosis/uterine fibroids)
Examination	Abdominal and pelvic	• An abdominal examination to exclude a pelvic mass (fibroid) • A vaginal and speculum examination to assess vulva, vagina and cervix to assess for a tumour, polyp or a foul smelly discharge • Cervical cytology (Pap smear) if indicated • A bimanual examination to elicit uterine enlargement

Source: Adapted from NICE 2018 guidance.

11.12 Nonsteroidal anti-inflammatory drugs

1. NSAIDs are prostaglandin synthetase inhibitors and act by inhibiting endometrial prostaglandin production leading to reduction in menstrual blood loss.
2. NSAIDs exert their anti-inflammatory effect through inhibition of cyclooxygenase enzyme, which catalyses the transformation of arachidonic acid to prostaglandins and thromboxanes.
3. Mefenamic acid (gastrointestinal effects are less likely compared to other NSAIDs) is the most commonly used NSAID for treatment of HMB and results in a reported blood loss reduction of 25%−50%.
4. NSAIDs and antifibrinolytic medications can be used together if they are beneficial
5. They can be used as long-term option.
6. They can also be used as adjuvant therapy with hormonal preparations.

11.13 Hormonal options

11.13.1 Levonorgestrel intrauterine system

1. The LNG-IUS contains an androgenic progestogen, levonorgestrel (LNG), which is slowly released to act on the local endometrial environment, preventing proliferation.
2. It may also impact on the frequency of ovulation.
3. Can decrease menstrual blood loss by up to 96% after 1 year of use and is even effective in women with fibroids without uterine cavity distortion.
4. A Cochrane review reported it to be significantly more effective at reducing HMB than oral treatment, and significant improvement in their quality of life.
5. The LNG IUS appears to be more acceptable for long-term use and is more cost effective than endometrial ablation techniques, but is associated with more minor adverse effects such as pelvic pain, breast tenderness and ovarian cysts.

11.13.2 Combined hormonal contraception

1. Acts by a regular shedding of a thinner endometrium and inhibiting ovulation, reducing blood loss by 50%, thus effectively treating both HMB and providing contraception.
2. An attractive option for low risk women experiencing frequent or irregular heavy bleeding, once an abnormal pathology has been excluded.
3. The COC may be used in a triphasic or tricycling manner to reduce the number of menstrual bleeds to 3–4 per year.
4. UKMEC Criteria may be used to assess suitability, especially risk profiling of individual women (http://fsrh.org/ukmec/).

The risks of the COCP are:

1. Thromboembolism, stroke, cardiovascular disease or breast cancer.
2. Contraindicated in women with a BMI >35 kg/m^2, smokers over 35 years, women with hypertension, vascular disease, migraine with aura, current/recent breast cancer, those with a personal or strong family history of venous thromboembolism or with a known thrombogenic mutation and breastfeeding women.

Progesterone only pill (POP):

1. POP is associated with irregular and unpredictable blood loss.
2. Not recommended as primary treatment for AUB.
3. Desogestrel containing POPs may provide effective treatment for some women.

Cyclical oral progestogens

1. Oral progestogens, taken three-times per day from day 5 to 26 of the menstrual cycle, have been shown to reduce blood loss by $>80\%$ although less acceptable choice than LNG-IUS.
2. More commonly used as a short-term measure to terminate a heavy bleed or regulate menstruation for a short duration.
3. Side effects include weight gain, bloating, breast tenderness, headache, acne and depression.

Injectable long acting progesterone

1. Intramuscular or subcutaneous injection of high dose depot medroxyprogesterone acetate given every 12 weeks (DMPA) can induce amenorrhea in up to 50% of women.
2. Women may have transient but reversible reduction in bone mineral density (osteopenia), if treatment is continuing beyond 5 years.
3. It is recommended that bone density assessment be considered after 5 years of treatment among long-term users.
4. The use of progesterone-only subdermal use as a treatment for HMB has not been studied.

Gonadotropin-releasing hormone agonists

1. Gonadotropin-releasing hormone (GnRH) agonists are synthetic peptides administered by an intramuscular, subcutaneous or intranasal route and induce a profound hypogonadal state (a medical menopause), with an amenorrhea rate of up to 90%.
2. Menopausal symptoms include: hot flushing, vaginal dryness, headaches, effect on bone density and decreased libido secondary to oestrogen deficiency.
3. Most of the side effects can be reduced with "add-back" hormone-replacement therapy (HRT).
4. "Add-back" HRT is introduced if the woman continues treatment for greater than 6 months, or sooner in symptomatic women.
5. GnRH agonists are particularly useful as a short-term option in the treatment of uterine fibroids (leiomyoma), which can reduce in size when ovarian hormone levels are suppressed.

Selective progesterone receptor modulators (SPRMs): Ulipristal Acetate (UPA)

1. SPRMs impart a tissue-specific partial progesterone antagonist effect.
2. Act upon progesterone receptors in the endometrium and the underlying myometrial tissue.
3. Oestradiol levels are maintained, thus hypoestrogenic side effects do not occur.
4. The mechanism by which these SPRMs reduce menstrual blood loss is poorly understood UPA is clinically proven to reduce size of uterine fibroids (leiomyomas and myomas), reduce menstrual blood loss or induce amenorrhea, and improve quality of life.
5. Short-term use of UPA is effective in treating HMB associated with uterine fibroids (3–10 cm in size).
6. UPA has the potential to provide a safe, fertility preserving, rapidly effective and convenient oral medical treatment for women with HMB whether associated with fibroids.
7. Liver function tests should be performed before starting treatment and at each month during the first two courses and then at the beginning of each course thereafter.
8. If liver function tests are abnormal, treatment should not be started.
9. Side effects include: headache (4%) and breast complaints (4%).
10. There are no publications to date on the clinical utility of SPRMs in the management of women with HMB who do not have fibroids or who have other conditions associated with HMB, such as adenomyosis.

Surgical treatment

Surgical management is only considered in women who have completed their family, with the exception of polypectomy and myomectomy where fertility can be retained.

Polypectomy

Endometrial polyps can be removed by hysteroscopic resection either under general anaesthesia, or in the outpatient setting.

Myomectomy

1. Uterine fibroids or leiomyomas can be identified by ultrasound in about 70%−80% of women by the time of the menopause.
2. In general, submucosal fibroids are more likely to cause AUB.
3. Some large intramural, and to some extent subserosal fibroids, that enlarge or distort the uterine cavity can cause bleeding problems.
4. Myomectomy is beneficial in selected cases and when medical interventions are less effective in the treatment of heavy menstrual bleeding for women who would like to preserve their uterus and fertility.
5. Transabdominal or laparoscopic surgical approaches are necessary to remove intramural and subserosal fibroids.
6. Transcervical hysteroscopic approach is utilised to remove submucosal fibroids.
7. Various techniques for reducing blood loss during myomectomy include: a preoperative course of GnRH analogue or SPRM.
8. Use of vasoconstriction agents (vasopressin), tranexamic acid and tourniquets during surgery should be considered.

Endometrial ablation

1. Endometrial ablation is a procedure that surgically destroys (ablates) the endometrium, to reduce menstrual flow, and some women may develop amenorrhea.
2. Newer techniques for endometrial ablation are associated with shorter operating times and can be performed under local rather than general anaesthesia.
3. Preablation endometrial histology should be obtained to exclude premalignant endometrial abnormalities,
4. Following endometrial ablation, there is a significant reduction in menstrual blood loss, and improved general and menstrual-related quality of life,
5. Up to 38% of women would have primary treatment failure requiring either a repeat procedure or a hysterectomy

Uterine artery embolisation

1. Uterine artery embolisation (UAE) is an established alternative treatment option for patients with fibroid-associated heavy menstrual bleeding.
2. It is minimally invasive, with reduced morbidity and faster recovery compared with surgery for women who wish to retain the uterus,
3. The procedure itself is not painful, but the ischemia to the uterus and the fibroids causes significant postprocedure pain,
4. There remains some concern as regards the effect of UAE on ovarian function,

MRI guided focused ultrasound treatment of uterine fibroids (MRgFUS)

1. MRgFUS is a noninvasive, outpatient procedure which uses magnetic resonance imaging (MRI) to locate the fibroids and direct high-intensity ultrasound energy to destroy fibroid tissue.
2. The procedure maintains fertility potential.
3. Long-term follow-up data are awaited.

Hysterectomy

Hysterectomy is the most definitive and effective treatment option for women who have no future fertility requirement and have heavy menstrual bleeding.

It has high satisfaction rate of up to 95%, 3 years following surgery.

It also provides a permanent relief from HMB when compared with other uterine sparing surgery such as endometrial ablation where up to 38% required further treatment because of ongoing excessive bleeding,

Hysterectomy can be carried by different routes: vaginal, open abdominal, laparoscopic assisted and robotic assisted.

The decision of which route and method should be used is influenced by the surgical skills, reasons for doing the procedure, the pelvic pathology that needs to be treated and the skills of the surgeon.

11.14 Conclusion

1. Heavy menstrual bleeding is a common gynaecological complaint and optimal management requires an understanding of the multiple aetiologies, diverse pathophysiological origins and how to diagnose the underlying pathologies.
2. Obesity per see is a risk factor for menstrual irregularities and precancer changed within the endometrium.
3. Women should have access to all the available treatment options and should have the opportunity to make informed choice with healthcare professional with an aim to improve her quality of life and optimise clinical outcomes.

Further reading

Bano R, Datta S, Mahmood TA. Heavy menstrual bleeding. *Obstetr Gynaecol Reprod Med.* 2016;26(6):167−174.

Bitzer J, et al. Medical management of heavy menstrual bleeding: a comprehensive review of the literature. *Obstet Gynecol Surv.* 2015;70(2):115−130.

Chodankar R, Harpur A, Mahmood TA. Heavy menstrual bleeding. *Obstetrics, Gynaecol Reprod Med.* 2018;28(7):196−202.

Heavy menstrual bleeding: assessment and management. NICE guideline [NG88]; 2018.

Maybin JA, Critchley HO. Medical management of heavy menstrual bleeding. *Womens Health (Lond).* 2016;12(1):27−34.

Munro MG, Critchley HOD, Fraser IS. The two FIGO systems for normal and abnormal uterine bleeding symptoms and classification of causes of abnormal uterine bleeding in the reproductive years: 2018 revisions. *Int J Gynaecol Obstet.* 2018;143(3):393−408.

Reavey JJ, Duncan WC, Brito-Mutunayagam S, Reynolds RM, Critchley HOD. Obesity and menstrual disorders. In: Mahmood T, Arulkumaran S, Chervenak F, eds. *Obesity and Gynaecology.* 2nd ed. Elsevier; 2020:171−175.

Obesity, incontinence, and pelvic floor dysfunction

12

Chu Chin Lim
Department of Obstetrics & Gynaecology, Victoria Hospital, Kirkcaldy, United Kingdom

12.1 Urinary incontinence

12.1.1 Introduction

1. Urinary incontinence (UI) is a common disorder that affects approximately 25% of the general population according to large epidemiological studies.
2. Several factors have been implicated in UI, including parity, operative vaginal delivery, length of labour, obesity, chronic cough, depression, anxiety, poor health status, lower urinary tract symptoms, previous hysterectomy, and smoking.
3. UI is a cause of significant morbidity and cost, estimated at over £500 million, representing 1% of the healthcare budget, in the United Kingdom, €400 billion in Europe, and between $25 and $50 billion in the United States.
4. Data related to quality of life (QoL) show that the impact of the disorder can be detrimental irrespective of the patients' age. Specifically, patients with UI suffer from depression, anxiety, and sexual dysfunction. The symptomatology is worse when obesity is also present as women with a body mass index (BMI) greater than $30 \, kg/m^2$ tend to have more severe symptoms.

12.1.2 Epidemiology

1. Norwegian EPICONT (Epidemiology of Incontinence in the County of Nord-Trøndelag) study, which included 34,755 women, researchers observed that obesity had a significant impact on UI (stress, urgency, and mixed).
2. It has also been estimated that 11% of the global population suffers from overactive bladder (OAB) and 8% from UI.
3. Known risk factors for UI include parity, vaginal childbirth, large babies, perineal trauma, operative delivery, increasing maternal age, prior hysterectomy, and BMI, which is another compounding factor.
4. A study has also shown that with each increase in BMI unit, the odds ratio (OR) of developing UI increases by 1.6 and women with BMI that exceeds $35 \, kg/m^2$ have a prevalence that peeks at 67.3%.
5. A recent meta-analysis of prospective cohort studies assessed the relationship between overweight/obesity and the risk of UI in young and middle-aged women; the risk of developing UI increased by about a third in women in the overweight category (35%) and nearly doubled in women with obesity (95%).
6. The effect of age on the prevalence of incontinence becomes minimal after the completion of the fifth decade of life; however, the severity of symptoms becomes more pronounced

Handbook of Obesity in Obstetrics and Gynecology. DOI: https://doi.org/10.1016/B978-0-323-89904-8.00017-2

after 70 years of age. Several factors contribute to this, including chronic ischemia of the lower urinary tract as well as the higher prevalence of abdominal obesity.

12.1.3   Pathophysiology urinary incontinence in the obese population

1. BMI correlates with intra-abdominal pressure, which increases intravesical pressure and exerts increased force on the pelvic floor, hence increasing the risk for stress urinary incontinence (SUI). Furthermore, increased intra-abdominal pressure can lead to the weakening of the pelvic floor innervations and musculature.
2. Chronically increased BMI is also associated with an elevated risk of UI in later life. Symptom severity also appears to worsen with the duration of increased BMI status, again confirming the detrimental effect of obesity on continence.
3. Functional disability and mobility problems often accompany gross obesity and are further risk factors for UI.
4. In women, an increase in adiposity is generally associated with a nearly linear increase in over active bladder (OAB) prevalence.
5. This contrasts with men, in whom as adiposity increases, the prevalence of OAB decreases to a certain point (BMI 27.5 kg/m^2), after which as adiposity increases, the OAB prevalence increases again.
6. In women, this relationship has been attributed to biomechanical (and neuroendocrine) factors, whereas men, who have greater pelvic floor strength, may be less susceptible to these forces with fewest symptoms in the overweight range.
7. The aetiology and pathophysiology of idiopathic overactive bladder is poorly understood even in the general population; hence, no clear mechanism of the association between an overactive bladder and obesity is available.
8. The systemic inflammatory state and oxidative stress associated with visceral obesity might play a role in development of OAB. Administration of an antioxidant agent has shown positive effects on LUTS in obese mice.

12.1.4   Nonsurgical treatment for urinary incontinence

1. A large randomised trial investigating the impact of a 6-month weight-loss program on outcomes of UI included 338 overweight and obese women with at least 10 UI episodes per week. This study found that weight reduction in the intervention group was approximately 8.0 kg (compared to1.5 kg in the control group), and this was accompanied by a significant reduction in incontinence episodes at 6 months (47% vs 28%). The difference was significant, however, only for cases with stress UI.
2. Outcomes concerning the impact of bariatric surgery on UI have shown the improvement or resolution of UI in approximately 55% of cases. Stress UI was less likely to be treated (47% of cases) compared to urgency UI (53% of cases). Worsening and new onset UI was observed in approximately 3% of cases.
3. Another study reporting on the impact of bariatric surgery in women with UI showed that incontinence-specific QoL scores were improved by 14%, while the proportion of women who were cured from any type of UI reached 58%.
4. There is a reported 73% reduction in overactive bladder symptoms in patients after surgically induced weight loss compared to their baseline symptoms. A similar reduction in

distressing overactive bladder symptoms has been described in studies with nonsurgical weight loss.

5. Conservative treatment of stress incontinence comprises pelvic floor exercises, supervised pelvic floor physiotherapy, electrical stimulation of pelvic floor muscles, and use of duloxetine.

6. Duloxetine was reported to be a useful alternative in obese patients with a good response in initial studies and has been described to have antiobesity and antibinge-eating properties, along with its known antidepressant effect in animal studies.

7. The dual effect of this drug may have a favourable impact on SUI in these patients by a variety of mechanisms including facilitation of weight loss.

12.2   Obesity and stress urinary incontinence

The efficacy of surgical procedures has been evaluated in nonobese as well as obese incontinent patients.

The majority of available evidence is based on the outcomes of patients treated with mid-urethral slings.

1. The impact of BMI on outcomes following midurethral sling placement is negative according to the findings of a recent systematic review that summarised data from 2846 women.

2. Objective cure rates are lower in overweight and obese patients compared to healthy controls, although subjective cure rates do not seem to differ.

3. The severity of obesity also negatively influences the results of the operation as morbidly obese patients are twice as likely to report a failure following a mid-urethral sling operation

4. In the long term, obese women undergoing sling procedures have worse outcomes according to the findings of a recent study that followed up patients for a period of 5 years.

5. Specifically, obese patients had worse objective cure rates (65.9% compared to 87.4% in nonobese) as well as subjective cure rates (53.6% vs 76.7%).

6. The incidence of urinary urgency incontinence was comparable in both groups; however, bothersome symptoms were more likely to persist in obese women (58.9% vs 42.1%).

7. In a case control study, compared with the nonobese population, the operating time was more prolonged with a slightly higher estimated blood loss in obese patients undergoing sling procedures for UI.

   a. Evidence concerning the efficacy of bulking agents in obese women is lacking.

   b. Evidence concerning bladder neck suspension procedures is limited to case reports; hence, definitive conclusions cannot be drawn to guide clinical practice.

12.2.1   Obesity and overactive bladder

1. Overactive bladder may be neurological or idiopathic in nature. The management of overactive bladder comprises bladder training, use of anticholinergics or more recently antimuscarinics, beta-3 agonists, neuromodulation, and cystoscopic injection of botulinum toxin.

2. Overactive bladder symptoms and their resulting distress are more common in obese patients compared to the general population.
3. A longitudinal study in changes in urinary function in parous women has shown that overactive bladder symptoms are more common in obese patients, similar to stress incontinence.
4. Obesity and poor lifestyle factors are markedly associated with incident or new-onset overactive bladder in a large database of >3000 patients. Such studies demonstrate a strong link with obesity and overactive bladder.
5. In a case control study, the distress caused by overactive bladder symptoms was considerably more in the obese group of patients. This may be due to coexisting mobility issues that prevent such patients from getting to the toilet in time when suffering urge incontinence and urgency.

The treatment of overactive bladder in the obese group of patients also comprises lifestyle changes and the use of medications.

12.2.2   Pharmacological treatment

1. Pooled data analysis of seven randomised placebo-controlled trials to evaluate the relationship between BMI and the efficacy or tolerability of solifenacin 5 and 10 mg has reported that baseline incidence of urge urinary incontinence (UUI) increased with increasing BMI and age. The treatment was effective for all OAB across all BMI categories, and between sexes.
2. Another trial and analysis had also shown a similar effect in another antimuscurinic and beta-3 agonist (mirabegron) and BMI was not associated with dose escalation.

12.2.3   Sacral neuromodulation

1. Only a few studies have explored predictors of treatment outcome with sacral neuromodulation.
2. Increased baseline BMI was found to decrease the chance of achieving $\geq 50\%$ UUI episodes reduction.
3. The study model estimated that a woman with a BMI of 25 would have a roughly 50% greater chance of achieving $\geq 50\%$ improvement in UUI compared with a BMI of 35.
4. In a prospective longitudinal study, investigators found that, among other factors, decreased BMI had a higher risk of reoperation.
5. In another study, the results suggested that lower BMI might predict reoperation; however, BMI was not significant when adjusted for other covariates.
6. A more recent retrospective review also failed to find any associations between obesity and reoperation.

12.2.4   Onabotulinum toxin A (BoNT-A)

1. Only one study had evaluated BMI as a predictive factor for treatment outcome for Onabotulinum toxin A in a cohort of women with idiopathic detrusor overactivity.
2. It is shown that besides smoking, higher BMI was associated with an increased risk of treatment failure.

12.3   Anal incontinence

1. Anal incontinence (AI) is the complaint of involuntary loss of flatus (flatus incontinence) or faeces (faecal incontinence FI).
2. Epidemiological data related to AI are more scarce compared to those available for UI. This is primarily owed to the emotional consequences of these disorders, which have a significant impact on the patient's self-esteem and QoL as well as to infrequent screening from healthcare providers.
3. The actual incidence of the disease is estimated to range from 7% to 15% in community-dwelling men and women.
4. AI is more prevalent among institutionalised individuals with data referring to nursing homes reporting an actual incidence between 50% and 70%.
5. This fact is primarily the result of the increased prevalence of dementia in these populations, a disorder that has a direct negative impact on anal sphincter control.
6. Systematic review of BS research (men and women) reported the rates of FI between 16% and 68%, in each case higher than the rates reported for nonobese individuals.
7. Several risk factors have been related to AI, including operative vaginal delivery, obstetric anal sphincter injury, advanced age, decreased physical ability, neurological diseases, obesity, and intestinal motility problems (primarily diarrhoea).
8. The impact of AI on patients' QoL can be catastrophic as the majority of them lack self-esteem and progressively diminish their social interactions to the minimum, together with hiding their problem from their relatives and in several occasions even from their general practitioner.

12.3.1   Pathophysiology anal incontinence in the obese population

Several pathophysiological mechanisms have been proposed for AI.

1. Altered stool consistency is among the proposed factors that contribute to FI as obese patients report altered bowel habits that are accompanied by unformed stools.
2. A recent prospective case-matched study that compared obese patients to age- and sex-matched nonobese patients with FI reported that the rates of FI were comparable between the two groups, although stool inconsistency seemed to be more prevalent among obese patients.
3. Significant differences were also observed in anorectal manometry results with obese patients having higher upper- and lower part resting pressures, higher intra-abdominal pressure during effort, and increased maximum tolerable volume.
4. The baseline anal resting and squeeze pressures were also shown to be increased in obese women with FI, suggesting that the threshold for leakage lowers as the pressure increases.

12.3.2   Weight loss and anal incontinence

1. Data on improvements in AI after weight loss are sparse. The impact of bariatric surgery on AI symptoms is rather disappointing.
2. The most recent systematic review summarised evidence from 20 BS studies revealed a modest effect with a reduction of the odds of FI by approximately 20%; however, the change was not significant (OR 0.80, 95% CI 0.53, 1.21).

3. This observation may be attributed to the relatively small number of enrolled patients.
4. A study that investigated functional anorectal parameters obese women with pelvic floor disorders revealed that bariatric surgery had no impact on internal and external anal sphincter size and on mean anorectal angle during squeeze and during defecation.

12.3.3 Obesity and treatment for anal incontinence

1. AI is primarily treated with behavioural treatment that aims to reduce stool inconsistency.
2. Various treatment alternatives have been proposed, including bowel training, biofeedback, antidiarrhoeal drugs, and bulk laxatives (in cases of chronic constipation).
3. Neither of these methods has been evaluated in obese populations.
4. Sphincteroplasty remains the cornerstone of treatment in cases of damaged anal sphincter.
5. A study that included 15 obese and 64 nonobese women, with a median follow-up period of 64 months, showed that, although the risk of complications was comparable between the two groups, improvement was less evident in obese patients.
6. Perianal bulking has also been used and showed promising results, however, data in obese populations are unavailable and they cannot be yet recommended for the treatment of FI.

12.4 Pelvic floor dysfunction

Pelvic organ prolapse seems to be the result of damage to the pelvic floor by different mechanisms of injury, such as obesity, childbirth, and menopause.

Several studies have shown that all the different types of pelvic floor prolapse are more common among obese women.

In one study, morbid obesity was associated with a significant increase in the occurrence of uterine prolapse (40%), rectocele (75%), and cystocele (57%).

The risk of prolapse progression in overweight and obese women compared with normal BMI increased by 32% and 48% for cystocele, by 37% and 58% for rectocele, and by 43% and 69% for uterine prolapse, respectively.

12.4.1 Weight loss and uterovaginal prolapse

Intensive nonsurgical and surgical methods of weight loss have not been shown to reverse the distressing nature or severity of symptoms of pelvic organ prolapse.

Weight loss in obese helps to stop the progression and worsening of symptoms of prolapse.

Weight loss may also help in reducing the postsurgical morbidity associated with obesity and prolapse surgery.

12.4.2 Pelvic organ prolapse surgery and obesity

Obesity is known to be associated with more complications in a postsurgical period due to restricted mobility and the resumption of activities of daily living.

Conservative measures such as the use of various types of pessaries should be considered as the first line of management.

The discussion about the route of surgery becomes more relevant in this group and it should be either vaginal or laparoscopic route to encourage early mobility.

The vaginal route has been reported to be associated with less febrile morbidity, postoperative ileus, and urinary infection.

There are no long-term follow-up studies reporting on the relationship between the success of prolapse surgery, or complications, and various classes of obesity.

12.5 Conclusion

1. Obese women are more prone to developing UI and AI compared to normal-weight women.
2. Increased intra-abdominal pressure seems to play a major role in the process of the development of UI with a direct negative impact on pelvic structures.
3. Evidence in the field of AI is less clear but suggests that stool inconsistency is more important than the increase in intra-abdominal pressure.
4. Weight loss results in important changes in urinary tract function, whereas its impact on anorectal manometry and functional parameters is minimal, if any.
5. Concerning continence procedures, long-term surgical outcomes for incontinence are less clear; however, obesity is associated with aggravated severity of symptoms postoperatively.
6. Evidence around the surgical treatment of FI is limited and does not suffice to form conclusions for clinical practice.

Obesity and pelvic organ prolapse 13

Chu Chin Lim
Department of Obstetrics & Gynaecology, Victoria Hospital, Kirkcaldy, United Kingdom

13.1 Introduction

1. Pelvic Organ Prolapse (POP) prevalence rates range from 10% in younger women and up to 50% in postmenopausal women.
2. Nearly 1 in 10 women will undergo surgical correction for POP in their lifetime.
3. This can be defined as descend into the vaginal space prolapse of >1 intrapelvic organ (uterus, bladder, rectum, and the urethra), presumably due to deficiencies in the pelvic support system that normally provides sustained support.
4. There are many known and unknown variables that affect the severity of POP and its symptoms.
5. Epidemiological studies have identified age, race, parity, size of infant, and body mass index (BMI) as independent risk factors for POP.
6. Aging and parity have been most consistently associated with POP; however, these factors are not modifiable. Obesity is a modifiable risk factor that may be influenced on a population level to reduce the public health and economic burden of POP.
7. As the population ages, the prevalence of POP is likely to increase, and more women will undergo surgical procedures to treat prolapse.
8. These two factors—increasing obesity rates and the aging population—will most likely increase the rates of POP beyond what is predicted.
9. Data from published cross-sectional and prospective studies suggest that being overweight or obese is associated with prevalent and incident POP as well as progression of POP; however, few studies have evaluated the impact of weight loss on subjective or objective POP or symptom severity

13.2 Obesity as risk factor for pelvic organ prolapse

1. POP is defined as the descent of the anterior vaginal wall, the posterior vaginal wall and/or the apex of the vagina (cervix or vault after hysterectomy).
2. For clinical and research purposes, the Pelvic Organ Prolapse Quantification (POP-Q) scale is used for an objective quantification of degree of the prolapse.
3. The prevalence of POP varies depending on the used definition. The subjective diagnosis of POP is mostly defined by the sensation of vaginal bulging. The reported prevalence range is 6%−11%. The subjective presence of a POP is strongly associated with a prolapse beyond the level of the hymen.

Handbook of Obesity in Obstetrics and Gynecology. DOI: https://doi.org/10.1016/B978-0-323-89904-8.00015-9

4. Risk factors for developing a POP can be divided into obstetric, lifestyle, comorbidity, nonmodifiable (e.g., age), social, pelvic floor, and surgical factors. Parity and aging are the strongest risk factors of POP. The most important lifestyle factor is a higher BMI.

5. The most probable mechanism of POP development among obese women is the increase in intraabdominal pressure that causes weakening of pelvic floor muscles and fascia. However, studies evaluating the association between obesity and POP have reported inconsistent conclusions.

6. Study showed symptomatic POP increased by 3% with each unit increase in current BMI. Recent published systematic review and meta-analysis showed that, compared with normal-weight women, women in the overweight and obese categories had risk ratios of at least 1.36 (95% CI, 1.20−1.53) and 1.47 (95% CI, 1.35−1.59), respectively, of developing POP.

7. A large US study that analysed data from 16,608 women showed progression of POP with increasing body weight. The excess risks for anterior vaginal prolapse were 32% and 48% in overweight and obese women, respectively, for posterior prolapse 37% and 58%, and for uterine prolapse it was 43% and 69%. However, weight loss did not significantly reduce degree of POP and suggested that damage to the pelvic floor associated with obesity may be irreversible.

13.3   Weight loss and the effects upon pelvic organ prolapse

1. Data on whether weight loss alters prolapse severity are also scarce. A large prospective study found that weight loss had only minimal effects upon anatomical prolapse.

2. The role of weight reduction on POP symptoms is an important clinical question as weight loss is an action that the patient can initiate through diet, exercise, and/or in some cases weight-loss surgery.

   a. Weight loss in an overweight or obese person is a positive action toward improving one's general health.

   b. Obesity is thought to cause increased intraabdominal pressure, which then transfers strain to supporting pelvic floor structures.

   c. Obesity may coexist with other comorbidities, such as diabetes, which are thought to contribute to poor tissue quality.

   d. Studies on this topic shed light on the complexity of POP as a disease process and how to appropriately counsel patients who are interested in weight loss as a treatment option.

3. There are limited studies on women who present specifically with symptomatic POP and underwent a weight loss program to treat their weight and prolapse symptoms.

4. Studies of the impact of obesity have focused on other pelvic floor disorders (PFDs) such as incontinence and overactive bladder, so weight loss has not directly been shown to decrease prolapse symptoms in these subset populations.

   a. One study looked at weight reduction on UI and pelvic floor anatomy, and found that in 378 women, the weight loss group with a mean weight reduction of 9.4%, only genital hiatus, perineal body, and Ap measurement were lower in the weight-loss group compared to the control group at 6 months. The authors concluded that there was little to no change in significant POP-Q variables after weight reduction.

b. A study from Egypt assessed 400 women and found vaginal prolapse in 65% of the sample, although symptoms were much less common.

c. In a group of women who underwent bariatric surgery, POP symptoms were reported by 56 women, and 15 women had documented anatomical prolapse. After surgery, 74% of the affected women had resolution of their prolapse symptoms.

13.4 Prolapse surgery in the obese woman

The obese woman with pelvic prolapse is a challenge for the pelvic surgeon.

For prolapse surgery, data on the relationship between obesity and the outcomes and complications are extremely limited.

Weight loss should be considered the primary option in obese women for its salutary effects on multiple organ systems and reducing PFD symptoms.

1. A very important question is how obesity influences the risk of intraoperative and postoperative complications as well as the outcome of surgical management of POP.
2. Laparoscopic or vaginal route may reduce the incidence of thromboembolic events associated with obesity.
3. The vaginal approach is also beneficial because it has been shown that obese women undergoing vaginal hysterectomy sustained fewer perioperative complications.
4. Irrespective of the type of surgery, the operation time in obese women is significantly longer than in healthy weight women.
5. Intraoperative surgery complications of the different vaginal surgeries (vaginal hysterectomy, anterior and posterior colporrhaphy, iliococcygeal hitch or posterior intravaginal sling), no differences are observed between obese and nonobese patients.
6. Complication rate of laparoscopic sacrocolpopexy is not different in obese and nonobese patient.
7. It seems that obese women are not at higher risk for perioperative and postoperative complications. Data from larger studies with longer follow-up are needed to be confident of this conclusion.

13.5 Recurrence of pelvic organ prolapse

Impact of obesity on surgical outcomes and the data regarding whether obesity is a risk factor for recurrence after POP surgery are controversial

1. The retrospective cohort study shows that after total vaginal hysterectomy with concurrent vaginal uterosacral ligament suspension, overweight or obese women have a similar overall risk of 20% prolapse recurrence (composite outcome definition of any anatomic prolapse beyond the hymen or pessary or repeat surgery).
2. Overweight and obese women are, however, more prone to recurrence in the anterior wall compared with normal-weight women.
   a. After anterior colporrhaphy, the risk of recurrence is relatively high in the short term and obesity is associated with increased odds of anatomic recurrence of anterior vaginal wall prolapse

    b. Five-year analysis of a prospective observational study where 376 women were followed after surgery for POP and/or UI showed no association was found with BMI in these surgical failures.

    c. A prospective study evaluated the development of POP in patients who underwent Burch colposuspensions. At 8-year follow-up, 38% had developed symptomatic prolapse and another 38% had asymptomatic prolapse but BMI was not found to have a significant association with surgical failure.

    d. In a secondary analysis at 2 years postoperatively of the Colpopexy and Urinary Reduction Efforts (CARE) trial, obese women were found to have significantly more prolapsed posterior vaginal wall, compared to healthy-weight women. The obese group reported more colorectal symptoms and related functional impact, but no differences were found in subjective prolapse symptoms and patient satisfaction outcomes.

Native tissue repair is the standard method for POP surgery, whereas the use of mesh in POP surgery has become controversial.

1. It is stated that the use of mesh should be reserved for high-risk individuals in whom the benefit of the use of mesh may justify the risks, such as individuals with recurrent POP.
2. In general, the population in studies on POP recurrence after mesh surgery often consist of a selected, high-risk group of women, which cannot be compared with the population in studies on POP recurrence after native tissue repair.
3. Looking at the outcomes of obese patients with the use of anterior trans-obturator mesh and vaginal sacrospinous ligament fixation, the surgical outcomes are not inferior compared to outcomes in nonobese women.

In an obese population with POP, no significant differences in the recurrence of POP after sacral colpopexy versus vaginal mesh colpopexy are noted, with better anatomical outcome of sacral colpopexy.

13.6   Conclusion

1. Worldwide, the number of women with overweight, obesity, or morbid obesity is impressive with prevalence of obesity ranging from 4% to 36%.
2. Overweight and obese women are more likely to have POP than normal weight women. This association is larger for clinically significant POP.
3. Weight loss (either by diet and exercise or by BS) is associated with large improvements in prolapse symptoms. Achieving a target weight loss between 5% and 10% of baseline weight will bring about complete resolution of prolapse symptoms in up to 70% of women.
4. Where surgery is deemed necessary, women should be advised that prolapse surgery appears equally safe in the obese patient but that the long-term failure rate after prolapse surgery is greater in the obese.
5. Obesity does not seem to be a strong risk factor for recurrence of POP in the short term. However, an increase in intra-abdominal pressure may have a negative impact on postoperative results in the long term.
6. Patients should be counselled on maintaining a healthy weight for their overall health and its impact on pelvic floor symptoms.
7. There are many gaps in the literature related to obesity and POP.

Key points

1. The most probable mechanism of POP development among obese women is the increase in intraabdominal pressure that causes weakening of pelvic floor muscles and fascia.
2. Obesity is associated with significant pelvic floor symptoms and impairment of QOL. Weight loss is likely not associated with anatomic improvement but may be associated with prolapse symptom improvement.
3. Weight loss should be considered a primary option in obese women for its beneficial effects on multiple organ systems and reducing PFD symptoms.
4. Although the operation time in obese women is longer than in healthy weight women, the complication rate of surgery has not been shown to be increased compared to nonobese patients, regardless of route of surgery.

Obesity and chronic pelvic pain

14

Chu Chin Lim
Department of Obstetrics & Gynaecology, Victoria Hospital, Kirkcaldy, United Kingdom

14.1   Introduction

1. Chronic pelvic pain is constant or intermittent pain persisting greater than 6 months in the pelvic or lower abdominal region.
2. This definition features wide variation worldwide and lacks a clear standard classification method even agreed among various scientific institutions worldwide.
3. Chronic pelvic pain is estimated to affect 4% of all women and 15% of reproductive age women.
4. Chronic pelvic pain is a symptom of a condition, not a diagnosable condition in itself.
5. Obesity is the excess accumulation of adipose tissue.
6. WHO 2016 estimated 40% of women worldwide were overweight with 15% classified as obese.
7. These numbers represent a tripling since 1975 from 650million to 1.9billion.
8. The most common measurement of obesity is a body mass index (BMI) value greater than 30 kg/m^2, however this does not effectively account for gender.
9. More female specific measures of obesity are
    a. Waist circumference greater than 88 cm (35inches)
    b. Waist-to-hip ratio greater than 0.85
    c. Greater than 30% body fat percentage.
10. There is a shortage of existing research in the field of obesity and chronic pelvic pain relationship.
11. In the passing comments which do mention the relationship, the implication is that increased BMI causes increased severity of chronic pelvic pain.
12. Most information regarding chronic pelvic pain and obesity is extrapolated from other chronic pain syndromes.

14.2   Obesity and pain physiology

1. Pain is the stimulation of nociceptors, transmitting impulses via spinal cord, thalamus, and limbic system to its perception portion of the cerebral cortex.
2. Chemical mediators are released in the offending area and can manipulate the signal all the way through the passage from receptor to brain.
3. Adipose tissue is a source of inflammatory proteins such as tumour necrosis factor alpha (TNF-α), interleukin-6 (IL-6), and IL-1 beta (IL-1β).
4. Common regulators of body homeostasis have shown involvement with inflammation and pain sign modulation.

Handbook of Obesity in Obstetrics and Gynecology. DOI: https://doi.org/10.1016/B978-0-323-89904-8.00037-8

5. Leptin, produced to inhibit hunger, may also increase C reactive protein (CRP) levels increasing pain perception.
6. Ghrelin is secreted in the stomach to increase hunger and glucose metabolism; its levels are decreased in obesity.
7. Ghrelin also inhibits IL-6, IL-1β, and TNF-α reducing inflammation.
8. Ghrelin increases nitric oxide synthase production which modulates μ-opioid receptors and produces an antinociceptive effect.
9. All these effects cause an increased level of inflammation and susceptibility to pain in obese people due to the reduction of ghrelin secretion.
10. Appetite is also stimulated by neuropeptide Y (NPY).
11. NPY stimulates appetite and reduces energy expenditure.
12. NPY also stimulates glucocorticosteroid production, gluconeogenesis, and glycogen storage.
13. Interactions with ghrelin, leptin, and insulin-like growth factor further contribute NPY to obesity.
14. Nociceptive effects of NPY appear to be linked to the site of the receptor, CNS NPY stimulate analgesia, peripheral postsynaptic receptors stimulate hyperalgesia.
15. Obese women have higher serum NPY levels.
16. The orexinergic system consisting of neuropeptides, orexin A and B stimulate appetite and reduce pain perception.
17. Orexin systems are stimulated in stress times, acute or chronic pain, to inhibit pain transmission and improve physical performance.
18. Due to this stimulation, chronic pain can contribute to weight gain.
19. Due to these associations, obesity is being considered a proinflammatory state.
20. Obese people are at greater risk of neuropathic pain development, for example, peripheral neuropathy.
21. The effect of obesity on pain is unclear with studies finding opposite results in terms of the relation between BMI and pain effects.
22. Obesity is more prevalent in chronic pain sufferers, pain limits physical activity.
23. This limited activity leads to weight gain and deconditioning.
24. This creates a cycle for pain and weight gain.
25. Weight can cause structural changes making development of pain more likely.
26. Quality of life surveys show that weight loss can significantly improve bodily pain scores.

14.3   Genetics of obesity and pelvic pain

1. Genetics is thought to play a role in obesity, pain perception, and sensitivity.
2. Studies indicate there is likely a heritable component to pelvic pain, through a heightened level of pain perception and sensitivity.
3. No current studies have been done to investigate the genetic link of obesity and chronic pelvic pain.
4. However, there is growing research showing a genetic link between obesity and pain in other areas, for example, lower back pain.
5. Melanocortin receptor 4 (MC4R) mutation cause appetite dysregulation and hyperphagia; incidence of MC4R deficiency is noticeably greater in obese populations compared to the general population.

6. Melanocortin receptors are also implicated in regulation of pain and chronic pain syndromes.
7. Another pathway may be glucocorticoid receptor gene polymorphisms.
8. Impaired receptor function creates glucocorticoid resistance.
9. Negative feedback because of this causes excessive production of glucocorticoids causing inflammatory cytokine stimulation such as CRP and visceral fat accumulation.
10. Many of the comorbidities of chronic pain and obesity are also implicated.

14.4   Psychological impact of obesity and chronic pain

1. There is a likely bidirectional effect between psychiatric disorders and both obesity and chronic pelvic pain.
2. Either reduce a person's quality of life and both likely cause an additive effect.
3. Both obesity and chronic pelvic pain demonstrate body image distortion elements.
4. There is minimal research investigating the psychological effect of both obesity and chronic pelvic pain in combination.
5. It has been suggested that obese women with chronic pelvic pain experience show an increase in depressive symptoms but not anxiety symptoms.
6. Chronic pain and obesity show a cumulative negative effect on psychological morbidity.
7. Anxiety and depression increase circulating levels of CRP and IL-6 in the obese individual.
8. Weight-loss treatments have been shown to improve quality of life and pain management in chronic pain management sufferers.
9. Pain treatments effect on weight loss has not been studied.
10. In developed countries, women from low socioeconomic backgrounds are more likely to be obese.
11. There is also a negative relationship between socioeconomic status and chronic pelvic pain.
12. Economic hardship has a negative effect on mental health.
13. In low socioeconomic groups there is potential for a compound effect of economic hardship, chronic pelvic pain, and obesity on mental health.

14.5   Impact of obesity on assessment of chronic pelvic pain

1. The excess adipose tissue in obesity increases the difficulty and limits the effectiveness of the pelvic examination which is necessary to distinguish between various aetiologies of pelvic pain.
2. This is compounded by the lack of appropriate equipment, such as larger instruments and bariatric examination couches.
3. The pitfalls and barriers present in obesity, chronic pelvic pain, and low socioeconomic groups have many similarities.
4. In groups with overlaps between these three conditions the barriers are raised due to the compound effect.
5. Due to difficulties in pelvic examination, imagining is becoming more prevalent. Ultrasound is the method of choice; however, alterations need to be made to minimise the effects of excess fat on the quality of images.

6. Experience in working with these images also presents a challenge.
7. Excess fat also presents an increased risk of muscular strain to the sonographer.
8. CT and MRI can produce better images, however there are restrictions on their use and there is an upper limit of capacity for weight and gantry diameter in these machines.
9. The supine position inside the machine can potentially cause hypoxia and hypotension secondary to aorto-caval compression by a large pannus.
10. Estimation of contrast dose for CT scan is challenging as it is based on lean body weight, and the increased radiation exposure, which is required for optimisation of image quality.

14.6   Impact of obesity on treatment of women with chronic pelvic pain

1. The oral contraceptive pill (OCP) is often used as an empirical treatment of chronic pelvic pain as oestrogen is a pain modulator.
2. Obesity and OCP both increase risk of venous thromboembolism (VTE) by two to three times, respectively.
3. Both obesity and OCP are likely to cumulatively increase the risk of VTE.
4. Obese OCP users are 10 times more likely to have a VTE event than normal nonusers.
5. Obese OCP users are also thought to potentially have an increased risk of acute myocardial infarction.
6. Hormonal treatment is the first line of common potential gynaecological conditions (endometriosis and adenomyosis),
7. The effect of OCP has not been adequately assessed in the treatment of chronic pelvic pain.
8. *Progesterone*-only subdermal implants and intrauterine devices are preferred methods of hormone suppression in obese patients.
9. Dosing recommendations of analgesic are based on total body weight from studies that do not include obese cohorts.
10. Obese people have a large magnitude of changes affecting clinical efficacy and drug metabolism, thought to be associated with lipophilicity of the drug.
11. Opioids and other sedating analgesics may exacerbate existing abnormal breathing patterns, thus compounding the risk of hypoxemia via central sleep apnoea, OSA, and ataxic breathing.
12. Neuropathic pain medications and hormonal suppressants may promote weight gain.
13. Psychological component of chronic pelvic pain in obese patients must be considered as pain management success is reduced in this combination.
14. Cognitive behavioural therapy, goal setting techniques, and development of coping strategies have been shown to be effective in obese chronic pelvic pain sufferers.

14.7   Impact of obesity on surgical management of women with chronic pelvic pain

1. Investigation and management of chronic pelvic pain may require diagnostic laparoscopy.
2. Laparoscopy requires a steep Trendelenburg position which can be difficult in obese patients, this is worsened by the excess adipose tissue viscerally as well as externally obscuring view and increasing the risk of injury.

3. The gastric changes in obese patients increases the risk of gastric aspiration during laparoscopy in this cohort.
4. Detrimental respiratory changes of obesity are compounded by the effects of the general anaesthesia, increasing the risk of hypoxemia and the degree of intrapulmonary shunting.
5. Pneumoperitoneum pressures necessary for laparoscopy further exacerbate the physiological changes of obesity.
6. Compromise between anaesthetist and surgeon regarding depth of incline and insufflation pressures are necessary to trade between surgeon fatigue and impact on respiratory physiology.
7. Laparoscopy is preferred to laparotomy due to the lower complication rate.
8. Robot-assisted laparoscopy may decrease operation time, blood loss, and length of hospital stay.
9. However, only literature for oncological problems is available and nothing specifically relating to chronic pelvic pain.

14.8   Obesity and endometriosis

1. Endometriosis is an oestrogen-dependent inflammatory condition and the most common gynaecological disease among reproductive age women with chronic pelvic pain.
2. Increased levels of adipose tissue in obese women results in increased levels of aromatase expression, causing higher oestrogen levels by the following mechanisms:
   a. the increased numbers of adipose cells result in increased aromatase enzyme activity resulting in increased circulating oestrogen;
   b. increased conversion of the less active oestrone to the more active oestradiol by the β-hydroxy steroid dehydrogenase type 1 enzyme;
   c. increased number of oestrogen receptors;
   d. decreased numbers of progesterone receptors;
   e. higher levels of leptin and lower levels of ghrelin in the peritoneal fluid of women with endometriosis show higher levels of CRP and IL-6.
3. These higher levels of oestrogen show the link between obesity and endometriosis, and other oestrogen-related cancers of ovary, breast, and endometrium are seen more often among obese women,
4. It is not yet proven if it is a causality or merely an association in this similar risk profile group.
5. However the correlation between higher BMI and the severity of endometriosis is controversial.
6. It is more likely that higher levels of oestradiol can result in neuromodulation with enhanced sensitivity of peripheral nerves and modulate CNS activity.
7. Minimal to mild disease can not be identified with imaging modalities, and MRI is more likely to identify endometriosis than transvaginal ultrasound.
8. Endometriosis requires long-term hormone suppression therapy, as previously discussed OCP increases VTE risk so therefore progesterone-only suppression is recommended.
9. Oophorectomy may be used as a last resort for treatment of endometriosis.
10. Recurrence of symptoms within 5 years may be due to increased aromatase enzymes in obese women.
11. Menopause increases risk of several other conditions also exacerbated by obesity (coronary heart disease, depression, anxiety, and all-cause mortality).

14.9　Obesity and adenomyosis

1. Adenomyosis is the presence of ectopic endometrial-like tissue within the myometrium.
2. Adenomyosis has a higher incidence in obese women.
3. Most symptoms of adenomyosis can be managed with hormone suppression or hysterectomy.
4. The challenges with hormonal suppression and surgical treatment are the same as endometriosis.
5. MRI preferable to ultrasound in cases of suspected adenomyosis because of greater sensitivity and positive predictive value.

14.10　Obesity and abdominal myofascial pain syndrome

1. Abdominal myofascial pain syndrome can affect up to 93% of women attending pain clinic.
2. The syndrome causes inflammation and intense pain in the pelvis activated by trigger points in muscle fascia, which can be palpated as taut bands or spasm of the rectus abdominus and pelvic floor muscles.
3. Pelvic organ prolapse, which obesity increases the risk of, may also be a cause of this pain.
4. Diagnosis of this condition is difficult due to the earlier challenges in pelvic examination as well as the thickening of subcutaneous layer.

14.11　Obesity and nongynecological causes of chronic pelvic pain

1. Non gynaecological causes can be urological, gastrointestinal, musculoskeletal, or psychological.
2. Interstitial cystitis and irritable bowel syndrome, although commonly seen in women with chronic pelvic pain and endometriosis, have questionable links to obesity as causes of pelvic pain.
3. Lower back pain is directly related to BMI and obesity and may refer to pelvic pain.
4. Victims of sexual, verbal, or physical abuse may be obese and have chronic pain including pelvic pain.

14.12　Conclusion

1. The inconsistency of the definition of chronic pelvic pain is problematic to coalescing ideas.
2. Although poorly researched, there are several probable links between obesity and chronic pelvic pain.

3. Multidisciplinary management of chronic pain results in reduced pain, somatisation, use of hospital resources, and increased mood and speed of return to work.
4. Obesity proves challenging for the assessment and management of chronic pelvic pain, as such adequate equipment is vital.
5. MRI is the preferred method of imaging.
6. Progesterone-only hormone suppression is more beneficial as it avoids complications.
7. Patient education, respectful communication, legitimisation of their pain, and multidisciplinary holistic care overcomes the barriers of management of chronic pelvic pain in obese women.

Further reading

American College of Obstetricians and Gynaecologists. Frequently asked questions: gynaecologic problems, FAQ099; 2011.

European Society of Human Reproduction and Embryology. Guidelines on the management of women with endometriosis; 2013.

Royal Australian and New Zealand College of Obstetricians and Gynaecologists. Chronic pelvic pain; 2017.

Royal College of Obstetricians and Gynaecologists. The initial management of chronic pelvic pain. Greentop guidelines, London, RCOG; 2005.

Seaman H, Ballard K, Wright JT, De Vries CS. Endometriosis and its co-existence with irritable bowel syndrome and pelvic inflammatory disease: findings from a national case control study-Part 2. *BJOG*. 2008;115(11):1392−1396.

Tan I-F, Horne AW. Obesity and chronic pelvic pain (In). In: Mahmood TA, Arulkumaran S, Chervenak FA, eds. *Obesity and Gynaecology*. 2nd ed. London: Elsevier; 2020:281−291. Available from: https://doi.org/10.1016/B978-0-12-817919-2.00031-0.

Obesity and clinical psychosomatic women's health

15

Chu Chin Lim

Department of Obstetrics & Gynaecology, Victoria Hospital, Kirkcaldy, United Kingdom

15.1 Introduction

1. The concept "clinical psychosomatic" brings together and emphasises the connection between mind and body as being relevant to clinical medicine when evaluating diseases that affect both physical and mental health concomitantly.
2. It accounts for the fact that the body and mind are not disparate entities. Rather, they are anatomically and physiologically linked via the neuroendocrine system, with their interplay influencing the maintenance of overall health and the generation of clinical psychosomatic disease conditions.
3. Obesity is often associated with diseases due to clinical psychosomatic interactions that can affect women's reproductive health, such as menstrual problems, metabolic disorders, infertility, gender-related violence, and cancer.
4. Both physical and mental illnesses in such despondent patients can lead to overeating and obesity. Early recognition and appropriate attention to relentless weight gain, often due to burgeoning psychosomatic issues, would likely prevent many cases of obesity.

15.2 Psychosomatic aspect of menstrual problems and obesity

15.2.1 Pathophysiology of psychosomatic menstrual issues

1. Normal menstrual bleeding lasts for about 5 days, accompanied by cramping abdominal pains that radiate to the thighs, hips, and lower back, though pains may start before menstruation begins.
2. These features are considered as normal by many who obtain symptomatic relief by rest, distractions, local heat, or analgesics (NSAIDs).
3. Irregular menstrual cycles may occur at menarche and when the pattern for a while changes to shorter premenopausal cycles associated with oligoovulation.
4. These phases can be associated with painful/heavy menstrual bleeding that affects psychosomatic welfare, more so in the obese.

15.3 Psychosomatic insights into menstrual problems in the obese

1. Obese woman can present with complaints of dysmenorrhoea, menorrhagia, menometrorrhagia, premenstrual syndrome (PMS), oligomenorrhea, amenorrhoea, or menopause. All of these are usually considered benign conditions.

2. They may cause dysphoria (anxiety and/or depression) in those experiencing these menstrual deviations, which can compel some women to seek relief by comfort eating, even if they might become overweight.
3. Nonmalignant growths, such as fibroids, ovarian cysts, and endometriosis, may be associated with heavy blood loss and dysmenorrhoea, which can promote overeating with an obesogenic body habitus.
4. Endometrial cancer or other malignancies of the female reproductive organs could present with menstrual problems, and these conditions occur more frequently in the obese. These cancers are affecting obese younger women as well.

15.4   Clinical psychosomatic approach to menstrual disorder

1. Menstrual problems may be perceived as ill-health, even if these problems are considered as normal by many, thereby encouraging some to gormandise in order to "feel good."
2. Such problems can be of significance in the obese teenager undergoing menarche, especially if discussions with parents/caregivers medicalise reasonable lifestyle restrictions as so-called menstrual abnormalities.
3. A medical referral that sometimes involves a gynaecological assessment could ensue. Gynaecologist may not be comfortable in assessing associated dysphoria that could lead to obesity.
a. Young patients may need gentle handling during the history-taking, examination, and investigations.
b. Relevant initiating/aggravating biopsychosocial factors deserve appropriate consideration when revealed at a medical consultation for menstrual irregularities; a clinician with psychosomatic expertise can often be successful in bringing about symptom relief with less invasive methods.
c. If a constitutional delay at menarche is diagnosed without any pathological contributing factors, reassurance and lifestyle alterations could limit the risk of overeating because of anxiety/depression, which could lead to becoming overweight/obese.
d. If endocrinological or chromosomal anomalies are confirmed, and complex hormonal/surgical treatments considered necessary to alleviate the menstrual problem, the patient may experience feelings of shock, grief, denial, or guilt, due to perceived loss of femininity.
4. Clear and empathetic communication can be enhanced by taking a psychosomatic viewpoint during discussions regarding the management of these health issues, even more so in the overweight/obese.
5. A greater appreciation for the psychosomatic management perspective could also benefit the gynaecologist facing such scenarios, by fostering better doctor–patient rapport, thereby making patients more likely to adhere to recommended lifestyle interventions with the goal of preventing obesity.
6. The problem warrants an evaluation by a gynaecologist with a clinical psychosomatic-oriented approach. Facilities for such evaluations are sparse, and there are limited numbers of trained staff who can deliver the necessary patient-cantered assessments.

15.5   Clinical psychosomatic approach to premenstrual syndrome

1. Premenstrual molimina (dysmenorrhoea, bloating, dysphoria, irritability, headaches, bowel symptoms, and breast tenderness) is perceived as normal and dealt with as trivial discomfort by many.

2. There is trend among many gynaecologists to classify such menstrual symptomatology as PMS.
3. A more severe form of premenstrual symptomatology, premenstrual dysphoric disorder (PMDD) is associated with tearfulness, sleeplessness, or a preference for being bed-bound, overeating or having food cravings—all factors that increase the risk of obesity, besides amplifying the psychological impact of menstrual molimina.
4. PMS and the more severe form PMDD have been considered as psychosomatic disorders related to endocrinological and autonomic alterations associated with changes in the woman's body and certain brain centres during the menstrual cycle.
5. These conditions can also promote weight gain from being sedentary and overeating, accordingly promoting obesity. In certain social groups, comfort eating is encouraged for women with PMS/PMDD as a coping strategy, often with little regard for potential weight increases.
6. Gynaecologists who are unfamiliar with psychosomatic issues usually prescribe hormonal medication, failing which surgical treatment is advised including hysterectomy, yet the problem often persists.
7. Consultation with a psychosomatically oriented clinician may persuade the patient to accept a trial of noninvasive management with risk of fewer potential side effects, including lifestyle modifications that could also remedy the ill effects of obesity.

15.6   Psychosomatic aspect of infertility and obesity

1. Infertility or subfertility is the inability to conceive despite regular unprotected intercourse for a defined period of time and can results in psychosomatic consequences that are often associated with obesity in women.
2. Woman has often been blamed for failures to conceive, which has persisted in many cultures even if investigations disclose that she is fertile.
3. Obesity can impact on fertility by impairing the development of ovarian follicles, causing defective oocyte maturation, and disrupting meiosis that causes abnormal embryo preimplantation.
4. Disappointing pregnancy outcomes can result in stress with guilt and self-blame because of unforgiving sociocultural attitudes. Thus infertile women facing failed attempts at conceiving, notably if obese, can acquire depression, phobic anxiety, and paranoid ideation.
5. Psychosocial distress can promote overeating and obesity, which further potentiates infertility due to ovulatory or unexplained causes in 50% of couples.
6. In vitro fertilisation is said to bring about lower live birth rates for couples with unexplained infertility, along with a higher risk of complications in the obese.
7. Endocrinological milieu associated with obesity also causes early miscarriages or pregnancy complications that result in the noncontinuation of pregnancy.
8. Couples can be steeped in dysphoria following repeated failed attempts to conceive and may welcome an obesogenic diet that is comforting.
9. Women undergoing assisted conception have high levels of stress. Stress associated with infertility may also cause sexual or marital conflicts, which have long-lasting effects, and comorbid dysphoria could manifest in couples.
10. Pregnancy rates may increase in obese infertile women after they undergo behavioural treatment for stress management and weight loss.
11. A gynaecologist with clinical psychosomatic skills could help resolve intense situations in the infertile couple by assessing specific needs, and appropriately counselling the obese who want to start a family.

15.7 Psychosomatic aspect of gender violence and obesity

1. Data from the United States confirms that Interpersonal violence (IPV) can lead to obesity. Although underreported for fear of reprisal from the male partner, IPV causes 4 million injuries annually, with 36% of women reporting rape, physical violence, or stalking, and 48% complaining of psychological aggression.
2. These can result in clinical psychosomatic issues and obesity in many affected women.
3. Women who experience both physical and nonphysical IPV are at an increased risk of experiencing numerous adverse health problems, such as chronic pain, gastrointestinal disorders, depression, anxiety, posttraumatic stress disorder, and sleep disorders, along with their overeating habits.
4. Besides, women may experience nonphysical forms of IPV, such as control through humiliation, verbal abuse, or threats of abuse to her or someone she loves, so that she lives in fear of her partner, thereby making her resort to comfort eating as a panacea for her dysphoria.
5. This is relevant to the practice of gynaecology worldwide as these women ignore weight gain even if it leads to obesity.
6. A large proportion of obesity in many populations would be prevented if IPV and linked overeating could be stopped.
7. The vital importance of ending violence against women to facilitate improved psychosomatic health internationally has been recognised by key international organisations such as the WHO and United Nations (UN).
8. The United Kingdom's National Institute for Health and Clinical Excellence developed relevant guidance to prevent/stop IPV.
9. Educating health professionals to deal effectively with women's clinical psychosomatic illnesses due to gender-related violence remains vital.

15.8 Psychosomatic aspect of severe pelvic/perineal dysfunction and obesity

1. Pelvic floor/perineal dysfunction, also referred to as pelvic/perineal dysfunction, relates to symptoms that bring about physical and mental ill-health.
2. The pelvic/perineum diaphragm comprises myofascial structures that support the pelvic organs and facilitate normal urogenital function by preserving their anatomical integrity and innervation from segments L3 5, S2 5 of the spinal cord.
3. Derangement of the nerve supply or injury to the muscles and ligaments of the pelvic floor or perineum can lead to the symptoms of pelvic/perineal dysfunction.
4. Bladder and bowel continence is a voluntarily acquired, socially appropriate behaviour learnt through a process of conditioning during childhood, so the loss of continence, especially in obese women, can impair physical and mental health.
5. Women who suffer from pelvic floor symptoms (which comprise urinary incontinence, anal incontinence, dyspareunia, prolapse, and haemorrhoids) will have severe biopsychosocial repercussions.
6. These symptoms can lead to severe physical and mental ill-health, yet studies of the impact on the relevant clinical psychosomatic health issues that are generated remain scarce. It is clinically important to evaluate the relationship of pelvic/perineal symptoms with physical and mental wellbeing that is compromised by urinary/faecal incontinence, too often in the obese.

7. The perceived psychosomatic misery borne by those with severe symptoms was not given enough attention. Incontinence causes anxiety/depression/phobias, which impact on overall welfare and health-seeking behaviour, thereby underscoring the importance of the psychosomatic approach toward symptom relief.

8. Clinical evidence for the sufferer's psychosomatic issues is scarce, particularly in obese/overweight patients.

9. Further research on obesity and the psychosomatic aspects of pelvic floor/perineal dysfunction in the obese would be clinically useful for improving patient-centred care as novel management insights accrue.

15.9 Psychosomatic aspect of gynaecological tumours and obesity

1. The incidence of endometrial cancer has risen over the last three decades to affect many at a younger age, mirroring the trend in obesity.

2. In utero foetal metabolic programming in the obese gravida can promote obese offspring, who would in turn be expected to be at increased risk of cancer.

3. There should be a multidisciplinary approach to support obesity gravida to lose weight during postnatal phase, as they are at increased risk of developing type 2 diabetes over the next 10 years,

4. Reduction of obesity would positively influence prevention of this cancer and other malignancies, thereby averting considerable clinical psychosomatic health burden.

5. Advising weight loss, ideally beginning with individualised lifestyle interventions may prevent endometrial cancer in those with a raised BMI.

6. Improvement in clinical psychosomatic health of younger women who wish to start a family is particularly relevant, as cancer treatment may permanently preclude natural childbearing.

7. Physical, mental, and social wellbeing in cancer survivors treated for early-stage endometrial cancer, showed improvement which is inversely related to BMI.

8. American Society for Clinical Oncology has been particularly active in promoting awareness among professionals and public alike of the links between obesity and cancer, including preparing a guide to aid physicians in advising weight reduction in obese women.

9. Primary and secondary preventative strategies for obesity and its complications seem sensible, such prudence may seem unacceptable to those who overeat due to familial/social pressures. Many patients who reach the obese habitus require tertiary management measures, such as weight-reducing or bariatric surgery.

15.10 Conclusions

1. A clinical psychosomatic approach aiming to reduce the urge to overeat seems reasonable for many facing these.

2. The trend for coping with menstrual irregularities using a calorie-rich diet seems unnecessary but is nevertheless followed by certain groups. Other treatments for menstrual problems, including PMS and PMDD, have ranged from pharmacotherapy to surgery, but these maladies could merit less-invasive management in many health issues.

3. Infertility/subfertility, with its psychosomatic implications, continues to be of great significance. Psychosomatic approaches that promote a healthy lifestyle could enable natural conception to childbirth after desired weight reduction in the obese.

4. Obesity can be promoted by inescapable IPV. This behaviour is associated with major clinical psychosomatic implications that promote overeating.
5. Pelvic floor dysfunction have a penchant for the obese. Psychosomatic aspect aggravated by the overweight habitus and symptom relief success is encouraged by weight loss in the obese.
6. Primary and secondary prevention of cancer and obesity could be a sound clinical psychosomatic approach. Prevention of obesity would reduce the disease burden of such cancers.
7. Health providers ought to consider the biological, psychological, social, and cultural factors that influence clinical psychosomatic interactions and promote obesogenic behaviour.

Further reading

Lal M, Sarhadi AHL. Obesity and clinical psychosomatic women's health. In: Mahmood TA, Arulkumaran S, Chervenak FA, eds. *Obesity and Gynaecology*. 2nd ed. 2020:293−312. Available from: https://doi.org/10.1016/B978-0-12-817919-2.00032-2.

Lal M, ed. *Clinical psychosomatic obstetrics and gynaecology: a patient-centred biopsychosocial practice*. Oxford: Oxford University Press; 2017.

Gonzalez-Mesa E. Obesity and psychosexual disorders. In: Mahmood TA, Arulkumaran S, Chervenak FA, eds. *Obesity and Gynaecology*. 2nd ed. 2020:313−318. Available from: https://doi.org/10.1016/B978-0-12-817919-2.00033-4.

Larsen SH, Wagner G, Heitmann BL. Sexual function and obesity. *Int J Obes*. 2007;31:1189−1198.

Obesity and cancer

16

Chu Chin Lim and Tahir Mahmood
Department of Obstetrics & Gynaecology, Victoria Hospital, Kirkcaldy, United Kingdom

16.1 Overview

1. Cancer development is stimulated by genetic components, environmental exposures, and lifestyle exposures; all impact the cellular microenvironment and initiate carcinogenesis.
2. Links between obesity and cancer have been documented for hormonally influenced cancers such as endometrial and breast cancer, and links to broader cancers are emerging.
3. Changes to systemic and microenvironment from obesity and hypernutrition not only cause significant risk for carcinogenesis but also impact the success and outcome of treatment.
4. BMI >40 kg/m^2 has a 1.6-fold higher risk of cancer death.
5. The growing epidemic of cancers in obese young people (24−29) that have previously predominantly been seen in only the elderly indicate that the need for control of obesity is of key importance to all nations

16.2 Epidemiological evidence for links between obesity and cancer

1. The following cancers have been linked to obesity with variable, but consistent relationships:
   a. oesophageal (with greater increases for higher BMI)
   b. gastric
   c. liver
   d. kidney (renal cell)
   e. pancreatic
   f. colorectal
   g. gallbladder
   h. multiple myeloma
   i. meningioma
   j. potentially some thyroid and ovarian cancers.
2. Study of colorectal cancer risk shows a link to increased risk for colon, but not rectal cancer, based on waist circumference.
3. Clear differences for risk of different cancer types exist in different genders.

Handbook of Obesity in Obstetrics and Gynecology. DOI: https://doi.org/10.1016/B978-0-323-89904-8.00021-4

4. Obesity as a unique risk factor for the development of cancer differs by site.
5. The International Agency for Research on Cancer working group on Body Fatness in 2016 noted sufficient evidence for links between obesity and postmenopausal breast, colon, endometrial, oesophageal, gallbladder, kidney, liver, meningioma, multiple myeloma, ovary, pancreas, stomach, and thyroid cancer.
6. Others have proposed links between advanced prostate cancer, mouth, pharynx, and larynx, although the relationship is less clear.

16.3   Cancers unique to or more common in women

1. 50% of endometrial cancer is associated with obesity, with also a striking dose/response curve from 1.5-fold increase for overweight women to 7.1-fold increase for class 3 obesity.
2. Primary mechanism likely linked to increase in aromatase, conversion of androgens to oestrogens, and then the hormonal influence on proliferation of endometrial cells.
3. Other mechanisms proposed for nongynecological cancers may also play a role, such as chronic inflammation and production of inflammatory cytokines, increases in insulin-like growth factors, and their associated cellular and genetic impacts.
4. Weight reduction decreases the risk as show by the use of bariatric surgery.
5. Breast cancer also shows a strong association with obesity.
6. Women's Health Initiative Clinical Study population shows increased risk for invasive breast cancer for overweight and above women compared to normal-weight women with the highest risks in those with grade 2 and 3 obesity.
7. Dose—response relationship was found again like endometrial cancer.
8. Also, primarily in hormone receptor positive disease cell types.
9. However, obese women who lost weight did not have a drop in risk unlike the impact seen with endometrial cancer and bariatric surgery.
10. When obesity most impacts the risk of breast cancer is unclear.
11. Obesity during trial may have been lifelong, and thus results reflect critical windows at reproductive age or earlier childhood as well.
12. Premenopausal obese women may actually have a protective effect from the body fat and decreased risk in that life stage.
13. Obesity in adolescence during breast development or premenopausal as an impact to risk factor compared to postmenopausal is difficult to study but important to understand for prevention.
14. High body fat proportions even in normal BMI may increase breast cancer risk.
15. Type and distribution of body fat is an area of intensive research and may hold keys to future prevention strategies.

16.4   Potential mechanisms for oncogenesis

1. Many potential pathways are engaged and could result in carcinogenesis with obesity.
2. Markers of disturbance of the microenvironment such as insulin resistance and inflammation may be more accurate risk predictors than BMI alone.
3. So too may waist-hip ratios as well as BMI.

4. Associated conditions, like metabolic syndrome and diabetes have significant impact on outcomes in breast cancer. However, mechanisms underpinning this relationship are still obscure.

5. Cross talk between adipocytes and macrophages is of particular interest as molecular mechanisms have been associated with adipocyte—macrophage breast cancer cell clusters.

6. Another component of the altered inflammatory environment may come from inhibition of natural killer cell activity. Further research on pathways in this area may provide additional avenues for cancer treatment specific to obesity-linked cancers.

7. Type and location of adipose deposits also play a role in hormones and cytokines secreted.

8. Association of increasing fat mass and increasing fasting insulin levels and hyperinsulinaemia implicates insulin as well as oestradiol as an important biologic link associating trunk adiposity with breast cancer risk.

9. Body composition rather than simply BMI has been highlighted as of great importance.

10. High body fat levels associated with elevated levels of insulin, CRP, IL-6, leptin, and triglycerides, as well as lowered high-density lipoprotein cholesterol and sex hormone-binding globulin.

11. Functional causal mechanisms supported by obesity may differ between locations with chronic secretion and inflammation being posited for the elevation of gallbladder cancer risks, reflux oesophagitis and inflammation for oesophageal cancer risk, associated hypertension with renal cancer risk, and endogenous oestrogen with breast and endometrial cancer risks.

12. Ultimately actions affecting oncogenesis are carried out at a cellular level with inflammatory mechanisms a common component.

13. One pathway comes through the downstream influence of increased aromatase activity and stimulation from circulating oestrogens, as noted previously with visceral fat.

14. Varying pathways interact and other downstream linkages between obesity and cancer are being explored to tease out critical interactions including those leading to insulin resistance.

15. One characteristic of the obesity—cancer relationship appears to be the significant ongoing cross talk between adipose tissue components, their products, and the cancer cells themselves, leading to a microenvironment that not only promotes carcinogenesis but also promotes metastasis.

16. Possibly the more disturbed the environment is with insulin resistance and the more exposure to growth-promoting secretome of adipocytes, the more aggressive and potentially metastatic the resulting tumour.

17. Genes are emerging as potential bridges between obesity and cancer; research is needed to fully explore the gene regulation of the various components involved.

16.5 Clinical implications for prevention and treatment of cancer in obese patients

1. Weight reduction seems a clear path for prevention, lifelong, given the risk windows where obesity interacts with oncogenesis may differ for various cancers from adolescence through postmenopausal life stages.

2. Prevention of disruption of the microenvironment and prevention of proinflammatory state locally and systemically will reduce certain cancers, such as endometrial, and is likely to reduce others as well, in addition to reduction in high blood pressure, metabolic syndrome, and diabetes.
3. A low-fat dietary program has shown promise for increase in overall survival with breast cancer, including potentially fewer deaths from cancers.
4. Medications addressing the role of insulin pathway, for example, metformin have been shown to have a positive role in lowering cancer incidences in diabetics.
5. Oncologists need to be mindful of changing recommendations for the treatment of obese patients, particularly accuracy, and the need for modifications of dosing of chemotherapeutic agents to treat the cancers.
6. There is a paucity of effective tools or interventions to address this health issue consistently and effectively at present other than bariatric surgery.
7. Finding a means to engage the population with effective and sufficient support will likely require engaging a broader health team with nutritionists and counsellors as well as the educational and motivational support from all levels of the healthcare team.
8. Individual health providers can advocate for policies within their practice, locally and nationally, to support more active lifestyles, availability of healthy food options, and healthy diets.
9. Evidence is emerging that encouragement of healthy physical activity lifelong and promotion of a balanced diet may be the most important actions organisations can take.
10. Focusing food production on healthy food and an environment that promotes physical activity may do more than any other mechanism for reduction in the burden of obesity-related cancers.
11. The overall reduction in costs to economies from the reduction of multiple diseases and disabilities linked to obesity makes it worth governmental interventions.

16.6   Ongoing needs for research

1. There is a critical need to understand the cellular cross talk that results in the oncogenic environment stimulating cancers in obese individuals.
2. This not only offers a mechanistic understanding but also opens doors to potential targets for prevention, treatment of the cancer, and reduction of recurrence and metastatic disease.
3. Increasing understanding of basic genetic and neuronal information about homeostasis, particularly weight regulation and energy balance, may provide additional means to prevent obesity epidemic and oncogenic outcomes.
4. The role of weight loss on the control of cancer is a difficult area to research but one that is very important to understand moving forward.
5. Public health advocacy and leadership globally must address the obesity epidemic.
6. Continued research into evidence-based pathways to reduce individual obesity and control appetite will additionally assist in the prevention of these cancers.
7. Expanding global knowledge of the benefits of addressing obesity and risks for individuals and populations in failing to do so, including cancer risks, is an important advocacy role for every health provider.

16.7   Endometrial cancer

1. Obesity accounts for about 40% of cases of endometrial cancer in the developed world.
2. There is a linear increase in risk of endometrial cancer with increasing weight and BMI.
3. Extensive evidence from studies suggests that overweight and obesity are strongly associated with type 1 endometrial cancer.
4. Overweight and obese have two to four times greater risk of developing endometrial cancer than healthy weight, regardless of menopausal status. Extremely obese are seven times more likely to develop type 1 endometrial cancer.
5. Obesity in menopause produces excess oestrogen production. Due to conversion of androgens into oestrone by aromatase.
6. Prolonged unopposed oestrogen exposure leads to a continuous spectrum of change from proliferative endometrium through endometrial hyperplasia/polyps to endometrial carcinoma.
7. Progesterone containing intrauterine contraceptive devices are good contraceptive choices for obese women.

16.8   Breast cancer

1. Obesity increases the risk of breast cancer only in postmenopausal women not using HRT.
2. Obese women at greater risk of death from breast cancer postmenopause.
3. Adult weight gain is the most consistent and strongest predictor of postmenopausal breast cancer risk and a large portion of deaths may be avoided if women could maintain a healthy BMI.
4. Central obesity presents a greater risk than distribution over the lower extremities.
5. Adult weight gain is associated with a higher risk of postmenopausal breast cancer than actual BMI.
6. No significant difference in risk in postmenopausal HRT users.
7. Premenopausal, obese women are at lower risk than healthy weight.
8. Oestrogen levels are 50%−100% higher in obese postmenopausal women compared to lean; tumour growth is therefore faster but detected later as breast tumour is more difficult in overweight women.

16.9   Ovarian cancer

1. Positive dose−response relationship between BMI and risk of epithelial ovarian cancer shown in research.
2. Higher BMI associated with a slight increase in risk of ovarian cancer, particularly in women who have never used HRT.
3. No association among users of HRT.
4. Obesity positively associated with clear cell tumours but less correlated with invasive endometrioid or mucinous tumours.
5. Association between obesity and ovarian cancer survival is not conclusive yet.

6. Weight adjustment for the dosing of carboplatin-based chemotherapy is challenging.
7. Incorporating a weight adjustment may cause greater grade 3 and 4 systemic side effects.
8. Not incorporating a weight adjustment may lead to an increased risk of disease progression if dose not high enough.
9. Women with obesity shown to have lower relative decrease in platelet counts and haemoglobin levels.
10. Trend toward increased risk for disease progression in women with BMI >30.

16.10 Cervical cancer

1. Associations between BMI and cervical cancer are limited and inconclusive.
2. Some studies reported cervical cancer to be associated with elevated BMI, others found a lower relative risk.
3. Increased risk among overweight and obese women was mainly for cervical adenocarcinoma with a smaller increased risk for squamous cell carcinoma.
4. Observed increased risk could be due to the decreased attendance for screening in the obese women population.
5. Retrospective study found BMI >35 with cervical cancer had a higher risk of both all-cause death and disease-specific death than normal weight counterparts. Treatment-related and biological factors may contribute to decreased disease-specific survival in morbidly obese patients with cervical cancer.

16.11 Mechanisms relating obesity to female malignancies

1. Obesity affects the production of peptides, sex hormone-binding globulin, and steroid hormones.
2. It is likely that the prolonged exposure to high levels of oestrogen and insulin associated with obesity may contribute to the development of female malignancies.
3. Oestrone and oestradiol levels are directly related to amount of adipose tissue in postmenopausal women. Androgens secreted from adrenal glands and ovaries are converted into oestrone by aromatase in the fat cells.
4. Lowered sex hormone-binding globulin results in higher circulating level of free active oestrogens.
5. Excess weight, increased plasma triglyceride levels and low levels of physical activity all raise circulating insulin levels causing chronic hyperinsulinaemia associated with breast and endometrium cancers.
6. Carcinogenic effects of hyperinsulinaemia could be directly mediated by insulin receptors in target cells or might be due to related changes in endogenous female sex hormone synthesis and bioavailability.
7. Insulin boosts IGF1 and both promotes cell proliferation and inhibits apoptosis.
8. Increased insulin and IGF1 blood levels result in reduced hepatic synthesis and blood concentrations of sex hormone-binding globulin, increasing bioavailability of oestradiol.

9. Proteins secreted by adipose tissue influence immune response and its regulation; vasculature and stromal interactions; angiogenesis; and extracellular matrix components.
10. Evidence of intention weight loss affecting cancer risk is limited, the risk reduction of breast cancer from daily physical activity was observed in an already healthy population, not the overweight or obese.

16.12 Effect of obesity on management of female malignancies

1. Obese patients have a poorer outcome compared to lean patients.
2. Obesity associated with both reduced likelihood of survival and increased likelihood of recurrence among patients with breast cancer regardless of menopausal status and after adjustment for stage and treatment.
3. Poorer outcomes reflect the biological effect of adiposity, delayed diagnosis, and higher treatment complication level.
4. Nonlocalised disease more common in high BMI self-detections.
5. Manual handling for examination of obese patients is more challenging and requires specific resources.
6. Dosage calculations and adjustments in treatment plans must be made in heavier patients.
7. A high BMI increases the risk of perioperative complications and mortality particularly in the presence of comorbidities.
8. Counselling should be given about these increased risks.
9. Obese patients should receive thorough cardiovascular and respiratory preoperative assessment as these patients are at a much higher risk of postoperative complications.
10. It is advisable to have planned admission to high dependency units for these patients.
11. Women with gynaecological malignancies can be managed in standard fashion in most instances, there is no need to compromise surgical treatment where indicated.
12. The route of surgery must be considered as abdominal procedures are more of an issue than vaginal.
13. Obesity may be a limiting factor in application of laparoscopic surgery, it may not allow a steep Trendelenburg and may prevent adequate mobilisation of small bowel out of the pelvis for proper visualisation.
14. However, laparoscopic, and robotic surgery has additional benefits for obese patients making it a desirable option.
15. Obesity presents a variety of problems in laparotomy incision placement and closure as well as the increased risk of wound infection and failure.
16. Possible aetiologies include decreased oxygen tension, immune impairment and tension, and secondary ischaemia along suture lines.
17. Pelvic access is challenging, and incidence levels of intraoperative complications are higher.
18. Experience, good assistance, retraction, and lighting are essential.
19. Regional anaesthesia (spinal or epidural) is encouraged as it can also help with postoperative pain control.
20. However, regional anaesthesia may be difficult or impossible with high BMI leading to difficulty locating the spine, an experienced anaesthetist is preferred.

21. Research shows bariatric surgery lowers risk of incident cancer, especially obesity-related cancers.
22. It was also found to result in significant beneficial changes in circulating biomarkers of insulin resistance, inflammation, and reproductive hormones in endometrial morphology and in molecular pathways that are implicated in endometrial carcinogenesis.
23. These results may have important implications for screening, prevention, and treatment of endometrial cancer.

16.13 Oncogenetics

1. At the tissue level, all cancers are the result of the cumulative effect of environmental insults and genetic changes leading to dysregulated, uncontrolled growth of abnormal clones of cells.
2. In a minority of cases, these tissue-specific genetic alterations leading to cancer are predisposed to by highly penetrant, inherited germline mutations.
3. Sporadic cancers are the result of detrimental genetic dysregulation of growth in the affected organ.
4. Whereas in the so-called inherited cancers, the underlying factors are genetic changes affecting all cells of the body.
5. In reality, there is a spectrum between true sporadic and inherited disease and many cancers may result from the cumulative effect of a number of lower penetrance familial genetic alterations.
6. The majority of gynaecological malignancies occur sporadically but both ovarian and endometrial are prominent features of key cancer predisposition syndromes.
7. The lifetime risk of women developing ovarian cancer is increased in BRCA hereditary breast and ovarian cancer syndrome and Lynch syndrome. Both syndromes are caused by mutations in tumour suppressor genes.
8. The BRCA hereditary breast and ovarian cancer syndrome is caused by mutations in the genes BRCA 1 and BRCA2, which are involved in postreplication DNA repair. The syndrome is associated with a significant increase in breast and ovarian cancer risks as well as the risk of other malignancies including prostate and pancreas.
9. Lynch syndrome is caused by mutations in a family of genes involved in DNA repair known as mismatch repair genes. The syndrome is associated with an increased risk of many cancers including colorectal, endometrial, gastric, and ovarian.
10. The women's age, previous personal history of cancer, and relevant family history of malignant disease can be pointers to a possible genetic predisposition and the genogram is useful.
11. A woman being diagnosed with ovarian cancer in her early 40s with a strong family history of breast cancer will need to be counselled for the possibility of an underlying BRCA mutation.
12. In a woman diagnosed with endometrial cancer in her 50s who has a past history of colorectal cancer, the possibility of Lynch syndrome will need to be considered.
13. Women and their families being offered genetic testing should be counselled comprehensively about the nature of the tests and the implications of their results,
14. Counselling should address the complex medical, ethical, and psychosocial aspects that are inherent to this process, especially the possibility of identifying variants of unknown mutations. Such discussion should take place at the multidisciplinary genetics clinic.

16.14   Future directions

1. Causal role of obesity in gynaecological malignancies must be researched and defined.
2. Dysregulation of adipokines is likely to contribute not only to tumorigenesis and tumour progression but also to metastatic potential.
3. Intervention strategies at individual and community levels are important for weight loss.
4. Future trials may study the effect of dietary changes on weight gain and cancer risk, the effect of patterns of physical activity in relation to weight gain and cancer risk. The combined effects of changes in diet and activity on obesity and female cancer risk.
5. Clinicians should be aware that preventing/treating obesity should be considered part of cancer prevention.
6. Further research must clarify the mechanism and role of bariatric surgery to lower the risk of incident cancer in severe obesity.

Further reading

Arulkumaran S, Denny L, Ledger W, Doumouchtsis S. *Oxford Textbook of Obstetrics and Gynaecology*. Oxford University Press; 2019 (ISBN 978-0-19-876636-0).
Gajjar KB, Shafi MI. Obesity and female malignancies. In: Mahmood TA, Arulkumaran S, Chervenak FA, eds. *Obesity and Gynaecology*, 2nd ed. 2020, London: Elsevier, 209−215. Available from: https://doi.org/10.1016/B978-0-12-817919-2.00024-3.

Obesity and breast cancer

17

Chu Chin Lim
Department of Obstetrics & Gynaecology, Victoria Hospital, Kirkcaldy, United Kingdom

17.1 Epidemiology

1. Breast cancer (BC) is the most prevalent female cancer, responsible for 15% of all cancer deaths in women worldwide.
2. 33% of BCs in post menopause are due to obesity.
3. Linear association between obesity and overall risk of BC in menopause has been reported with a hazard ratio of 1.05 (99% Confidence interval, CI 103−1.07), for each body mass index 5 kg/m^2 increase.
4. Inverse correlation between obesity and BC in premenopause, however level of effect varies depending on many other anthropometric parameters (Hazard ratio, HR 0.89, 99% CI 0.86−0.92).
5. In premenopause women, for each 5 kg/m^2 increase in BMI, there was a 7% and 5% reduction in BC risk in Caucasian and African women, respectively, while there was a 5% increase in Asian women. These risks were very sensitive to waist hip ratio and height of the women.
6. Obesity increases risk of developing *hormone receptor positive BC* (both oestrogen and progesterone) in women greater than or equal to 65 years old (HR 1.25, 99% CI 1.16−1.34).
7. But obesity does not increase hormone receptor positive BC in women less than or equal to 49 years old (HR 0.79, 99% CI 0.68−0.91).
8. Relationship between obesity and *hormone receptor negative BC* is more complex:
   a. Obesity increases risk in premenopausal women (Relative risk, RR 1.06, 95% CI 0.71−1.60)
   b. Decreases risk of hormone receptor positive BC (RR0.78, 95% CI 0.67−0.92)
   c. Hormone receptor negative BC risk increased in postmenopausal women who have never used HRT (multivariate HR 1.59, CI 1.08−2.34).
9. Overall risk of obesity-dependent BC is lower for women on HRT, suggesting HRT is confounding factor in obesity-cancer relationship.
10. No association between obesity and risk of specific BC subtypes demonstrated to date.
11. Intentional weight loss is associated with lower BC risk.
12. In bariatric surgery observational trials, a weight loss of approximately 30% was associated with a reduction in BC risk of up to 80%.

17.2 Pathogenic mechanisms

1. Development of BC in obese women may be influenced by various factors including:
   a. endogenous sex hormones
   b. hyperinsulinaemia

Handbook of Obesity in Obstetrics and Gynecology. DOI: https://doi.org/10.1016/B978-0-323-89904-8.00018-4

    c. insulin-like growth factor 1
    d. hyperglycaemia
    e. adipokines
    f. chronic inflammation
    g. microbiome

17.2.1   Sex hormones

1. Oestrogen levels are higher in obese women due to peripheral conversion of circulating androgens to oestradiol by aromatase enzyme.
2. Obese women have reduced sex hormone binding globulin, causing greater bioavailability of oestradiol and testosterone.
3. Oestrogens have mitogenic and mutagenic effects to promote proliferation, genetic instability, and DNA damage in both normal and neoplastic mammary epithelial cells.
4. The risk of developing BC is not only due to an increase in oestrogen levels, but higher levels of androgens in both premenopausal and postmenopausal obese women also play a role in the pathogenesis.

17.2.2   Hyperinsulinaemia

1. Obesity is closely related to metabolic syndrome, insulin resistance and hyperinsulinaemia.
2. 80% diabetic women are obese.
3. Raised waist circumference or waist−hip ratio also predicts T2DM risk, irrespective of BMI.
4. Hyperinsulinaemia promotes carcinogenesis by (a) direct promotion of cell growth or (b) indirect use of IGF-1 axis.
5. Overexpression of insulin and IGF-1 receptors in cancer cells may also create expression of hybrid receptors capable of binding to both molecules.
6. Hyperinsulinaemia causes increased IGF-1 concentration due to suppression of the binding proteins 1 and 2. Also due to activation of GH receptor increasing secretion of GH stimulating IGF-1.
7. Insulin and IGF-1 binding triggers various mechanisms which promote carcinogenesis and neoplastic spread.
8. There is a direct relationship between higher levels of circulating IGF-1 and the risk of developing BC, specifically ER + tumours and the risk of developing chemotherapy resistance.
9. Excess insulin acts synergistically with IGF-1 and increases aromatase enzyme activity via sex hormone route.
10. Hyperglycaemia is also linked to visceral fat and influences tumour development.
11. Elevated glucose levels promote metastasis and increased invasiveness due to the epithelial to mesenchymal transition process.
12. Hyperglycaemia also acts indirectly on BC cells by increasing insulin and IGF levels, inflammatory cytokines such as IL-6 and TNFα, oxidative stress, and platelet activation.
13. Hyperglycaemia also alters the epigenetic regulation of neoplastic cells,"hyperglycaemic memory" to activate oncogenic pathways even if blood glucose levels return within normal range.

17.2.3   Adipokines

1. Family of polypeptides synthesised by adipocytes including over 100 different molecules —common studied molecules are leptin and adiponectin, which have opposite biological effects
2. Leptin is a potent proinflammatory agent and its concentrations are proportional to total body fat levels.
3. Leptin has several activities: mitogenic, antiapoptotic, immunosuppressive, proangiogenic alone, and acts in synergy with vascular endothelial growth factor expression, all relating to carcinogenesis.
4. Binding of leptin to long form receptor activates several signalling pathways involved in the control of cell survival, proliferation, differentiation, migration, and invasion.
5. Meta-analysis demonstrated a positive association between leptin levels and BC risk.
6. Leptin and receptor are associated with more severe BC cases and potentially act as BC risk biomarkers.
7. Adiponectin is secreted by the visceral adipose tissue, has potent antiinflammatory activity, and levels are inversely correlated with body fat.
8. Adiponectin has various influences on carcinogenesis as it reduces fatty acid and protein synthesis cellular growth, proliferation, DNA mutagenesis, and increases apoptosis.
9. All these effects are indirectly achieved by sensitising cells to insulin and inhibiting inflammation.
10. They are directly achieved by sequestering growth factors at pre-receptor level or activating and inhibiting pathways.
11. Low adiponectin concentrations are associated with increased BC risk and an inverse association between adiponectin concentrations and BC recurrence in ER/PR patients has been reported.

17.2.4   Chronic inflammation

1. Obesity is a state of chronic low-grade inflammation which plays an important role in tumour development and progression.
2. Visceral fat, leptin, and oestrogen levels are associated with an increase in proinflammatory molecules in obese women, promoting carcinogenesis.
3. IL-1β, IL-6, and TNFα are all increased in obese women and promote T-regulatory lymphocytes chemotaxis and inhibit the cytotoxic activity of CD8 [+] T cells. This mechanism is associated with poor BC prognosis.

17.2.5   Microbiome

1. Human microbiome is proven to play a fundamental role in some diseases including cancer.
2. Greater microbial alterations are observed in BC patients than healthy women.
3. Increased caloric intake leads to dysbiosis, creating alterations in the carbohydrate and lipid metabolism, insulin resistance, and perturbations in endocrine systems.
4. The gut microbiota may induce the transformation of chemical compounds derived from the host diet into obesogenic and diabetogenic molecules that play a role in carcinogenesis.

5. Alterations in the gut microbiota may also influence the production of oestrogen metabolites and the circulating levels of oestradiol.
6. Dysbiosis, obesity, and increased oestrogen levels may act synergistically to increase the risk of BC.

17.3   Diagnosis

1. Obesity may negatively impact BC diagnosis as obese women are usually less aware of the importance of a healthy life style.
2. Obese women less likely to access mammography; the main reason listed was pain during the procedure.
3. Psychosocial factors and a low socioeconomic status contribute to barriers between patient and physician and hamper their medical care.
4. Societal stigma can affect patients; mental resolve may impact their perception and decision-making.

17.4   Therapy

17.4.1   Surgery

1. More difficult to ventilate and intubate obese patients making anaesthesia more risky.
2. There is a higher risk of deep venous thrombosis, pulmonary embolism, urinary tract infections, myocardial infarction, pneumonia, and reoperation.
3. Obesity is linked to an increase in major and minor surgical complications in primary breast surgery even without reconstruction.
4. Cosmetic outcome of breast reconstructive surgery in both implant-based and autologous-based is poorer in obese than healthy women.
5. Complications include wound dehiscence, haematoma, seroma, and flap failure or necrosis.
6. A BMI $>40\,\text{kg/m}^2$ has been identified as the threshold at which complication rates become prohibitively high.
7. The cosmetic outcome with breast conservation (lumpectomy and radiation) is poorer in obese than in nonobese women, as women with BMI $>30\,\text{kg/m}^2$ had more postoperative breast asymmetry and deformity.
8. Obese patients have higher incidence of surgical site infections, return to emergency department after discharge, and hospital readmission within 30 days of surgery.
9. Sentinel node mapping is more difficult in obese women as node identification rates are lower.

17.4.2   Radiotherapy

1. Obese patients may receive increased doses to critical organs such as heart and/or lungs especially in supine position. Prone whole-breast radiation and hypofractionated radiotherapy minimise toxicity.

2. High BMI and large breast size are associated with increased risk of dermatitis after whole-breast radiotherapy.

17.4.3 Chemotherapy

1. Challenges present due to the presence of comorbidities and balance between efficacy and toxicity.
2. Obese patients are more likely to receive insufficient chemotherapy doses compared to normal weight which can have a negative impact on disease-free and overall survival.
3. American Society of Clinical Oncology recommends full weight-based chemotherapy doses are used in the treatment of obese cancer patients and toxicity managed as in nonobese women.
4. Obesity has been associated with a higher risk of cardiotoxicity after treatment with Trastuzumab in women with HER-2 positive BC requiring close monitoring and risk factors management.
5. Obesity is considered a factor of resistance to anticancer therapy.
6. Obesity modifies the pharmacokinetics of chemotherapy drugs and is hypothesised to induce biological modifications of adipose tissue promoting resistance to the drugs used.

17.4.4 Endocrine

1. Endocrine therapy may be less effective as obesity is associated with elevated aromatase activity and serum oestrogen levels in postmenopause.
2. Current literature reports that Anastrozole is associated with worse outcomes than Tamoxifen.
3. For obese postmenopausal women, Letrozole, as a more potent inhibitor of aromatase, should be used.
4. Currently for obese premenopausal women, it is recommended to use Exemestane plus ovarian suppression when indicated.

17.5 Prognosis

1. Obese women with BC have worse overall and disease-free survival statistics than nonobese women regardless of menopausal stage or therapy.
2. The relative risk for total mortality is 1.41 (95% CI, 1.29−1.53) and the RR for BC-specific mortality is 1.35 (95% CI, 1.24−1.47) for obese versus normal weight patients.
3. For each 5 kg/m^2 increment of BMI, total mortality rises by upto 17%.
4. It is unclear if postdiagnosis weight loss has any impact on BC survival outcomes.
5. Reduction in fat intake reduces both risks of developing BC and also death postdiagnosis
6. Evidence suggests metformin may have antitumour activity in BC as it improves many potential physiologic mediators of obesity effects on BC.
7. Metformin is also associated with modest weight loss.
8. There are no data on the safety and the impact of BC outcome of bariatric surgery and approved weight-loss medications.

Further reading

Benedetto C, Canuto EM, Borella F. Obesity and breast cancer. In: Mahmood TA, Arulkumaran S, Chervenak FA, eds. *Obesity and Gynaecology.* 2nd ed. Elsevier; 2020:201−208. Available from: https://doi.org/10.1016/B978-0-12-817919-2.00023-1.

Female obesity and osteoporosis

18

Chu Chin Lim

Department of Obstetrics & Gynaecology, Victoria Hospital, Kirkcaldy, United Kingdom

18.1 Introduction

1. Obesity and osteoporosis are two important problems affecting global health with a high impact on both mortality and morbidity.
2. Both have multifactorial aetiologies, including genetic and environmental components, with potential interactions between them.
3. Obesity is a condition of excessive body fat due to an imbalance when energy intake exceeds energy expenditure over a prolonged period.
    a. In healthy adults, body weight is tightly regulated by several environmental, nutritional, and hormonal factors.
    b. Postmenopausal women often show increased body weight, likely due to a decrease in basal metabolism, alteration of hormonal levels, and reduced physical activity.
    c. They are often affected by hypertension, dyslipidaemia, diabetes mellitus (DM), and cardiovascular disease and also have an increased risk of developing some cancers.
1. *Osteoporosis* is a metabolic bone disease that is characterised by excessive skeletal fragility (due to a reduction in both bone quantity and quality) leading to an increased risk of developing spontaneous and traumatic bone fractures and even death.
    a. It is characterised by a reduction in bone mass; it is typically defined in an individual with a bone mineral density (BMD) T-score that is 2.5 or more SD values below normal (T-score -2.5 or less).
    b. Normal aging is associated with a high incidence of osteoporosis and bone marrow adiposity. Bone remodelling and adiposity are both regulated through the hypothalamus and sympathetic nervous system. Adipocytes (the cell for storing energy) and osteoblasts (the bone from a common progenitor)—the mesenchymal stem cell.
2. Interestingly, obese women have always been considered protected against osteoporosis.
    a. Increased mechanical loading, associated with increased body weight, contributes to increases in bone mass.
    b. One potential problem with these phenotypic studies is that the correlation between body weight (or BMI) and bone mass may not necessarily represent a correlation between obesity per se and osteoporosis, because it is excessive fat mass rather than total body weight that defines obesity.
    c. Body weight is a heterogeneous phenotype consisting of fat, lean muscle, and bone mass. Fat mass accounts for approximately 16% and 25% of total body weight in normal-weight men and women, respectively; the majority of the remaining body composition is lean mass.
    d. Conclusions from studies about the relationship between obesity and bone mass may be confounded by the mechanical loading effects of total body weight on the skeletal system.

Handbook of Obesity in Obstetrics and Gynecology. DOI: https://doi.org/10.1016/B978-0-323-89904-8.00043-3

18.2   Normal bone metabolism

Osteoblasts, osteoclasts, and osteocytes are the main cells of the bone.

1. Osteoblasts are nonproliferative bone building cells that originate from osteoblast progenitor cells and aid in the formation of the bone matrix by secreting osteoid, a substance responsible for bone mineralisation. It plays the most important role in bone formation
2. Osteocytes (mature osteoblast) are unable to divide and no longer secrete matrix components.
3. Osteoclasts originate from macrophage monocyte cell lineage and participate in bone resorption, ultimately leading to decreased bone mass.
4. Preosteoblasts express receptors for different types of growth factors, pro-inflammatory cytokines, and hormones, including bone morphogenic proteins, Wnt, transforming growth factor-beta, parathyroid hormone (PTH), interleukin-6 (IL-6), 5-HT, insulin/insulin-like growth factor, and tumour necrosis factor (TNF).
5. Binding of these ligands with their receptors induces the activation of different types of transcription factors responsible for osteoblast differentiation, maturation, and survival.

18.2.1   Interaction of various hormones

1. Parathormone stimulates osteoclasts and releases calcium and phosphate in the blood.
2. Calcitonin inhibits osteoclasts and increases deposition of the calcium in the bone.
3. Calcitriol stimulates the absorption of calcium and phosphate from the small intestine and ensures availability in the bone. It also activates the osteoblasts to synthesise collagen.
4. Oestrogen inhibits bone resorption.
5. Growth hormone stimulates bone formation.
6. Insulin increases synthetic activity of osteoblasts while glucocorticoids inhibit osteoblasts.

18.3   Epidemiology of osteoporosis and obesity

1. Osteoporosis has become a significant health problem as approximately 200 million people worldwide are estimated to have osteoporosis.
   a. There are estimated 10 million aged more than 50 years in the United States with osteoporosis.
   b. In a Brazilian study, the prevalence of fragility fracture in women and men aged higher than 40 years was 15.1% and 12.8%, respectively was reported.
2. The WHO's World Health Statistics (2015) shows the obesity rate among adults in Europe is 21.5% in males and 24.5% in females.
3. It is projected that 60% of the world's population, that is, 3.3 billion people, could be overweight (2.2 billion) or obese (1.1billion) by 2030 if recent trends continue.
4. Age and female gender increase the risk of developing both obesity and osteoporosis.
   a. Age-related changes in body composition, metabolic factors, and hormonal levels after menopause, accompanied by a decline in physical activity, may all provide mechanisms for the propensity to gain weight.
   b. Increase in fat mass often characterised by replacement of lean mass (LM) by adipose tissue.

c. The process of bone loss begins soon after menopause due to increased bone resorption by osteoclasts that exceed bone formation by osteoblasts.

d. Osteoblast function declines with aging, determining the imbalance between bone resorption and bone formation.

5. Premenopausal women with increased central adiposity had poorer bone quality and stiffness and markedly lower bone formation.

6. Most recent studies have reported that BMD appears to be a better long-term predictor of death than blood pressure or cholesterol.

18.4 Relationship between fat and bone: epidemiologic and clinical observations

1. The most powerful determinant of fracture risk is the amount of bone in the skeleton, as defined by either BMD or BMC

a. Extensive data have shown that high body weight or BMI is correlated with high BMD or BMC and a decrease in body weight leads to bone loss.

b. There is also evidence to support the view that fat mass, a component of total body weight and one of the most important indices of obesity, has a similar beneficial effect on increasing bone mass, thereby reducing the risk of osteoporosis.

c. In normal pre- and postmenopausal women, total body fat was positively related to BMD throughout the skeleton, and this effect was found in both White and Japanese subjects

d. Study showed that "rapid" bone losers had significantly lower fat mass than the "slow" bone losers.

2. However, in contrast to the abovementioned reported results, other independent groups have shown that excessive fat mass may not protect against decreases in bone mass.

a. In a large-scale study of Chinese and White subjects, when the mechanical loading effect of total body weight was statistically removed, then fat mass was negatively correlated with bone mass, thus suggesting that fat mass actually has a detrimental effect on bone.

b. A study conducted on evaluation of BMD in individuals with high body mass index has shown that obese patients have a significant reduction in bone mineral mass for age and BMI. It also showed the evidence that morbid obesity may not be considered a protective factor against osteoporosis in both female and male population.

3. Evidence from environmental factors and medical interventions also support an inverse correlation between fat mass and bone mass.

a. Physical exercise increases bone mass while reducing fat mass.

b. Consumption of milk which is a good source of highly absorbable calcium has been shown to increase peak bone mass at puberty, slow bone loss, and reduce the incidence of osteoporotic fractures in the elderly.

c. Menopause has also been associated with increased bone loss, increased fat mass, and decreased LM.

d. Hormone replacement therapy is an effective means of attenuating loss of LM and bone and reversing menopause related obesity in postmenopausal women.

e. Osteoporosis and obesity are side effects of treatment with gonadotropin-releasing hormone agonists.

f. The clinical use of glucocorticoids has been shown to cause decreased bone mass and an increase in central obesity.

4. The finding that all of these interventions have opposite effects on fat vs. bone mass supports the concept that there is an inverse correlation between fat and bone mass and that fat does not have a protective effect on the bone.

5. LM is the strongest predictor of BMD at all sites. It is important that LM should also be the target for improvement when considering prevention and/or management of osteoporosis.

18.5   Hormonal effect of obesity and osteoporosis

18.5.1   Beta-cell hormones (pancreatic hormones)

1. Insulin resistance is highly correlated with obesity.

2. Insulin is a potential regulator of bone metabolism as osteoblasts have been shown to have insulin receptors as well as IGF-1 receptor. Insulin is cosecreted with Amylin that directly stimulates osteoblasts proliferation in vitro and in vivo.

3. Hyperinsulinaemic patients develop a cluster of abnormalities, including androgen and oestrogen overproduction in the ovary and reduced production of sex hormone binding globulins in the liver.

4. There is increased free concentration of sex hormones resulting in reduced osteoclasts activity and possibly increased osteoblasts activity, leading to increased bone mass.

5. Fasting insulin levels were significantly and positively associated with BMD of the radius and spine in middle-aged women.

6. The complex effects of insulin on the bone are similar to the complicated relationship between fat and bone.

18.6   Adipocyte hormones

1. Adipocyte is not just an inert organ for energy storage but it expresses and secretes a variety of biologically active molecules, such as oestrogen, resistin, leptin, adiponectin, and IL-6 that affect human energy homeostasis and bone metabolism.

2. Adipocyte is an oestrogen-producing cell particularly in postmenopausal women. Oestrogen inhibits bone turnover by reducing osteoclast-mediated bone resorption and by stimulating osteoblasts-mediated bone formation.

3. In postmenopausal women extragonadal oestrogen synthesis in fat tissue becomes the dominant oestrogen source. The role of adipocyte as oestrogen producers may become more important for the bone metabolism in postmenopausal women.

4. Women with premature menopause who lose bone rapidly have lower levels of both oestrone and oestradiol than slow losers, and this may be accounted for by their lower body fat.

18.6.1   Leptin

1. The effect of leptin on obesity is mediated by a series of integrated neuronal pathways, including the catabolic pathway represented by proopiomelanocortin (POMC) neurons and the anabolic pathway represented by neuropeptide Y (NPY).

2. Leptin stimulates POMC neurons, and it results in reduced food intake and increased energy expenditure.
3. Leptin controls bone formation through inhibiting NPY gene expression in the hypothalamus hypothalamic relay, thereby suggesting a central mechanism also to be involved in its action on the bone modelling.

18.6.2 Adiponectin

1. Adiponectin regulates energy homeostasis and has antiinflammatory and antiatherogenic effects.
2. It increases insulin sensitivity and its circulating levels are reduced in obesity and diabetes.
3. Adiponectin and corresponding receptors are expressed in primary human osteoblasts, suggesting a link between adiponectin and bone. There is an inverse relationship between serum adiponectin level and BMD.

18.6.3 Resistin

1. Resistin is expressed in mesenchymal bone marrow stem cells, osteoblasts, and osteoclasts and increases osteoblasts proliferation and cytokine release, as well as osteoclast differentiation.

18.6.4 IL-6

1. IL-6, a pluripotent inflammatory cytokine, is released from adipocytes, the adipose tissue matrix, and elsewhere.
2. Adipose tissue accounts for one-third of the circulating levels of IL-6. Just like leptin, overweight and obese children and adults generally have elevated serum levels of IL-6.
3. Proinflammatory cytokines, including TNF-a, IL-1, and IL-6, are key mediators in the process of osteoclast differentiation and bone resorption and IL-6 antagonises osteoblasts differentiation.
4. Chronic inflammation and increased proinflammatory cytokines induce bone resorption and bone loss in patients with periodontitis, pancreatitis, inflammatory bowel disease, and rheumatoid arthritis.
5. The accelerated bone loss at menopause is linked to increased production of proinflammatory cytokines, including TNF-a, IL-1, and IL-6.
6. The significant increase in the development of osteoarthritis in obese human subjects provides evidence that chronic inflammation influences bone metabolism.

18.6.5 Adipsin

1. Adipsin similarly has been shown to negatively affect osteoporosis, particularly DM induces osteoporosis.
2. Increased expression is seen among DM and obese patients, and it has been shown to decrease bone formation.

18.7   Obesity of the bone

1. Bone marrow mesenchymal stromal cells are the common precursors for both osteoblasts and adipocytes.
2. Aging may shift the composition of bone marrow by increasing adipocytes, osteoclast activity, and decreasing osteoblasts activity, resulting in osteoporosis.

18.8   Clinical and diagnostic implication of the concept obesity of the bone

1. Errors in BMD determinations commonly seen in markedly obese individuals are because of fat deposition in bone marrow.
2. Dual-energy X-ray absorptiometry measurements may be falsely elevated by increased body fat, whereas measurements of trabecular BMD by quantitative computed tomography may be decreased by greater marrow fat.
3. Increase in bone marrow adiposity may explain secondary causes of osteoporosis such as Cushing's syndrome, DM, glucocorticoids levels, and immobility, which are associated with obesity.

18.9   Treatment implications of the concept obesity of the bone

1. The present treatment options for osteoporosis primarily are either antiosteoblastogenesis or proosteoblastogenesis in nature.
2. Decreased levels of vitamin D are a hallmark of osteoporosis and bone fractures.
3. Vitamin D has been shown to act by:
   a. inhibiting bone marrow adipogenesis as an additional mechanism beside its known actions on bone;
   b. affecting body fat mass by inhibiting adipogenic transcription factors and lipid accumulation during adipocyte differentiation; and
   c. influencing adipokine production and the inflammatory response in adipose tissue.
4. *High-fat diet* has a pivotal role in bone formation because it markedly reduces the rate of Ca21 absorption by the intestine and thereby decreases the availability of Ca21 required for osteogenesis.
5. *Vitamin D deficiency* in the serum prevents intestinal uptake of Ca21 from the diet and hereby signals the parathyroid gland to secret increased levels of PTH.
6. *Increased secretion of PTH* induces osteolysis and prevents osteogenesis by supplying adequate levels of calcium and phosphorus in the blood necessary for metabolic processes and neuromuscular function.
7. *Alendronate* is a widely used bisphosphonate that stimulates osteoblastic differentiation while inhibiting adipogenesis in vitro, thereby suggesting an anabolic effect on bone through the differentiation of mesenchymal stem cells.
8. *PTH* also has been shown in the past to induce osteoblasts differentiation, inhibit adipogenesis, and suppress osteoclasts apoptosis.

9. *PTH-related protein* has been shown to induce a mild osteogenic effect and inhibit the adipocytic effect in human mesenchymal stem cells, thereby helping in osteoporosis beside its known action to modulate bone formation through promoting osteoblasts differentiation.

10. *Strontium ranelate* has both antiresorptive and anabolic effects on bone. However, it has been recently shown that adipogenesis is negatively affected in the presence of strontium ranelate with a concomitant dose-dependent decrease in the expression of adipogenic markers and changes in adipokine profile, thereby generating a favourable osteogenic effect within the bone marrow milieu.

11. *Dietary relevant mixtures* of isoflavones and their metabolites, lignans and their metabolites, coumestrol, and a mixture containing all of them have been shown to inhibit adipocyte differentiation as their additional mechanism of action in preventing osteoporosis independent of their concentration.

18.10 Bariatric surgery and bone health

1. Bariatric surgery adversely affects bone health. The skeletal effects of bariatric surgery are presumably multifactorial, and mechanisms may involve nutritional factors, mechanical unloading, hormonal factors, and changes in body composition and bone marrow fat.

2. The Roux-en-Y gastric bypass (RYGB) procedure combines restriction and malabsorption techniques and involves creating both a small gastric pouch and a deviation of a segment of the small intestine.

3. Metabolic bone disease (osteomalacia and osteoporosis) is a well-documented long-term complication of obesity surgery.

4. Abnormalities in calcium and vitamin D metabolism begin shortly after gastrointestinal bypass operations but clinical and biochemical evidence of metabolic bone disease may not be detected until many years later.

5. RYGB has significant impact on calcium and vitamin D metabolism. Diet restriction reduces the exogenous load of calcium and vitamin D and decreases intake of macronutrients that positively affect their absorption. In RYGB the proximal jejunum is bypassed, excluding an important site of calcium absorption, which contributes to the decreased calcium load.

6. In addition, the reduction in food intake leads to increased release of cortisol and decrease in IGF-I serum levels, both adaptations potentially impair calcium absorption.

18.11 Management after Roux-en-Y gastric bypass

1. Follow-up includes careful examination to detect subclinical fracture.
2. Patient stature should be measured before and at regular intervals after surgery.
3. Patients undergoing RYBG should be screened for osteoporosis with bone density measurement.
4. Laboratory evaluation includes calcium, albumin, magnesium, PTH, and 25(OH) D.
5. Operated patients should be encouraged to perform regular weight-bearing physical exercise. Physical activity is part of the strategy to reduce fracture risk.

6. Calcium and vitamin D supplementation should be prescribed in all bariatric patients. Patients should be advised to take slightly higher daily doses of vitamin D and Calcium than RDI recommendations: 1500 mg calcium and 2000 IU vitamin D.
7. 25(OH) D serum levels should be checked every 2 months to ensure adequate levels. One should consider pharmacological treatment in patients if BMD is below 21.5.

18.12   Pragmatic approach for obese women

18.12.1   What should a physician advise to an obese woman?

1. Physicians have a responsibility to recognise obesity as a gateway disease and help patients with appropriate prevention and treatment pathways for obesity and its comorbidities, including osteoporosis.
2. Treatment should be based on good clinical care and evidence-based interventions, and it should be individualised, multidisciplinary, and focused on realistic goals of prevention of weight regain and weight maintenance.
3. Advice, treatment, care, and the information given to the patients should be nondiscriminatory and culturally appropriate.
4. It should also be accessible to people with additional needs such as physical, sensory, or learning disabilities and to people who do not speak or read English.
5. The main requirement of a dietary approach to weight loss and osteoporosis is that total energy intake should be less than energy expenditure, and they should eat plenty of fibre-rich foods, including five portions of a variety of fruit and vegetables each day.
6. Interventions to increase physical activity should focus on activities that fit easily into people's everyday life, such as walking, and should be tailored to people's individual preferences and circumstances.
7. Attention should be paid to women who are at risk of developing obesity just like pregnant and menopausal women.
8. Women should be encouraged to increase their physical activity to lose weight, as evidence suggests that physical activity can reduce the risk of type 2 diabetes, cardiovascular disease, sudden death, cancer (especially cancer of the endometrium), depression due to body image, and osteoporosis.
9. Adults should be encouraged to do at least 30 minutes of at least moderate-intensity physical activity on 5 or more days a week.
10. Obese women should take higher dose folic acid and vitamin D along with additional calcium.
11. Pharmacological and surgical treatment should be initiated based on clinical assessment.

18.13   Conclusion

1. The relationship between fat mass and bone is confounded by complex genetic backgrounds and by interactions between metabolic factors and regulatory pathways influencing both obesity and osteoporosis.
2. The previous concept that obesity is protective for osteoporosis may not stand up to careful scrutiny as the new concept of bone marrow fat deposition seen in obesity has

emerged, supporting the detrimental effect of obesity for bone health. Thus obesity, especially central obesity, may not be considered protective for osteoporosis.

3. Considering that obesity can be associated with fracture and that obesity treatment also can damage the skeleton, it is reasonable to conclude that the primary target should be obesity prevention.

Further reading

Bano R., Mahmood T.A. Female obesity and osteoporosis (In) Obesity and Gynaecology, (eds) Mahmood TA, Arulkumaran S, Chervenak FA, 2020, 2nd ed., pages 265−272 <https://doi.org/10.1016/B978-0-12-817919-2.00029-2>.

Berarducci A, Murr MM, Haines K. Risk and incidence of falls and skeletal fragility following Roux-en-Y gastric bypass surgery for morbid obesity. *Osteoporos Int.* 2007;18 (S):201.

National institute of clinical excellence (NICE). Osteoporosis: assessing the risk of fragility fracture. Clinical guideline (CG146) http://www.nice.org.uk.

Shapses SA, Riedt CS. Bone, body weight and weight reduction—what are the concern. *J Nutr.* 2006;136(6):1453−1456.

Zhao LJ, Liu YJ, Liu PY, et al. Relationship of obesity with osteoporosis. *J Clin Endocrinol Metab.* 2007;92:1640−1646.

Menopause and hormone replacement therapy

19

Chu Chin Lim

Department of Obstetrics & Gynaecology, Victoria Hospital, Kirkcaldy, United Kingdom

19.1 Introduction

Menopause is an inevitable component of ageing and encompasses the loss of ovarian reproductive function, either occurring spontaneously or secondary to other conditions. This has a significant influence in women's quality of life and the likelihood of healthy ageing:

1. Temporal changes in health and quality of life (vasomotor symptoms (VMS), sleep disturbance, and depression).
2. Longer term changes in several health outcomes (urogenital symptoms, bone, and lipids).
3. The loss of sex hormones during ageing contributes to changes in body mass, musculoskeletal integrity, sexual dysfunction, and long-term risks of health and disease.
4. Metabolic syndrome (insulin resistance syndrome) increases in prevalence after menopause and consists of insulin resistance, abdominal obesity, dyslipidaemia, elevated blood pressure, and proinflammatory and prothrombotic states.
5. The prevalence of obesity [body mass index (BMI) of >30 kg/m^2] is higher in postmenopausal women than in premenopausal women.
6. This is a multifactorial process consequence of reduced energy expenditure due to physical inactivity, compounded by depression, muscle atrophy, and a lower basal metabolic rate.
7. Increase of total body fat and a redistribution of body fat from the periphery to the trunk, which results in visceral adiposity. Postmenopausal women had 36% more trunk fat, 49% greater intraabdominal fat area, and 22% greater subcutaneous abdominal fat area than premenopausal women in studies where computed tomography and magnetic resonance imaging have been used.

19.2 Oestrogens and menopausal obesity

1. Recent studies with oestrogen receptors (ER) knockout mice have helped to unravel the role of the ERs in brain degeneration, osteoporosis, cardiovascular diseases (CVDs), and obesity.
2. Sex hormones help integrate metabolic interaction among major organs that are essential for metabolically intensive activities such as reproduction and metabolic function.
3. Sex steroids are also required to regulate adipocytes' metabolism and also influence the sex specific remodelling of particular adipose depots.

Handbook of Obesity in Obstetrics and Gynecology. DOI: https://doi.org/10.1016/B978-0-323-89904-8.00011-1

4. The function of oestrogens is mediated by nuclear receptors (ER α and ER β) that are transcription factors that belong to the superfamily of nuclear receptors. ER α and ER β are expressed by human subcutaneous and visceral adipose tissue, whereas only ER α mRNA has been identified in brown adipose tissue.

5. ER α plays a major role in the activity of adipocytes and sexual dimorphism of fat distribution. Polymorphism of ER α in humans is associated with risk factors for cardiovascular disease.

6. Body weight and BMI were significantly higher in perimenopausal and postmenopausal than in premenopausal women.

7. SWAN cross-sectional analysis showed no correlation between obesity and age at natural menopause, but obesity was associated with a likelihood of surgical menopause.

8. Concentrations of sex hormones partially control fat distribution: men have less total body fat but more central/intraabdominal adipose tissue, whereas women tend to have more total fat in gluteal/femoral and subcutaneous depots.

9. Fat tissue and regional fat tissue as a percentage of total fat tissue were higher in the trunk and arms in perimenopausal and postmenopausal than in premenopausal women. The shift to a central, android fat distribution can be counteracted by HRT.

10. The mean oestradiol level E2 change in transition from premenopausal to postmenopausal was less pronounced in obese women when compared with nonobese women. Obese women had lower premenopausal mean E2 levels but higher postmenopausal mean E2 levels.

11. Large observational studies have reported that obesity is a key factor for perimenopausal VMS but not postmenopausal VMS.

12. Data from WHI have shown that obesity is an important correlate for multiple urogenital symptoms, and obese women were twice as likely to report severe vaginal discharge and almost four times more likely to report severe itching/irritation compared with low, normal-weight women, controlling for diabetes.

13. It has been hypothesised that adipose tissue functions as an insulator and interferes with normal thermoregulatory mechanisms of heat dissipation.

14. A study using computed tomography demonstrated an increase in subcutaneous adipose tissue with age with age, independent of menopausal status, whereas visceral and total body fat increased only in women who became postmenopausal during the 4 years of follow-up. The change in visceral obesity was accompanied by a decrease in visceral circulating oestradiol and increase in FSH.

15. Ultrasound scanning has shown no difference in antral follicle count between obese and nonobese women in late reproductive age.

16. Follicular dysfunction and alterations in central nervous system regulation of hormonal levels among obese women may be contributory factors.

17. Change occurs in the primary source of circulating E2 as the menopause transition progresses; the primary source of circulating E2 premenopausally is the ovary, whereas in postmenopause the primary source of circulating E2 is the aromatisation of androgens within the adipose tissue.

19.3 Lifestyle intervention and hormone replacement therapy

Hot flushes and menopausal symptoms, in general, are more frequent in obese women compared to women with normal BMI.

1. Women who gain weight during the menopausal transition are more prone to have menopausal symptoms.
2. Obesity is associated with a greater likelihood of VMS, although women who are overweight (BMI from 25 to 30 kg/m^2), as opposed to obese (BMI >30 kg/m^2), are more likely to have severe symptoms.

Obese postmenopausal women are at increased risk of developing coronary heart disease (CHD).

1. According to the Nurses' Health Study, a 5 kg/m^2 increase in BMI is associated with a 30% increase in the incidence of CHD in women, independently of other CHD risk factors, such as age, smoking, physical activity, alcohol intake, or family history of CHD.
2. Stroke risk increases linearly with increasing BMI, independently of sex and race.
3. Women with BMI >32 kg/m^2 have a relative risk of 2.37 of developing ischaemic stroke.
4. Women who gain 10−20 kg during their adult life have a 69% increase in the risk of ischaemic stroke.

Obese postmenopausal women are at increased risk of developing breast cancer with a relative risk of breast cancer ranging between 1.26 and 2.52.

A 5 kg/m^2 increase in BMI is associated with a 12% increase in the incidence of breast cancer.

1. Possible explanations are the higher endogenous oestrogens produced by the aromatisation of precursor adrenal and ovarian androgens in adipose tissue and mitogenic IGF-1 activity associated with insulin resistance.
2. Apart from absolute body weight, the weight gained after 30th−40th year of age and especially perimenopausally appears to constitute an extra risk for breast cancer.

Obesity is associated with the increased risk of venous thromboembolism (VTE).

VTE is rare in premenopausal and young postmenopausal women and its incidence increases with age, BMI, and the presence of prothrombotic mutations (factor V-Leiden and prothrombin G20210A).

1. Obesity is a biologically plausible risk factor for VTE, but the mechanisms underlying the relation of obesity with VTE are not totally understood.
2. A strong positive correlation between plasminogen activator inhibitor-1 (PAI-1) level and BMI has been reported.
3. PAI-1 is the main fibrinolytic inhibitor, and reduced plasma fibrinolytic potential may be a risk factor for venous thrombosis.
4. Decreased fibrinolysis because of a high level of PAI-1 could explain in part the association of VTE with overweight and obesity.
5. Increased BMI was associated with higher levels of prothrombotic factors such as fibrinogen and factor (F) VII.
6. Both oral oestrogen and obesity may have synergistic effects on the imbalance between procoagulant factors and antithrombotic mechanisms.
7. By contrast, transdermal oestrogen appears to have little or no effect on haemostasis. Alternatively, increased C-reactive protein levels have been reported in obese individuals with a history of VTE, and low-grade inflammation could explain in part these findings.
8. In addition to the effects on haemostasis and inflammation, obesity may also have direct mechanic effects on the venous area.

9. An increased BMI may result in a higher VTE risk through an increased intra-abdominal pressure and a decreased venous return.
10. These effects may result in venous hypertension, varicose veins, and venous stasis, which promote the development of VTE.

19.4   Healthy lifestyle

1. All women at midlife should be encouraged to maintain or achieve a normal body weight, be physically active, adopt a healthy diet, limit alcohol consumption, and not smoke.
2. For obese women, weight loss may lessen VMS as well as reduce the risks of CVD, diabetes, urinary incontinence, breast, pancreatic, and endometrial cancers, and dementia.
1. HRT administration to postmenopausal women is associated with a significant decrease in the incidence of type 2 diabetes by the following mechanism:
   a. Oestrogens seem to influence glucose homeostasis through increased glucose transport into the cells.
   b. Lack of oestrogens has been associated with a progressive decrease in glucose-stimulated insulin secretion and insulin sensitivity as well as with insulin resistance.
   c. In the systemic circulation, E2 and oestrone are partly bound to sex hormone binding globulin (SHBG), as well as to albumin, as is testosterone.
   d. Increasing or decreasing SHBG levels will affect the amount of unbound oestrogen and testosterone in the circulation.
   e. The aim of additional progesterone in women with intact uteri is to protect against the development of endometrial pathologies, including hyperplasia and cancer.

For those who require pharmacological therapies, average-dose HRT is the most effective treatment for VMS with reductions in both frequency and severity in the order of 75%, and HRT may improve quality of life in symptomatic women.

1. Obese postmenopausal women requiring HRT should be thoroughly evaluated at baseline and the severity of symptoms and risk of fracture should be weighed against individual risks of breast cancer, CVD, and VTE.
2. There is a rationale to use the lowest effective dose—oral conjugated equine oestrogens (CEEs) 0.300−0.400 mg or oestradiol 0.5−1 mg orally or 25−50 μg transdermally, and it may be preferable using the transdermal route.
3. HRT should be avoided in those with unexplained vaginal bleeding, active liver disease, previous breast cancer, CHD, stroke, personal history of thromboembolic disease, or known high inherited risk.
4. CVD risk factors do not automatically preclude HRT but should be taken into account.
5. Upregulation of the hepatic synthesis of procoagulants is another known effect of oral oestrogens.
6. Transdermal oestradiol does not seem to increase the risk of venous thromboembolic events.
7. Evidence shows that transdermal oestrogen ($< = 50$ μg) is associated with a lower risk of deep vein thrombosis, stroke, and myocardial infarction compared to oral therapy and may be the preferred mode of treatment in women with an increased thrombosis risk, such as obese women and smokers.

Genitourinary syndrome (GSM) is a relatively new terminology describing vulvovaginal changes at menopause, as well as urinary symptoms of frequency, urgency, nocturia, dysuria, and recurrent urinary tract infections.

1. Vaginal dryness is common after menopause and unlike VMS usually persists and may worsen with time.
2. Urogenital symptoms are effectively treated with either local (vaginal) or systemic oestrogen therapy.
3. Oestrogen therapy restores normal vaginal flora, lowers the pH, and thickens and revascularises the vaginal lining.
4. The number of superficial epithelial cells is increased, and symptoms of atrophy are alleviated. Low-dose vaginal oestrogen improves vaginal atrophy without causing proliferation of the endometrium.
5. Initiation of HRT is usually contraindicated in women with a personal history of breast cancer or VTE, or those with a high risk for breast cancer, thrombosis, or stroke.
6. Transdermal oestrogen therapy may be considered and preferred when highly symptomatic women with type 2 DM or obesity, or those at high risk of CVD, do not respond to nonhormonal therapies.
7. Commencement of systemic hormone therapy is not recommended for women who are aged >60 years.

19.5 Progesterone

The combination of the oestrogen with a progestogen is used to avoid undue chronic stimulatory effects on the endometrium.

Endometrial cancer is the most common gynaecologic cancer, it is estimated that risk of endometrial cancer increases about 59% for every 5 unit increase in BMI (kg/m^2), and overweight and obesity are responsible for 57% of all cases of endometrial cancer in the United States.

Obesity increases on exposure to oestrogen unopposed by progesterone in pre- and postmenopausal women. The inclusion of progesterone appears to increase breast cancer risk, but progestogens are still indicated to prevent endometrial hyperplasia and cancer risk.

1. Progesterone is naturally produced in the ovaries (particularly the corpus luteum), in the placenta, and, to a certain extent, in the adrenals, and there are a variety of synthetic progestogens.
2. One of these progestogens, dydrogesterone, is a retroprogesterone and, another, drospirenone (DRSP), is a spironolactone derivative.
3. The "newer" progestogens belong to different classes based on their structure. For each of them the progestogenic, as well as the antioestrogenic action, is common.
4. The antiandrogenic effect is relevant for dienogest and DRSP and minor for nomegestrol acetate. None of them have a glucocorticoid effect.
5. DRSP is different due to its strong antimineralocorticoid action and has a favourable effect on blood pressure. In addition, these progestogens do not interfere with the positive effect of oestrogens on lipid and carbohydrate metabolism, do not augment haemostasis

processes as monotherapy, and avoid induction of abnormal proliferation of the endometrium in doses clinically tested.

Therefore all three progestogens appear to be suitable for the treatment of menopausal women.

The most recent Position Statement of the North American Menopause Society on HRT suggests that HRT may help attenuate abdominal adipose accumulation and the weight gains that are often associated with the menopause transition, and it significantly reduces the diagnosis of new-onset type 2 DM.

19.6 Emerging therapies

HRT could create important health risks; it is highly desirable to discover new alternatives in the menopause-related symptoms management, with minor side effects.

1. Dehydroepiandrosterone (DHEA) have been available over the counter. DHEA serves as a precursor for oestrogens and androgens from foetal life to post menopause.
2. DHEA may be an inactive precursor pool for the formation of bioactive steroid hormones.
3. DHEA-sulfate (DHEAS) represents the most abundant sex steroid in plasma in humans (more than 1000 times higher than oestradiol and testosterone levels), but its serum concentration goes down to 10%−20% of its maximum level by around the age of 70 years. The large difference between low and high serum DHEA levels has a major clinical impact.
4. Among postmenopausal women with coronary risk factors, lower DHEA levels were linked with higher mortality from CVD and all-cause mortality.
5. DHEA 10 mg daily administration in symptomatic postmenopausal women with lower (5th percentile) baseline DHEAS levels, may improve climacteric and sexual symptoms and directly reverse some age-related changes in adrenal enzymatic pathways, including adrenal DHEA and progesterone synthesis.

An oral selective ER modulator (SERM), ospemifene, has been approved for the treatment of moderate-to-severe pain during intercourse associated with vulvovaginal atrophy.

A tissue-specific SERM−oestrogen complex (a combination of oral CEE and bazedoxifene (BZA) (a SERM)) has been approved for the management of moderate-to-severe VMS in women with an intact uterus.

1. Tissue selectivity is achieved through the concurrent use of oestrogen and a SERM, which replaces a progestogen and selectively blocks the undesirable actions of oestrogen.
2. In the case of CEE-BZA, the proliferative effects of oestrogen are blocked in the uterus and possibly also the breast, whereas the bone-sparing actions of oestrogen are preserved.
3. The rationale for combining oestrogens with a SERM (T-SEC, combination of CEE and BZA (SERM)) is to retain the beneficial effects of oestrogens on VMS, VVA, and bone while incorporating the antiestrogenic effects of the SERM on the breast and endometrium to improve the overall safety profile.
4. The tissue-selective oestrogen complex (combination of 0.45 mg of oral CEE and 20 mg BZA (a SERM)) has been approved for the management of moderate-to-severe VMS in the United States and Europe.

Tibolone is a synthetic steroid that is rapidly converted to two metabolites with estrogenic activity and to a third metabolite characterised by a mixed progestogenic/androgenic activity.

1. Tibolone controls hot flushes, sweating, and mood symptoms and is effective in improving libido, due to its androgenic component.
2. Randomised, controlled studies show that tibolone increases bone mineral density and reduces fracture risk. These beneficial effects are seen over long-term treatments (over 10 years) and both in early and late postmenopausal women as well as in women with established osteoporosis.
3. The combined analysis of randomised clinical studies on tibolone indicates no increase in risk of breast cancer development compared with placebo.
4. Tibolone treatment is associated with a reduction of proliferation and a stimulation of apoptosis in normal breast cells that are possibly attributable to the impact of this compound on the activity of oestrogen-metabolising breast enzymes.
5. The metabolisation of tibolone is tissue selective, and the conversion to the progestogenic metabolite is particularly active in the endometrium.
6. Investigation of endometrial histology in women treated with tibolone shows no hyperplasia and a high level of atrophic endometrium, indicating no proliferative effect of this molecule.

A number of nonhormonal therapies are efficacious against menopausal VMS and should be considered for women who do not wish to take oestrogen or those with contraindications.

1. For VMS, many drugs have demonstrated efficacy in several studies: paroxetine, fluoxetine, and citalopram (which are selective serotonin reuptake inhibitors); venlafaxine and desvenlafaxine (selective noradrenaline reuptake inhibitors); clonidine (α2-adrenergic receptor agonist); and anticonvulsants (gabapentin and pregabalin).
2. Paroxetine and fluoxetine are potent cytochrome P4502D6 (CYP2D6) inhibitors and as they decrease the metabolism of Tamoxifen (a SERM used in the treatment of breast cancer)—which may reduce its anticancer effects—these drugs should be avoided in tamoxifen users.
3. Consistency of treatment response and efficacy of the various alternative options remain questionable.

19.7 Conclusion

1. The decision to start HRT in a women transitioning towards menopause requires a personalised discussion on the unique balance of risks and benefits.
2. During counselling, the importance of improving lifestyle, dietary habits, and implementing physical activity, especially their preventive role in cardiovascular disease, should be reiterated.
3. Evidence suggests that the balance of benefits and risks for HRT is more favourable within the first 10 years of menopause.
4. Based on the WHI data, the greatest risk appears to be associated with combined oestrogen—progestin therapy, therefore newer preparations may have a useful role in future practice.

5. Menopausal hormone therapy remains the most effective treatment of VMS and is also indicated for GSM and bone protection.

6. The duration of use for HRT should be individualised to take account of each patient's requirements and risk-benefit profile.

7. The ultimate aim is to relieve bothersome menopausal symptoms and reduce the risk of osteoporosis and cardiovascular disease, without increasing the risk of endometrial or breast cancer.

Further reading

Al-Safi ZA, Polotsky AJ. Obesity and menopause. *Best Pract Res Clin Obstet Gynaecol.* 2015;29:548−553.

Caretto M, Giannini A, Simoncini T, Genazzani AR. Obesity, menopause, and hormone replacement therapy. In: Mahmood TA, Arulkumaran S, Chervenak FA, eds. *Obesity and Gynaecology.* 2nd ed. Elsevier; 2020:273−279. Available from: https://doi.org/10.1016/B978-0-12-817919-2.00030-9.

European Menopause and Androgenic Society. EMAS position statement: managing obese postmenopausal women. *Maturitas.* 2010;66(3):323−326.

Genazzani AR, Komm BS, Picker JH. Emerging hormonal treatments for menopausal symptoms. *Expert Opin Emerg Drugs.* 2015;20(1):31−46.

Roberts H, Hicke M. Managing the menopause: an update. *Maturitas.* 2016;86:53−58.

The NAMS 2017. Hormone therapy position statement. The 2017 hormone therapy position statement of the North American Menopause Society. *Menopause.* 2017;24(7):728−775.

Intraoperative care during gynaecology surgery

Chu Chin Lim and Tahir Mahmood
Department of Obstetrics & Gynaecology, Victoria Hospital, Kirkcaldy, United Kingdom

20.1 Introduction

Obesity is associated with various conditions, including diabetes mellitus, hypertension, hypercholesterolemia, heart disease, asthma, and arthritis. All these conditions contribute to increased morbidity and mortality in gynaecology surgery. Obese women with metabolic syndrome (specifically hypertension and diabetes) who underwent general, vascular, and orthopaedic surgery are at increased risk of perioperative morbidity and mortality compared with normal-weight patients.

Based on the data, the American College of Obstetricians and Gynecologists made the following recommendations in a recent committee opinion paper:

1. Gynaecological surgeons should have the knowledge to counsel obese women on the risks specific to this group.
2. As with all patients, evidence demonstrates that, in general, vaginal hysterectomy is associated with better outcomes and fewer complications than laparoscopic or abdominal hysterectomy.
3. Wound complications, surgical site infections, and venous thromboembolism are the main cause of morbidity in obese women who underwent gynaecology surgery.
4. Every effort should be made to offer all patients, regardless of BMI, the least invasive procedure in order to decrease complications, length of hospital stay, and postoperative recovery time.

20.2 Decision-making about surgery

1. Ensure that surgery is appropriate and there are no other alternative nonsurgical management options to deal with immediate issues until weight management has been addressed.
2. Consider conservative therapies, such as the Levonorgestrel-releasing intrauterine system for menstrual dysfunction, bladder retraining and physiotherapy for urinary problems, and pessaries for prolapse, should be considered as an option.
3. Obese women and their families should receive careful counselling about the increased risk of complications during surgery, possible technical challenges which may be encountered during surgery, and specific issues related to postoperative recovery
4. It is the clinician's duty to help them understand the problem from a medical point of view, and how risks related to surgery can be reduced. There is a case to offer bariatric

Handbook of Obesity in Obstetrics and Gynecology. DOI: https://doi.org/10.1016/B978-0-323-89904-8.00031-7

surgery for morbidly obese woman, if conservative treatment had failed, and she has other significant comorbidities.

20.3   Physiological changes in obese women

1. Central (visceral) obesity leads to several cardiovascular and haemodynamic changes associated with physiological abnormalities. A Scottish survey reported that the prevalence of cardiovascular disease was 37% in adults with a BMI >30 compared to only 10% in adults with a BMI of <25.
2. Hypertension is common in obese patients with 60% of obese patients having mild—moderate hypertension and 5%—10% having severe hypertension.
3. Cardiac arrhythmias are more common in obese patients and may be caused by a number of factors, including hypoxia, electrolyte imbalance, myocardial hypertrophy, and myocardial infiltration of the conducting system.
4. Autopsy studies have shown that there is an association between obesity and cardiomyopathy with a 20%—55% increase in cardiac diameter, ventricle size, and cardiac weight for the obese patients compared to the nonobese patient.
5. Class III obesity is associated with a decrease in functional residual capacity, shallow breathing pattern, an increase in peak inspiratory pressure, decreased expiratory reserve volume, and total lung capacity.
6. The functional residual capacity (FRC) is reduced in the obese patients when lying in supine position with an impaired tolerance for the Trendelenburg position for the laparoscopic surgery.
7. Steep Trendelenburg position along with CO_2 Pneumoperitoneum results in a greater arterial partial pressure of CO_2 ($paCO_2$) during Laparoscopic and robotic surgery. The end-tidal CO_2 remains constant and therefore leads to an elevated $PaCO_2$-$EtCO_2$ gradient (hypercapnia), which in turn reflects increased dead space, "Obesity hypoventilation syndrome." The FRC is further compromised by anaesthesia to levels lower than closing capacity resulting in airway closure and hypoxemia.
8. Obesity is a well-established risk factor for developing obstructive sleep apnoea; the higher the BMI, the higher the risk.
9. As the FRC and expiratory reserve volume drops, the mismatch in ventilation perfusion promotes alveolar collapse and atelectasis at the lung bases. The decrease in chest wall compliance can be as high as 60% after pneumoperitoneum is created.
10. Morbidly (Class III) obese patients are at a higher risk of developing hypercapnia and acidosis which can cause cardiac arrhythmias and vasoconstriction of pulmonary vessels, depressive effect on cardiac myocardial contractility, and tachycardia.
11. Obese patients are at increased risk of gastric acid aspiration, especially during minimal access/or invasive abdominopelvic surgery, because of increased intragastric pressure, large gastric volume, altered secretion of adipokines, predisposition to reflux, lower gastric pH, and delayed gastric emptying.
12. Increased intraabdominal pressure during minimally invasive surgery can reduce the peak femoral systolic velocity and increase the femoral vein cross-sectional area. Sequential compression devices should be used to reverse this effect along with the use of prophylactic antithrombotic agents to prevent deep venous thrombosis.

20.4 Preoperative evaluation

1. Preoperative detailed assessment by the anaesthetic team should be considered.
2. The anaesthetist will consider whether tracheal intubation and airway management will be difficult due to adipose tissue in the neck and limited neck/cervical spine movement.
3. Obese patients with metabolic syndrome undergoing noncardiac surgery are at increased risk of cardiovascular complications. A 12-lead cardiogram is recommended at preoperative evaluation and other tests, such as echocardiogram, based on the class of obesity and findings of physical examination, should also be considered.
4. In patients with diabetes mellitus, blood glucose evaluation and counselling the woman on the importance of euglycemia to improve postoperative wound healing are important.
5. Specialist investigations are required if obstructive sleep apnoea is suspected from a history of daytime somnolence, morning headaches, nocturnal wakening, and partner reports of loud snoring and apnoeic episodes during sleep.
6. Thorough abdominal and pelvic assessment should be carried out to decide upon the best route for surgery with the help of ultrasound scanning and magnetic resonance imaging to determine the best route of surgery. Even an examination under anaesthesia may provide more guidance.

20.5 Equipment and general considerations

1. Careful surgical planning is required in terms of personnel and availability of equipment.
2. There should be clear and early communication among members of the multidisciplinary team involved in the care of obese patients to agree on the plan of management and the availability of resources (appropriate equipment).
3. Risk to the patients in receiving suboptimal care due to lack of appropriate equipment can be a potential medico-legal issue.
4. Risk assessment and training for staff is required for appropriate manual patient handling, to protect themselves and patient.
5. Bariatric operating table that can handle 300 kg with appropriate extension is needed together with "Obesity Packs" as recommended by the Royal College of Anaesthetists (United Kingdom).
6. Obese patients are at risk of slipping off the table during position changes and therefore they must be secured to the table. All pressure points should be well padded, as there is a risk of nerve injury and of rhabdomyolysis of the gluteal muscles leading to renal failure among class 111 obese.
7. Special hospital beds should be available that can accommodate the weight and enable movement of the patient without manual handling.

20.6 Anaesthetic challenges

1. As a good practice, two experienced anaesthetists may pair up to support each other during complex procedures for class 11–111 obese.

2. During anaesthesia, obese patients in supine position require a 15% higher minute ventilation to maintain normocarbia.
3. In Trendelenburg position, the steeper the head-down position, and higher the pneumoperitoneum pressures, the greater the problem becomes, requiring higher minute ventilation.
4. An imbalance between perfusion and ventilation within the lung tissue results is increasing difficulty for the anaesthetist to maintain oxygenation for these patients especially in prolonged and complex surgery.

20.7 Thromboprophylaxis

1. The risk of perioperative deep vein thrombosis and pulmonary embolism is higher among obese people than among those of normal weight and it occurs in 5%−12% of obese patients who undergo surgery.
2. Obese people are at increased risk of venous stasis pre- and postoperatively.
3. Obesity is associated with increased levels of fibrinogen and factor VIII. Therefore, if they are on any oestrogen-containing treatment, then it should be discontinued at least 4 weeks before surgery.
4. Appropriately sized thromboembolic stockings should be used along with mechanical devices such as intermittent pneumatic compressions during surgery and also during the postoperative period, especially for class III obese until mobile.
5. There should be a departmental evidence-based protocol about the use of low-molecular-weight heparin (LMWH), starting a minimum of 2 hours postoperatively, unless there is a contraindication (it should be recorded in the case notes).
6. Treatment with LMWH should be extended to 4 weeks in cases of pelvic surgery for malignancy.

20.8 Sepsis prophylaxis

1. Obesity has been shown to impair immune response leading to impaired chemotaxis and macrophage differentiation.
2. Studies have also shown that obesity is associated with an increase in the risk of urinary tract infection, and pyelonephritis has been reported to be almost five times more common in obese female patients compared to nonobese.
3. Several studies have also reported an increased risk of skin and soft tissue infection in obese women.
4. For that reason broad spectrum antibiotics prophylaxis should be administered to patients especially to those with class II−III obesity.

20.9 Intraoperative challenges

20.9.1 Laparoscopic surgery

1. Gaining access into the intraperitoneal cavity can be challenging in the obese patient.
2. Laparoscopic surgery could be challenging in Class II−III patients because of the higher rate of failed entry, hindered manipulation, and poor views.

3. Increased distant between skin and peritoneum makes placement of Veress needle more difficult.
4. The longer Veress needle (150 mm) may be required to avoid preperitoneal insufflations.
5. Longer ancillary trocars (up to 150 mm) may also be useful.
6. Transumbilical open technique or entry at Palmer's point is recommended by the Royal College of Obstetricians and gynaecologists in morbidly obese women.
7. Ancillary port placement can be challenging due to poor visualisation of inferior epigastric vessels.
8. It is recommended that incision is made at the extreme lateral edge of the rectus sheath for ancillary ports to avoid injury to the pelvic sidewall vessels.
9. All the trocars should be cuffed ports to avoid displacement during the procedure.
10. Exposure is often compromised due to omental fat and limited manipulation of instruments.
11. This is compound be limited head down tilt (position due to ventilation difficulty).
12. A higher operating pneumoperitoneum pressure is often required and this further prevents satisfactory ventilation and positioning of patient.
13. Port-site hernia is more common in obese patient. It is vital to close any port size of 10 mm or more in layers.
14. Port closure techniques that affords laparoscopic visualisation is recommended (e.g., Endoclose).
15. To minimise the risk of port site herniation, smallest ports feasible should be used.

20.10 Open abdominal surgery

1. Abdominal surgery is challenging and should be the last resort if other routes prove impossible.
2. Access and adequate exposure is difficult due to amount of subcutaneous adipose tissue.
3. There are higher incidences of intraoperative complications due to limited access and/or distorted anatomy.
4. To overcome this, it is vital to have good assistance, appropriate instruments, retraction and lighting.
5. Obese women are at increased risk of wound infection; therefore perioperative administration of adequate amount of antibiotics is vital for reduction in wound infection rates.
6. Lowest infection is observed in patients with antibiotics administration before incision is made
7. A higher dose or weight-dependent dose of antibiotics should be considered as a standard antibiotics prophylactic regime failed to achieve adequate tissue concentration in obese women.
8. Meticulous operating technique with minimal tissue handling and good haemostasis is vital to prevent postoperative complications.

20.11 Postoperative issues

1. Class II—III obese individuals should be admitted to a high dependency unit for postoperative care.

2. All patients should be fitted with thromboembolic-deterrent stockings, and advised on rehydration and early mobilisation.
3. Adequate dose of LMWH should be administered based upon weight of the individual and the risk factors, and most patients may require extended duration of venous thrombo-embolic prophylaxis.
4. Adequate analgesia is crucial to allow early mobilisation.
5. Respiratory morbidity (postoperative hypoxemia) is more common in obese patients, due to reduced FRC and atelactasis. It can be improved with supplemental oxygen, semire-cumbent positioning and chest physiotherapy.

20.12   Conclusion

1. Obese patients commonly have comorbid conditions that can complicate intraoperative care.
2. A thorough assessment of the risk benefit should be discussed with women.
3. Appropriate planning of infrastructure upgrading to allow safe management of morbidly obese patients in an appropriate, safe, and adequately equipped environment is vital in the current upward trend of obesity in the world.

Further reading

American College of Obstetricians and Gynaecologists: Committee Opinion Number 619; January 2015 (Reaffirmed 2019): Committee on Gynaecologic Practice. American College of Obstetricians and Gynaecologists; 2019.

Booth CM, Moore CE, Eddleston J, et al. Patient safety incidents associated with obesity: a review of reports to the National Patient Safety Agency and recommendations for hospital practice. *Postgrad Med J.* 2011;87:6949.

Demaria EJ, Carmody BJ. Perioperative management of special populations' obesity. *Surg Clin North Am.* 2005;85:1283−1289. Available from: https://doi.org/10.1016/j.suc.2005.09.002.

Lim C, Mahmood TA. Challenges in gynaecological surgery in obese women. In: Mahmood TA, Arulkumaran S, Chervenak FA, eds. *Obesity and Gynaecology.* 2nd ed. London: Elsevier; 2020:217−222. Available from: https://doi.org/10.1016/B978-0-12-817919-2.00025-5.

Nieboer TE, Johnson N, Lethaby A, et al. Surgical approach to hysterectomy for benign gynaecological disease. *Cochrane Database Syst Rev.* 2009;(3). Available from. Available from: https://doi.org/10.1002/14651858.CD003677.pub4. Art. No: CD003677.

Royal College of Obstetricians and Gynaecologists. *Preventing Entry-Related Gynaecological Laparoscopic Injuries_RCOG Green Top Guideline No49.* London: RCOG; 2008.

Shaw RW. The effects of patient obesity in gynaecological practice. *Curr Opin Obstet Gynecol.* 2003;13:1798.

Section 2

Obstetrics

Pathological basis of effects of obesity on pregnancy outcome

21

Tahir Mahmood
Department of Obstetrics & Gynaecology, Victoria Hospital, Kirkcaldy, United Kingdom

21.1 Introduction

1. Obesity has reached epidemic proportions globally and nearly tripled worldwide between 1975 and 2016.
2. According to the World Health Organisation (WHO) in 2016, more than 1.9 billion adults aged 18 years and older were overweight, and of those over 650 million adults were obese.
3. Overall approximately 13% of world's adult population (11% of men and 15% of women) was obese in 2016.
4. A dramatic rise in overweight and obesity has been reported among children and adolescents aged 5−19 years over the last four decades.
5. Obesity is a complex condition with serious pathophysiological, social, and psychological implications that affects virtually all ages and socioeconomic groups.
6. Obesity has a negative impact on fertility as well as causing increased rates of congenital malformations and adverse obstetric outcomes.
7. Obese women have more saturated subcutaneous fat stores and tend to accumulate fat more centrally than lean women.
8. Obesity is a state of chronic inflammation, and it is this that is thought to result in the increase in insulin resistance (IR), via modulation of insulin signalling.
9. Central obesity is associated with metabolic syndrome, including gestational diabetes mellitus (GDM), gestational hypertension, and preeclampsia.
10. The incidence of preeclampsia has increased (25% rise in the United States) between 1987 and 2004, and in particular there is an increase in severe preeclampsia reported in the United States in parallel with a threefold rise in obesity.

21.2 Classification of body mass index

1. The prevalence of overweight and obesity is commonly assessed by using body mass index (BMI), defined as the weight in kilograms divided by the square of the height in metres (kg/m^2) (Table 21.1).

BMI is a relatively simple anthropometric index of total adiposity that does not discriminate between muscle and fat mass.

Handbook of Obesity in Obstetrics and Gynecology. DOI: https://doi.org/10.1016/B978-0-323-89904-8.00027-5

Table 21.1 The international classification of BMI in adults (WHO classification).

Weight group	BMI range	Additional cut-off
Underweight	<18.50	<18.50
Severe thinness	16.00−	<16.00
Moderate thinness	16.00−17.99	16.00−16.99
Mild thinners	17.00−18.49	17.00−18.49
Normal	18.5−24.9	18.50−22.99
		23.00−24.99
Overweight	=/>25.00	=/>25.00
Preobese	25.00−29.99	25.0−27.99
		27.50−29.99
Obese class 1	>30.00	>30.00
	30.00−34.99	30.00−32.99
		32.50−34.99
Obese class 2	35.00−39.99	35.00−37.49
		37.50−39.99
Obese class 3	=/>40.00	=/>40.00

There is a linear relationship between BMI and some of the adverse metabolic effects on:

1. Blood pressure (essential hypertension and PET)
2. Cholesterol and triglycerides
3. IR (GDM and diabetes mellitus T2)

    However markers of absolute and relative accumulation of abdominal fat, such as increased waist circumference and waist-to-hip ratio are more sensitive.

21.2.1 Waist circumference

Central obesity has been described as waist circumference >102 cm in men and >88 cm in women, however it is believed that waist and hip ratio are more sensitive, the normal ratio being <1.

1. *It is well-recognised that increased waist−hip ratio is associated with:*
    a. Increased risk of MI
    b. Hypertension
    c. Heart failure
    d. Total mortality with cardiovascular disease
    e. Disturbance of the renin angiotensin system
    f. Activation of the coagulation cascade plays a role
2. *For waist circumference associated with increased risk of cardiovascular disease:*
    a. with an increase of male waist from an average of 94 cm to 102 cm, there is a substantial increased risk of cardiovascular disease
    b. with an increase of female waist from an average of 80 cm to 88 cm, there is a substantial increased risk of cardiovascular disease
    c. An increase in waist circumference of 2% and a 0.01% increase in waist hip ratio is associated with 5% increased risk of future cardiovascular disease events

21.2.2 Visceral obesity is linked with elevated oxidative stress and systemic inflammation

1. There is activation of the coagulation cascade and disturbance of the renin−angiotensin system.
2. More importantly there is enhanced lipid and protein oxidation.
3. There is generation of oxidative low density lipoproteins

21.3 Increased disease burden secondary to obesity

21.3.1 Obesity was linked in up to 20% of all cancer-related deaths

There is overall increased risk of cancer in women for the following cancers:

1. postmenopausal breast cancer 9%
2. colon cancer 11%
3. renal cancer 25%
4. oesophageal cancer 37%
5. endometrial cancer 39%

21.3.2 US data show that the risk of developing obesity-related cancer appears to increase progressively in successively younger birth cohorts

1. The 25−49-year-old age group had a significant increase in six of the 12 obesity-related cancers:
   a. multiple myeloma;
   b. colorectal;
   c. uterine corpus;
   d. gall bladder;
   e. kidney; and
   f. pancreatic cancer.

21.3.3 Gynaecology

1. Early menarche
2. Menstrual disorders with increased risk of developing uterine fibroids and adenomyosis
3. Anovulation and infertility
4. Polycystic ovaries (> 50% of women are overweight or obese)
5. Suboptimal response to infertility treatment
6. Increased risk of recurrent miscarriages
7. Pelvic floor dysfunction

21.3.4   Obstetrics

1. Increased risk of recurrent miscarriage
2. Congenital anomalies
3. Prematurity
4. Abnormal foetal growth
5. Shoulder dystocia
6. GDM
7. Pregnancy-induced hypertension (PET)
8. Caesarean section/operative deliveries/perineal tears
9. Anaesthetic complications
10. Stillbirth
11. Neonatal death rates
12. Thromboembolic disease
13. Infant hypoglycaemia
14. Neonatal jaundice
15. Low Apgar scores of 7 at 5 minutes
16. Increased risk of maternal death
17. Wound infection
18. Endometritis

21.4   Other comorbidities

1. Ischaemic heart disease
2. Cerebrovascular disease (stroke, neurodegenerative disease, and cognitive impairment)
3. Gallstones (nonalcoholic steatohepatitis)
4. Osteoarthritis
5. Sleep apnoea
6. Psychological illness
7. Impaired physical functioning

21.5   Physiological changes during pregnancy in normal weight women

In normal pregnancy, the first and second trimester is a state of anabolism and in the anabolic phase there is hyperphagia, reduced IR, and increased fat storage.

By late pregnancy there is a catabolic state. IR facilitates increased lipolysis and gluconeogenesis to allow for foetal growth and weight gain.

21.6   First and second trimester of pregnancy

1. All women increase maternal fat stores in early pregnancy irrespective of prepregnancy adiposity.

2. Total fat appears to increase to a peak towards the end of the second trimester.
3. During the early to mid-trimester stage of pregnancy, there is an anabolic state where foetal demands are limited, maternal fat stores increase in part to maternal behaviour, hyperphagia, and increased adipose tissue lipogenesis.
4. Insulin sensitivity is normal or even slightly improved with peripheral sensitivity to insulin and hepatic glucose production.
5. Lipogenesis and fat accumulation is favoured by pregnancy-related endocrine changes including increasing levels of oestrogen, progesterone, and cortisol.

21.7 Third trimester

1. During late pregnancy, there is a switch to a state of catabolism with a marked increase in lipolysis rates and a corresponding rise in maternal free fatty acids and glycerol.
2. This change is enhanced by the increased production and activity of hormone-sensitive lipase and a concomitant decrease in lipoprotein lipase activity.
3. Exaggerated catecholamine release in response to even modest hypoglycaemia contributes to this switch as well.
4. Insulin's effect on lipolysis and fat oxidation in liver and muscle are significantly impaired.
5. Reduced expression of peroxisome proliferative-activated receptor γ (PPAR γ) and its target genes may also contribute to accelerated fat metabolism.
6. This primarily lipid-based metabolism in the mother increases availability of glucose and amino acids for the foetus
7. With advancing gestation, plasma cholesterol, plasma cortisol, and triglyceride concentrations rise by 25%−50%, 100%−160% (upto 1.6-fold increase), and 200%−400%, respectively.
8. The increase in triglyceride concentration is mainly due to VLDL triglycerides that show a threefold increase from 14 weeks gestation to late pregnancy.
9. There is significant triglyceride enrichment of the HDL, intermediate density lipoprotein, and LDL fractions, compared to the accompanying increase in phospholipids and cholesterol in these fractions.

21.8 In normal weight women fat distribution

1. The majority of fat is accumulated centrally in the subcutaneous component of the trunk and upper thigh,
2. In the later stages of pregnancy, there is an increase in the thickness of preperitoneal fat (visceral) and the ratio of preperitoneal to subcutaneous fat.

21.9 Amino acid metabolism in normal weight women

1. There is an increase in protein synthesis during the second and third trimester of 15% and 25%, respectively, in maternal tissue, including the liver, breast, and uterus.

2. A grater maternal protein synthesis in the second trimester is associated with an increase in birth length and accounts for 26% of its overall variance.

21.10   Pathological changes in obese women during pregnancy

BMI is a relatively simple anthropometric index of total adiposity which does not discriminate between muscle and fat mass.

1. Visceral adiposity in early pregnancy appears to correlate better than subcutaneous fat or BMI with metabolic risk factors.
2. In the third trimester, severely obese women have significantly greater abdominal and visceral fat stores compared with lean women.
3. In obese women, there is excess fat which accumulates in the visceral tissue and organs.
4. This happens when the adipose tissue has reached the maximum capacity, a spillover of lipid from adipocytes resulting in the increase of circulating free fatty acids.
1. The accumulation of excess fat happens at the following ectopic sites:
   a. Visceral adipose tissue
   b. Intrahepatic
   c. Intramuscular
   d. Renal sinuses
   e. Pericardial
   f. Myocardial
   g. Perivascular
2. The presence of macrophages together with expression of some inflammatory factors is more frequent in omental fat (visceral) than in subcutaneous peripheral fat.

21.11   There is increased mass of metabolic reactive visceral adipose tissue

1. In visceral fat, there is a higher turnover of lipids due to its greater sensitivity to catecholamine-induced lipolysis and decreased sensitivity to insulin.
2. Liver is exposed to chronic elevation of nonessential fatty acids, which produce an alteration in liver metabolism and promote hepatic IR.
3. As obese pregnant women have more saturated subcutaneous fat stores, they tend to accumulate fat more centrally than lean women.

21.12   Hyperlipidaemia

1. Physiological hyperlipidaemia of pregnancy is exaggerated in obese women with higher serum TG, VLDL, cholesterol, and FFA concentration than lean women.
2. There are lower levels of HDL-cholesterol, although LDL-cholesterol and total cholesterol appear similar.

3. The ability of insulin to suppress lipolysis is also reduced during pregnancy, leading to a greater postprandial increase in FFA, increased gluconeogenesis, and IR.

21.13 Low-grade chronic inflammation

1. Obesity generally is regarded as a state of low-grade chronic inflammation. This inflammation occurring in metabolically important organs, such as the liver and adipose tissue, has been referred to as "meta-inflammation".
2. The mechanism that links inflammation and IR is through activation of the kinase JUN-N terminal kinase which occurs in response to a variety of stress signals, including FFAs, proinflammatory cytokines, and reactive oxygen species.
3. Recently there has been focus on the unfolded protein response (UPR) to ER stress, which activates apoptotic pathways in stressed cells.
4. The downstream signalling cascade triggered by UPR has three main arms, which can induce inflammation, influence insulin receptor signalling, and activate apoptosis.
5. The activities of ER stress pathways in liver, adipose tissue, and skeletal muscle are key factors in the development of IR in these organs.
6. Energy imbalance leads to adipocytes hypertrophy and hyperplasia, thus causing adipocytes dysfunction.
1. *Adipocyte dysfunction leads to elevated levels of adipokines:*
   a. Low adiponectin
   b. Increased leptin levels
   c. Increased IR
   d. Increased androgen
   e. Increased cortisol
   f. Raised free fatty acids
   g. Increased oxidative stress
   h. Reduced immune cell recruitment
   i. Increased inflammatory cytokines

Table 21.1 *changes secondary to the chronic inflammatory milieu due to visceral obesity.*

1. *There is an exaggerated inflammatory state which leads to:*
   a. endothelial and vascular dysfunction;
   b. leading to excess lipolysis;
   c. increased triglycerides;
   d. increased free fatty acids; and
   e. low high-density lipoprotein and raised LDL.
2. *They all contribute to:*
   a. hypertension and atherosclerosis;
   b. cardiac muscle dysfunction;
   c. low levels of apoptosis;
   d. increased cell proliferation; and
   e. carcinogenesis,

(Continued)

(cont'd)

3. *IR leads to:*
   a. low sex hormone binding globulin synthesis;
   b. increased circulatory bio available E2, testosterone and free androgens; and
   c. increased insulin-like growth factor (IgF) leading to cell proliferation and cancer.

21.14   Adiposity and pregnancy-specific insulin resistance

1. Placentally derived growth hormone and placental lactogen both modulate maternal insulin sensitivity.
2. Placenta is also a source of a number of other hormones which influence metabolism including leptin, adiponectin, corticosteroids, and inflammatory mediators, such as TNF-α, and proinflammatory cytokines (IL1, IL8, MCP-1).
3. Thus placental and preexisting maternal pathways interact to adapt maternal metabolism over the course of pregnancy.
4. Preexisting obesity appears to drive oxidative stress and lipotoxicity within the placenta, and thus leads to dysregulation of normal placental function.

21.15   Glucose metabolism in obese women

1. In obese women, there is a loss of the reduction in fasting glucose in early pregnancy.
2. In late pregnancy, insulin-mediated glucose utilisation worsens by 40%−60% and insulin secretion increase several fold.
3. Skeletal muscle is the primary site of glucose disposal, and, along with AT, becomes severely IR during the latter half of pregnancy.
4. There is significant enhancement of peripheral and hepatic IR.
5. There is postprandial obesity-related IR, leading to an increase in metabolic fuels (glucose, lipids, and amino acids), exposing the foetus to increased availability of these nutrients.

21.15.1   The mechanism of the link between obesity and IR is multifactorial, and includes effects on

1. insulin signalling
2. Subclinical inflammation;
3. increased release of inflammatory mediators (TNF-α etc.);
4. adaptation of adipose function, low levels of circulating adiponectin; and
5. reduction in the lipogenic transcription factor (PPAR-γ).

Thus overall reduction in IR and increased lipolysis leads to:

1. haemostasis plasminogen activator inhibitor 1, adiponectin, and Leptin; and

2. chronic positive energy balance in obesity leads to: Raised TG levels leading to adipocyte hypertrophy and subsequently hyperplasia through adipogenesis.

21.15.2 Ectopic fat deposition leads to

1. Excessive fat in skeletal muscle is thought to promote peripheral IR by a reduction in insulin-mediated glucose uptake.
2. The exposure of the liver to elevated FFA leads to reduced hepatic insulin extraction, leading to systemic hyperinsulinaemia and accelerated gluconeogenesis.
3. Excess FFA can also lead to decreased function and apoptosis of pancreatic beta cells, contributing to the state of relative insulin deficiency.

21.16 Amino acid metabolism in obese women

1. A greater maternal protein synthesis in the second trimester is associated with an increase in birth length and accounts for 26% of its overall variance.
2. Obesity is associated with a greater supply of gluconeogenic amino acids to the liver with preference of their use over glycogen for glucose production.

21.17 Visceral adiposity as modulator for proinflammatory–prothrombotic state

A complex adipocyte-induced proinflammatory changes, and prothrombotic state:

1. raises serum leptin;
2. raises IL-6, TNF-α and tissue factor (leads to impaired chemotaxis and macrophage differentiation);
3. alters secretion of adipokines;
4. increases levels of plasminogen activator inhibitor type 1(PAI-1);
5. raises von Willebrand factor;
6. raises fibrinogen and prothrombin levels;
7. shows evidence of increased coagulation and platelet activation;
8. increases/worsens venous stasis; and
9. results in endothelial injury and dysfunction.

21.18 Raised serum leptin

1. Leptin levels are increased by the insulin stimulation of adipocytes. Their levels are elevated with food intake and lower in fasting state.
2. Lower levels of adiponectin leads to increased circulatory insulin that can be followed by hyperandrogenaemia that inhibits gonadotrophin secretion.
3. Leptin has a direct peripheral action on many tissues, including adipose tissue, stimulating lipolysis and fatty acid oxidation.

4. Leptin is also expressed in placenta. It binds to at least six leptin receptors.
5. Visceral fat is the main determinant of circulating maternal leptin in the first trimester of pregnancy; leptin will therefore be higher from early pregnancy in obese women.
6. Absolute leptin levels rises linearly with worsening obesity in pregnancy.
7. Leptin is an important proinflammatory agent and its systematic concentrations are proportional to the amount of body fat.
8. Leptin is mitogenic, antiapoptotic, immunosuppressive, and proangiogenic.

21.19   InterLeukin-6

1. InterLeukin-6 (IL6) is expressed by adipocytes.
2. IL6 and TNF-α may also increase oestrogen levels systematically.

21.19.1   Adiponectin

1. Adiponectin has potent antiinflammatory activity.
2. Adiponectin is predominantly secreted by visceral adipose tissue.
3. Its circulating levels are inversely correlated with body fat.
4. It is only produced by mature adipocytes, which in obese individuals represent only about 20% of the total number of cells.
5. Adiponectin appears to be protective against vascular disease.
6. Adiponectin is positively associated with HDL cholesterol and negatively with triglycerides and LDL cholesterol.
7. Adiponectin concentrations fall in obesity and are inversely related to glucose and insulin concentrations.
8. Adiponectin levels tend to be lower in women with preeclampsia than in normotensive pregnancies and lowest in the severe obese.
9. Low adiponectin concentrations are associated with increased risk of breast cancer.

21.20   Plasminogen activator inhibitor-1

1. PAI-1 levels are more closely related to fat accumulation in the liver, suggesting that in IR individuals, the fatty liver is an important site for PAI-1 production.
2. PAI-1 levels tend to be elevated in obese individuals with metabolic syndrome.
3. PAI-1 is expressed in stromal cells, including monocytes, smooth muscle cells, and preadipocytes.
4. Visceral adipose tissue seems to have up to five times the number of plasminogen activator inhibitor 1 (PAI-1)-producing stromal cells compared with subcutaneous adipose tissue.
5. There is increased concentration of small, dense LDL, hepatic overproduction of VLDL.
6. There is an inability to suppress hepatic glucose production.
7. There is impaired glucose uptake and oxidation.
8. There is inability to suppress release of nonesterified fatty acids from adipose tissue.

21.21 TNF-α

1. TNF-α may contribute to IR by release of FFA from adipocytes or by blocking the synthesis of adiponectin.
2. It also activates nuclear factor kB and hence increases the expression of adhesion molecules on endothelial cells and vascular smooth muscle cells.
3. This contributes to endothelial dysfunction, and ultimately atherogenesis.
4. Serum concentrations of TNF-α and placental expression TNF-α receptors are increased in preeclamptic pregnancies.

21.22 Angiotensinogen

1. Angiotensinogen is mainly synthesised in liver but also by adipocytes that contribute to circulating concentrations.
2. Adipocytes also synthesise the other major components of the renin−angiotensin system (RSA).
3. The RSA is increasingly being implicated in the pathogenesis of IR, probably via the generation of reactive oxygen species.
4. It is possible that some diagnoses of PET in the obese in fact relate to the unmasking of underlying hypertension by pregnancy.
5. There are significant trends to increasing blood pressure at booking with higher BMI category throughout pregnancy.
6. PET is associated with an increased risk of cardiovascular, renal, and metabolic disease in later life.
7. Being born of a preeclamptic pregnancy is also associated with a greater metabolic and CVD risk throughout life.
8. All these changes hold many parallels with metabolic syndrome in the nonpregnant and generally are in keeping with the metabolic and vascular phenotype of obesity.

21.23 Ghrelin

1. It is a neuropeptide hormone secreted from the stomach to increase hunger and is involved in glucose metabolism.
2. Plasma levels are decreased in obesity.
3. It also increases the production of nitric oxide synthase, which, in turn, modulates μ-opioid receptors to produce an antinociceptive effect.
4. There is an increased level of inflammation and heightened susceptibility to pain in obesity due to reduction of the protective antinociceptive and anti-inflammatory effect of ghrelin.
   a. CNS NPY receptors stimulate analgesia, while peripherally located postsynaptic receptor trigger hyperalgesia.
   b. Obese women are known to have higher circulating serum levels of NPY.
   c. The orexinergic system consists of neuropeptides: Orexin A, and Orexin B that stimulates the appetite and can reduce the perception of pain.

    d. In times of stress such as acute or chronic pain, the orexin system is stimulated.

    e. Obesity also appears to contribute to the risk of developing neuropathic pain disorders.

    f. In both diabetic and nondiabetic population, obese individuals have a higher incidence of developing peripheral neuropathy.

21.23.1 Unhealthy adipokines are upregulated in obesity: elevated levels of the following are also noted

1. Adipocyte fatty acid-binding protein
2. Lipocalin-2
3. Chemerin
4. Visfatin
5. Vaspin
6. Resistin
7. 1 β and monocyte-chemoattractant protein-1

    An increased production of SAA (lipolytic adipokine serum amyloid A) by enlarged adipocytes may contribute towards IR. RBP4, an adipocyte-secreted molecule, is elevated before the development of diabetes, and seems to signal the presence of IR.

21.24 Bone health

1. Insulin is a potential regulator of bone growth and metabolism, as osteoblasts have insulin receptors as well as IGF-1 receptor.
2. Leptin controls bone formation through a hypothalamic relay, by inhibiting NPY gene expression.
3. Adiponectin and corresponding receptors are expressed in primary human osteoblasts.
4. Resistin plays a role in bone remodelling and is expressed in mesenchymal bone marrow stem cells, osteoblasts, and osteoclasts, and increases osteoblasts proliferation and cytokinase release, as well as osteoclasts differentiation.
5. Increased production of IL-6, IL-1, and TNF- α leads to the development of osteoarthritis in obese women.
6. Adipsin-raised levels are seen in DM and obese patients.
7. There is increased fat deposition in bone marrow as well.
8. High-fat diet markedly reduces the rate of calcium absorption by the intestine, hence decreased availability of the Ca required for osteogenesis.
9. Metabolically, vitamin D levels are low, as women with a BMI >30 are at increased risk of vitamin D deficiency compared to women with a BMI of <25.
10. Vitamin D deficiency in the serum prevents intestinal uptake of calcium from the diet, and thereby signals the parathyroid gland to secrete increased levels of PTH.
11. Increased PTH induces osteolysis and prevents osteogenesis by supplying adequate levels of calcium and phosphorus in the blood necessary for metabolic process and neuromuscular function.
12. RYGB bariatric surgery leads to abnormalities in calcium and vitamin D absorption but its effects on bone health may not be apparent until many years later.

21.24.1 Cardiac disease and cardiovascular alterations during obese pregnancy

1. Increased blood volume
2. Increased heart rate
3. Increased cardiac output
4. Increased stroke volume
5. Reduced peripheral resistance
6. Increased arterial pressure
7. Increased preload (shifting the frank starling curve to the left)
8. Left ventricular wall stress
9. Cardiovascular hypertrophy
10. Pulmonary hypertension
11. Significantly increased risk of ischaemic heart disease, heart failure and arrhythmias

21.25 Long-term changes in cardiac function induce

1. ventricular modelling with enlargement of the cardiac cavities; and
2. increased wall tension;
3. leading to ventricular hypertrophy,
4. decreased diastolic chamber compliance; and
5. increase in left ventricular filling pressure
   a. Left ventricular diastolic dysfunction, systolic and diastolic dysfunction is the main precursors of heart failure in obesity.
   b. These cardiac adaptations are also modulated by the duration of the obesity.
   c. The cumulative effects of smoking, high low-density lipoproteins, high cholesterol, and IR increase the risk of cardiovascular disease, and obesity acts as a compounding factor.

21.25.1 Increased risk of arterial thrombosis in obesity is due to

1. Central obesity
2. Hypertension
3. Diabetes mellitus type 2
4. Dyslipidaemia

21.25.2 Increased risk of venous thromboembolism in obesity is due to

1. Increased procoagulant activity
2. Impaired fibrinolysis
3. Increased inflammation
4. Endothelial dysfunction
5. Altered lipid metabolism
6. Altered glucose metabolism

7. Adipose tissue produces several cytokines (adipokines):
   a. raised leptin;
   b. low levels of adiponectin;
   c. tumour necrosis factors (TNF-α); and
   d. plasminogen activator inhibitor PAI-1.
8. There is Oxidative stress which also leads to:
   a. platelet activation; and
   b. endothelial damage.
9. Shedding of thrombogenic endothelial cell derived microparticles
10. Increased plasma oestrogen levels
11. Local factors such as large size deep veins with reduced flow velocities and valvular dysfunction also contribute.

21.26   Smoking, obesity, and cardiovascular disease risk

1. Obese chronic smokers have increased levels of TNF-α, decreased levels of adiponectin, and increased IR.
2. Endothelial dysfunction, which is an early marker of atherosclerosis, is present in obese individuals as well as in chronic smokers. Therefore smoking among obese people would compound endothelial dysfunction and increase the risk of cardiovascular disease.

21.27   Preeclampsia

1. A recent review of studies linking obesity has reported a summary risk estimate of 4.14 (95% CI 3.61–4.75) for the development of preeclampsia in women with BMI >35 compared with BMI < 25.
2. There is a clear gradation and risk of developing preeclampsia rises incrementally as the BMI increases.
3. It appears that women with the greatest load of metabolic abnormalities have the highest incidence of PET.
4. Visceral fat mass is strongly correlated with metabolic risk factors such as hypertension and preeclampsia.
5. Basal fat oxidation increases by at least 50% during pregnancy (physiological hyperlipidaemia) but this is exaggerated in obese pregnant women.
6. They have greater increases in total and VLDL triglycerides and cholesterol and small LDLs and lower HDL.
7. The small-density LDL particles are increased in both obesity and preeclampsia.
8. Preeclampsia is now generally regarded as being a state of oxidative stress.

It is likely that dyslipidaemia triggers the development of placental bed atherosclerosis and preeclampsia.

1. There is a rise in proinflammatory cytokines (IL-6, TNF Alpha), raised CRP, raised PAI-1, endothelial dysfunction, IR, and dyslipidaemia.

21.28   Maternal obesity and in-utero programming

21.28.1   Animal studies have reported that

1. *In-utero environment of chronic inflammation leads to:*
   a. vasoconstriction;
   b. increased platelet aggregation;
   c. increased lipid storage in placenta;
   d. increased foetal hypoxia and poor placental function;
   e. development of atherosclerotic plaques in placental arterioles;
   f. and an adaptive shift in blood supply away from the kidney and heart; and
   g. reduced proliferation and maturation of cardiomyocytes leading to myocardial hypertrophy and fibrosis resulting in premature cardiac dysfunction.

21.29   Human studies have shown

1. Maternal obesity is associated with onset of early cardiovascular disease in the offspring.
2. Premature death of offspring in adolescents and adulthood.
3. Maternal in-utero environment has an impact on foetal programming from neonatal health to-long term adult health.
4. Human maternal obesity is associated with early cardiovascular disease and premature death of offspring in adolescence and adulthood.

21.29.1   Epigenetics

1. Changes mediated by the maternal environment can be transduced to the next generation via epigenetic changes, that is, DNA modifications other than changes in the DNA sequence itself which can be heritable.
2. Apart from DNA methylation, such epigenetic changes include histone modifications and small noncoding RNAs.
3. Consistent methylation changes in metabolic genes are closely associated with later obesity.
4. It has been estimated that neonatal epigenetic marks could explain a significant proportion of the variance in childhood obesity.

21.29.2   Evidence

1. In a large study (Lee et al.) of 28,540 women and their 37,709 offspring were followed up and it was shown that obese women were at significant risk of major cardiovascular events and hospitalisation for a cardiac event compared to a mother with a normal BMI.
2. Children exposed to in utero maternal environments secondary to obesity-related metabolic syndrome are also at increased risk of transmitting intergenerational risk of obesity on to future generations.

21.30   Various hypotheses of transmission of intergenerational obesity are as follows

1. *The developmental over nutrition hypothesis* proposes that high maternal glucose and high free fatty acids and amino acids plasma concentrations result in permanent change in appetite control, neuroendocrine function, and allergy metabolism in the foetus, thus increasing the risk of IR and glucose intolerance.
2. *Longitudinal studies* have shown a plausible link between maternal over nutrition during intra-uterine, breast feeding period and foetal programming?
3. *Foetal skeletal development* is influenced by shifting mesenchymal stem cells differentiation from myogenesis towards adipogenesis. This shift permanently impairs the physiological function of the offspring's skeletal muscles.
4. *Epigenetic modification* may influence the phenotype later in life, thus impaired glucose tolerance during pregnancy leads to adaptation in leptin gene DNA methylation.
5. *In utero exposure to very high levels of leptin* by the offspring leads to a hyperphagic and obese phenotype in adulthood.

21.30.1   Intergenerational obesity

1. Maternal obesity is associated with increased risk of adiposity and noncommunicable diseases in the offspring.
2. This effect is independent of the shared genetic and environmental factors between the mother and child.
3. Epidemiological studies have revealed stronger correlation between offspring and maternal BMI than with parental BMI.
4. Maternal BMI was associated with offspring adiposity at 12 and 24 months while maternal glycaemia was correlated with offspring adiposity at birth but not at 12 and 24 months.
5. Foetuses of obese mothers have greater percentage body fat, elevated cord blood, leptin, and IL-6 and they are more insulin resistant at birth.
6. Southampton study showed that chicken born to nondiabetic mother with a higher BMI had greater fatness at 9 years of age.
7. "Growing up today study" (United States): children born to mothers with gestational diabetes had approximately 40% increased risk of adolescent obesity at 9–14 years after birth.
8. A systematic review of 45 studies concluded that prepregnancy overweight and obese are associated with approximately two- to threefold increased risk of offspring being overweight/obese.
9. Both children who are large for gestational age (LGA) and exposed to either maternal obesity or gestational diabetes are at increased risk of developing metabolic syndrome in childhood.
10. Avon longitudinal study has reported that women whose weight gain during pregnancy exceeded the IOM criteria are more likely to have children with greater BMI, waist circumference, leptin, systolic blood pressure, elevated inflammatory factors, including CRP and IL-6, and lower HDL cholesterol.
11. Offspring of mothers with DM-T1 have been found to be at increased risk of glucose intolerance and impaired insulin secretion, suggesting that it is the exposure to

intrauterine hyperglycaemia or associated metabolic derangement that is responsible for this transgenerational effect of maternal diabetes.

12. LGA offspring of mothers with normal glucose tolerance, and LGA offspring of mothers with GDM have increased fat mass and decreased lean body mass.

13. GDM was twice more common in subjects with a diabetic mother, compared with those with a diabetic father.

14. EPOCH study has demonstrated that GDM was associated with increased overall and abdominal obesity, and a more central fat distribution at 6−13 years old.

15. HAPO study have shown that offspring exposed to untreated GDM had increased risk of being overweight, obese, and increased adiposity at age 7, independent of maternal BMI during pregnancy. GDM was associated with a 51% increase in risk of being overweight/obesity in the offspring.

16. There is a linear relationship between the level of maternal glycaemia during pregnancy and offspring adiposity. The occurrence of macrosomia was associated with higher glucose concentration at screening. HAPO study reported an increase in fasting, 1 and 2 hours glucose at 24−32 weeks gestation which was associated with approximately 1−1.5-fold increased risk for neonatal macrosomia (BW > 90th centile) and neonatal hyperinsulinaemia (cord blood C peptide >90th centile).

17. GDM has been associated with increased cardiometabolic risk in the offspring. GDM mothers' offspring were found to have:
    a. significantly higher systolic and diastolic blood pressures;
    b. lower HDL Cholesterol; and
    c. elevated C-peptide predicted glucose intolerance in the offspring at 8 and 15 years of age, as well as metabolic syndrome at 15.

18. In one study, offspring of mothers with GDM had increased IR and central obesity 15 years later.

21.30.2   Offspring of diabetic mothers had

1. significantly raised levels of markers of endothelial dysfunction;
2. raised E-selection;
3. vascular cell adhesion molecule 1;
4. raised leptin;
5. increased waist circumference;
6. raised BMI;
7. raised systolic blood pressure; and
8. reduced adiponectin levels.

The effect of physical activity:

1. Physical activity improves glucose intolerance and sensitivity in improving noninsulin-dependent glucose intake;
2. improves the ratio between HDL and LDL cholesterol by increasing activity of lipoprotein lipase;
3. decreases triglycerides;
4. increases fibrinolysis and decreases platelet aggregation;
5. improves oxygen uptake in the heart and in peripheral tissues;
6. lowers the resting heart rate by increasing vagal tone; and
7. lowers blood pressure.

21.31 Summary

Maternal obesity is rising incrementally in all age groups globally. It is associated with significant maternal and foetal morbidity and mortality. Maternal obesity during pregnancy is a strong risk factor for intergenerational metabolic disorders, especially among females. Visceral obesity is associated with a state of low-grade chronic inflammation, IR, increased lipolysis, and prothrombotic changes.

Further reading

Mahmood et al., 2020Mahmood TA, Arulkumaran S, Chervenak FA, eds. *Obesity and Gynaecology*. 2nd ed. London: Elsevier; 2020.

Preconception care for obese women

Gamal Sayed[1] and Tahir Mahmood[2]
[1]Women's Wellness and Research Centre and Clinical Department, College of Medicine, Qatar University, Doha, Qatar, [2]Department of Obstetrics & Gynaecology, Victoria Hospital, Kirkcaldy, United Kingdom

Obesity poses increased health risks for women not only before pregnancy but also puts these women and their babies at various adverse outcomes during pregnancy.

These women and their offspring are at increased long-term risks of weight retention, with significantly increased likelihood of long-term obesity and metabolic syndrome when compared with normal BMI mothers.

According to Public Health England, adult overweight and obesity is predicted to reach 70% by 2034.

Health professionals involved in the care of pregnant women should receive education on maternal nutrition and its impact on maternal foetal and child health.

Women (and health practitioners) may not be aware of all the risks. These risks also include effects on future offspring and should be appropriately shared in a sensitive manner during the prepregnancy counselling consultation by an appropriately trained health professional. Women should be supported to lose weight before pregnancy.

- Women should be counselled as regards increased risks related to their reproductive performance.
- Information should be provide about the effects of maternal obesity on the offspring.
- They should be informed that working towards a healthy weight will significantly reduce these risks:

Prepregnancy risks:

- Infertility and prolongation of the time to get pregnant (due to hyperandrogenaemia causing decrease in gonadotrophin secretion as well as the association of obesity with polycystic ovarian disease and anovulation).

Metabolic syndrome and Type 2 diabetes:

- This risk incrementally increases with higher BMI.
- Women with obesity have increased insulin resistance as well as their dietary patterns may increase that risk.
- More than 80% of type 2 diabetes can be associated with obesity.
- Metabolic syndrome during pregnancy is associated with higher risk of gestational diabetes, hypertension of pregnancy and thromboembolic disease.

Handbook of Obesity in Obstetrics and Gynecology. DOI: https://doi.org/10.1016/B978-0-323-89904-8.00044-5

Disorders of lipid metabolism:

- Elevated serum cholesterol, low density lipoprotein, and very low density lipoprotein, as well as reduction in serum high density lipoprotein (HDL).
- These effects and especially the lower levels of serum HDL cholesterol may increase the risk of coronary heart disease.

Folic acid deficiency:

- There are low serum folate levels in obese women.
- Additionally, obese women are less likely to have adequate folate intake in their diet or take nutritional supplements containing folate.
- Folic acid needs to be started 1−3 months before pregnancy and aspirin is usually started at 12 weeks.

Vitamin D deficiency:

- Obese women have lower levels of serum vitamin D concentrations.
- It is unlikely that the requirements in pregnancy are met by diet alone.
- Furthermore, in countries with less exposure to sunlight at the appropriate wavelength, this deficiency can be more marked, as vitamin D synthesis depends on skin exposure to sunlight.
- COVID-19: Obesity increases the morbidity and mortality from Coronavirus-19 disease and this seems independent of other risk factors.

Antenatal risks

- Miscarriage.
- Increased congenital abnormalities as neural tube defects, spina bifida, cleft lip and palate, anorectal atresia, hydrocephaly, cardiovascular, septal anomalies, and limb reduction abnormalities.
- Hypertensive disorders of pregnancy as preeclampsia. Difficulty in measuring blood pressure, therefore use of an appropriate sized cuff is recommended.
- Metabolic syndrome including gestational diabetes.
- Venous thromboembolism (VTE). The risk increases with BMI >30 kg/m^2 and is higher with increasing obesity. This risk is throughout the pregnancy and lasts 6 weeks postpartum.
- Foetal macrosomia.
- Increased mental health problems as depression, antenatal and postpartum anxiety.
- Increased induction of labour.
- Death. In the UK 60% of pulmonary embolism-related deaths (2003−2008) occurred in women who had BMI >30 kg/m^2 as opposed to 20% prevalence.
- Difficulty in assessing foetal size and external foetal heart tracing

Intrapartum risks

- Dysfunctional and prolonged labour.
- Foetal macrosomia increases shoulder dystocia with its sequalae as brachial plexus injury.
- Difficulty in palpating the presenting part, foetal size and external foetal heart rate tracing.
- Caesarean section.
- Less rates of successful vaginal birth after caesarean section.
- Difficult venous access.

Anaesthetic complications:

- Obese women are at higher risk of anaesthesia-related complications. The higher the BMI, the greater the complications.
- Obesity is associated with higher initial failure rate of epidural insertion, increased resite rates and failed intubation.
- Obesity is considered a significant risk factor to maternal mortality from anaesthetic complications.

Postpartum risks

- Postpartum haemorrhage
- Wound infections and wound separation
- VTE
- Lesser rates of initiation and maintenance of breast feeding
- Postpartum weight retention for the mother

Short- and long-term risks for foetuses and infants include:

- Stillbirth
- Prematurity
- Macrosomia
- More prone to develop obesity in childhood. Infants of obese mothers tend to have more body fat than mothers with normal BMI
- Increased risk of childhood asthma
- Metabolic disorders in childhood and later life
- Possible increased risk of autism spectrum disorders and developmental delays in childhood

Preconceptional counselling:

- Health behaviours stem before pregnancy. There will be limited potential to impact unhealthy eating or lifestyle habits to cause significant change after pregnancy has already occurred.
- Preconceptional counselling and contraceptive consultations provides an excellent opportunity to address obesity in the reproductive age women.
- A holistic approach is ideal and women planning to get pregnant should also be informed of risks of smoking, drugs, and alcohol, which may not only affect a healthy lifestyle but can hinder them from achieving it.

Smoking

- Smoking with obesity may amplify various health issues.
- The risk of VTE and coronary heart disease is increased.
- Women should be advised to stop smoking before embarking on pregnancy.

Alcohol

- Women should be informed that there is no "low safe limit" during pregnancy. The UK's chief medical officer guideline advises not to drink at all during pregnancy.
- It is particularly important to limit alcohol intake while trying to lose weight for various reasons:
    o Firstly, the calories in alcohol may offset any dietary caloric restrictions.
    o Secondly drinking is associated with increased hunger and in turn will lead to more food consumption.

o There is an association with various nutritional deficiencies.
- Binge drinking can be associated with unprotected intercourse leading to unplanned pregnancy.
- Alcohol intake during pregnancy can cause various birth defects.
- Foetal alcohol spectrum disorder (FASD) is a blanket term comprising various abnormalities due to alcohol consumption in pregnancy.
- These include physical, mental or behavioural problems and these include lifelong learning disabilities.
- Foetal alcohol syndrome is the most severe form of FASD and includes facial abnormalities, growth restriction, and lifelong learning disabilities
- Chronic alcohol consumption may lead to various nutritional deficiencies: this includes deficiency in vitamins. Of particular importance is vitamin B9 (folic acid) which can occur in up to 80% of alcoholics.
- A combined effect of alcohol, folic acid deficiency and obesity may confound folate deficiency-related birth defects as microcephaly, neural tube defects, and facial malformations

Unplanned pregnancies

- 45% of pregnancies are unplanned.
- Health education and losing weight before pregnancy are essential steps in managing obesity in pregnancy and the overall approach to the obesity pandemic.
- An unplanned pregnancy can represent various missed opportunities and has several effects on women and children including:
  o Late antenatal care
  o Late testing for various abnormalities, such as gestational diabetes
  o Late start to implement various dietary modifications
  o Late starting of various medications and supplements as folic acid and aspirin at the appropriate time. Missed opportunity to risk assess for thromboembolism: Some women with multiple risk factors need to start LMWH in the first trimester based on their risk scoring
  o Reducing alcohol and giving up smoking before pregnancy
  o Not being up to date with vaccinations before pregnancy.
  o Performing sexual health checks including cervical screening before pregnancy.
- Antenatal and postpartum depression

Diagnostic challenges

- Difficulty in assessing the nuchal translucency measurements due to obesity.
- Less clarity in imaging of foetal structures at ultrasound scan to screen for structural anomalies. This can decrease accuracy of detection of structural anomalies in obese women.
- Amniocentesis and chorionic villus sampling can be technically challenging and associated with higher miscarriage rates.
- Noninvasive prenatal testing for trisomies can be less effective in obese women. It is worth noting that other serum biomarkers are not affected by obesity as they are adjusted by weight.

Weight loss before pregnancy can carry health benefits into pregnancy and future pregnancies. Weight management strategies *before* pregnancy include:

Diet

Various approaches to diet should be considered to achieve weight loss. Some of the basic principles are as follows:

- Involve a dietician.
- Health education about various types of diet and calories.
- A healthy diet does not necessarily mean it would be helpful in reducing weight.
- Women should be educated about both healthy and low-in-calories diets which would aid in weight loss.
- Calculate the total energy expenditure. This can be done by using the World Health Organization equations for calculating energy expenditure.
- Overall caloric intake should be reduced and less than expenditure, and this should be the basis for any weight loss intervention.
- Aim to reduce caloric intake by approximately 500 kcal/day. This should reduce the weight by approximately 0.5 kg/week
- In calculating the macronutrients in the reduced caloric diet, aim for a balanced diet for protein, carbohydrates, and healthy fats. It has been recommended that diet should include higher protein and lower carbohydrates.
- Avoid or reduce alcohol, sugary beverages, and simple sugars in diet.
- Aim for consistency as it will take time

Exercise

- Physical activity is very important in mental and physical overall health. This, however, is different from the exercise effects on losing weight.
- Contrary to common belief, on its own without diet modifications it can only lead to modest reductions in weight. In other words, it will be difficult to lose weight with exercise alone. *It is difficult to "outrun" a bad diet.*
- According to the World Health Organization guidelines, updated in 2020, there is a reaffirmation that physical activity is better than none.
- A combined aerobic and resistance training or muscle strengthening in addition to diet seems to be a reasonable approach.
- All adults should undertake weekly 150−300 minutes of moderate intensity exercise or 75−150 vigorous intensity exercise or an equivalents combination.

Behavioural therapy

- The statement from US Preventive Services Task Force did not make specific recommendations on initiating behavioural counselling in primary healthcare settings. The health benefits seem to be small.
- However, clinicians can selectively counsel patients' readiness to embrace changes, explore barriers to change, and direct to social support services that can aid in behavioural changes.
- This however is different from behavioural counselling that is usually done by dieticians to promote nutritional change in behaviour by proper education and nutritional counselling with regard to the different types of diet and food.

Medications

- A number of medications for management of obesity have been used; these include Orlistat, Liraglutide, Phentermine-topiramate, and Bupropion-naltrexone.

- Little data exist on their effects on pregnant women and the manufacturers recommend that these medications are contraindicated in pregnancy.
- The British national formulary however suggests that Orlistat can be used with caution in pregnancy and is to be avoided during breast feeding.

Bariatric surgery

National institute of clinic excellence in the UK recommends bariatric surgery for women with BMI $>40\,\text{kg/m}^2$ or >35 in the presence of other comorbidities. There are numerous bariatric surgeries. They decrease weight by three main ways:

- Restriction of caloric intake by decreasing stomach capacity (example sleeve gastrectomy)
- Malabsorption of nutrients by decreasing the absorption length of small intestine (example jujeno-ileal bypass)
- Combination of above (example Reux-en-Y gastric bypass)

Care of women after bariatric surgery:

- Inform women that following weigh loss they are at higher risk to get pregnant if they are not using contraception. Cycles may become regular and rate of anovulation decreases, thus improving fertility.
- Women should be advised to wait for 12−18 months before getting pregnant as this is the time where these women are usually actively losing most weight, so this time is ideal to optimise the success of weight loss. Additionally, it gives time to address nutritional deficiencies.
- Risk of various nutritional deficiencies including iron, folate, calcium, B12, vitamin D, fat-soluble vitamins, fats, and proteins, especially in procedures that lead to malabsorption such as gastric bypass. It is thus important to know which type of surgery was performed.

Breast feeding advice

- Women should be advised about the potential benefits of breast feeding.
- Women who breast feed can lose weight faster than those who do not breast feed.
- Breast feeding also reduces risk of developing Type 2 diabetes mellitus.
- Obese women may find it difficult to initiate and maintain breast feeding for various reasons.
- Delay in lactogenesis in obese women.
- Impaired prolactin response to suckling.

Suggested prepregnancy counselling session:

Measure weight, height, blood pressure, and calculate BMI.

Develop risk-based list of issues which should be specifically addressed now and at subsequent visits.

Consider screening for hyperglycaemia.

Advice on weight reduction and lifestyle modifications including stopping smoking and reducing or stopping alcohol.

Referral to dietician (or this can be done in primary health care services) for guidance on the effective interventions to lose weight, healthy eating before and during future pregnancy, and in between pregnancies.

Clinical evaluation of concurrent medical conditions and treatments which may increase risks for the mother and the offspring.

Health education on the risks before and during pregnancy and to the offspring. Explain that women who are healthy before pregnancy have a higher chance of getting pregnant, being healthy in pregnancy, safer delivery, and a healthy baby thereafter.

Initiate discussion on benefits of breast feeding, including its potential to aid weight loss.

Start folic acid 5 mg, 1−3 months before conception to continue throughout the first trimester.

Consider giving vitamin D 10 μg daily during pregnancy and in breast feeding.

Further reading

Care of women with obesity in pregnancy RCOG; 2018.

Obesity in pregnancy ACOG; 2015. (re-affirmed 2020).

Making the case for preconception care. Planning and preparation for pregnancy to improve maternal and child health outcomes. Public Health England; 2018.

Maternal obesity in the UK: findings from a national project. Centre for maternal and child enquiries; 2010.

Obesity and pregnancy. Royal College of Physicians of Ireland; 2013.

Obesity in pregnancy. SOGC (Canada); 2010.

Moyer V.A. Screening for and management of obesity in adults: U.S. Preventive Services Task Force recommendation statement. U.S. Preventive Services Task Force. *Ann Intern Med.* 2012.

Weight optimisation strategies in pregnant obese women

Alasdair Hardie[1] and Tahir Mahmood[2]
[1]Department of Obstetrics and Gynaecology, Royal Infirmary of Edinburgh, Edinburgh, Scotland,
[2]Department of Obstetrics & Gynaecology, Victoria Hospital, Kirkcaldy, United Kingdom

Roughly one in five pregnant women are obese and this number is increasing. High maternal weight gain in pregnancy is associated with adverse maternal and foetal outcomes. These risks are increased in mothers who are already obese at booking. Strategies aiming to optimise maternal weight during pregnancy have been shown to be effective in reducing gestational weight gain and improving maternal health outcomes.

23.1 Guidelines on maternal weight gain during pregnancy

In 2009 the US Institute of Medicine (IoM) set out guidelines advising the optimal weight gain during singleton pregnancies. The primary goal of this strategy was to increase the proportion of babies born with a birth weight of between 3000 and 4000 g. This guidance also takes into account the long-term maternal and child health consequences of weight gain in pregnancy. FIGO advises that 'pregnant women with a BMI ≥ 30 should be advised to avoid high gestational weight gain. Weight gain should be limited to 5−9 kg' (Table 23.1).

23.2 Antenatal consequences of increased gestational weight gain

1. Increased risk of developing gestational hypertension/preeclampsia
2. Increased risk of developing gestational diabetes
3. Increased risk of complications in labour and delivery:
   a. Increased risk of induction of labour and failure of induction labour
   b. Prolonged duration of labour
   c. Increased risk of caesarean section.

Handbook of Obesity in Obstetrics and Gynecology. DOI: https://doi.org/10.1016/B978-0-323-89904-8.00026-3

Table 23.1 IOM guidance 2009 about total weight gain and rates of weigh gain during pregnancy.

Prepregnancy weight status (body mass index category)	Recommended total weight gain ranges		Recommended rates of weight gain in the second and third trimester*	
	Pounds	Kilograms	Pounds/week	Kilograms/week
Underweight (<18.5 kg/m²)	28–40	12.5–18	1.0 (1.0–1.3)	0.51 (0.44–0.58)
Normal (18.5–24.9 kg/m²)	25–35	11.5–16	1.0 (0.8–1.0)	0.42 (0.35–0.50)
Overweight (25–29.9 kg/m²)	15–25	7–11.5	0.6 (0.5–0.7)	0.28 (0.23–0.33)
Obese (>30 kg/m²)	11–20	5–9	0.5 (0.4–0.6)	0.22 (0.17–0.27)

*Calculations assume a 0.5-2 kg (1.1-4.4 lbs) weight gain in the first trimester.

23.3 Postpartum consequences of increased gestational weight gain

• Increased risk of long-term obesity
• Increased risk of postpartum depression
• Small increase in the risk of developing postmenopausal breast cancer
• Reduced rates of initiation and reduced duration of breastfeeding.

23.4 Consequences of increased gestational weight gain for the child

• Increased neonatal and infant mortality
• Increased risk of intrauterine death
• Increased risk of large for gestational age (LGA) baby
• Increased risk of childhood obesity
• There is weak evidence linking increased gestational weight gain to acute lymphoblastic anaemia, attention deficit hyperactivity disorder, and breast cancer.

23.5 Consequences of low gestational weight gain

• Increased risk of small for gestational age infant
• Increased risk of preterm birth
• Limited evidence shows an association with childhood asthma.

Currently only a third of pregnant women's weight changes in line with the IoM guidance. Obese women are twice as likely to exceed the IoM's weight gain guidelines compared to mothers with a normal BMI. Similarly, mothers from lower socioeconomic groups are more likely to gain weight during pregnancy. Weight gain in pregnancy is influenced by modifiable behaviours such as dietary intake and exercise, and weight optimisation strategies can be used to target and ameliorate these behaviours. There is evidence that interventions during pregnancy using strategies which focus on diet, exercise, and lifestyle management can reduce:

- the incidence of LGA babies;
- the incidence of gestational hypertensive disorders;
- the risk of Caesarean delivery and birth trauma;
- the risk of respiratory distress;
- the risk of intrauterine death; and
- rates of future childhood obesity.

23.6   Strategies for weight management

- Preconception weight optimisation is a key strategy in both reducing prepregnancy obesity and increased gestational weight gain. The RCOG advises that all women of childbearing potential should be informed of the risks of obesity in pregnancy and "supported to lose weight before and between pregnancies."
- The postpartum period offers an opportunity for interaction with health services and lifestyle intervention. Women with a BMI >30 should be offered a structured weight loss programme following delivery. This weight loss programme should:
  a. Be tailored to the woman's needs individually.
  b. Consideration should be given to barriers to weight loss in post-natal period.
  c. Combine healthy eating and physical exercise (moderate exercise for at least 30 minutes 5 days per week).
  d. Provide professional support for an adequate period to consolidate lifestyle change.
- All pregnant women with a BMI >40 should be referred to a dietitian for specialist consultation as early as possible in pregnancy. Dietetic advice should take into account preexisting sociocultural beliefs.
- Obese women have a higher rate of nutritional deficiencies compared to women with normal BMIs. Diets prescribed to pregnant women should encourage the intake of nutrient rich foods, such as:
  i. Fruits and vegetables
  ii. Wholegrains
  iii. Nuts and seeds
  iv. Legumes
  v. Pregnant women should be advised to reduce intake of red or processed meats and fried foods.
- Pregnancy-based diets which have been proven to be effective in reducing maternal weight gain and improve obstetric outcome advocate:
  i. Low glycaemic index diet composed of fresh fruit, vegetables, and unprocessed wholegrains and legumes.
  ii. Composition of fat (30%), protein (15%−20%), and carbohydrate (55%).
- Bariatric surgery is contraindicated during pregnancy and a minimum interval of 12 months is recommended between weight-loss surgery and conception. Following bariatric surgery patients are at increased risk of nutritional deficiencies and require enhanced surveillance during pregnancy.

- The efficacy of weight optimisation programmes is increased by education of the risks of obesity in pregnancy in programmes as well as regular professional follow-up to ensure sustained dietary/lifestyle modification. Technological solutions such as phone apps can be helpful in achieving this.
- Weight-loss drugs are not recommended during pregnancy.
  - **i.** Orlistat is a lipase inhibitor which reduces the absorption of dietary fat. Studies of foetuses exposed to orlistat have not shown any increase in congenital malformation.
  - **ii.** Phentermine/topiramate reduces appetite, increases satiety, and reduces calorific intake. It is associated with orofacial clefts in the newborn. Both drugs are expressed in breast milk and therefore not recommended during lactation.
  - **iii.** Lorcaserin hydrochloride is a serotonin receptor agonist which acts on the 5HTc receptor to reduce appetite and increase satiety. It has been shown to reduce birth-weight of offspring in animal studies. Lorcaserin is contraindicated in pregnancy.

23.7   Exercise in pregnancy

- ANZCOG advise that "women without contraindications should participate in regular aerobic and strength conditioning exercise during pregnancy." Evidence has shown that exercise:
  - **i.** Is helpful in the management and prevention of gestational diabetes and hypertensive disorders.
  - **ii.** May be associated with shorter and less complicated labours.
  - **iii.** May improve neonatal outcomes and is associated with improved neonatal body composition and infant cardiac autonomic control.
- Specific obstetric contraindications to exercise to consider are persistent bleeding; abnormal placentation; hypertensive disorders; conditions which may increase the risk of spontaneous preterm birth (shortened cervical length, PPROM, multiple pregnancy); and foetal growth restriction.
- ANZCOG advise for pregnant women to be physically active on most if not all days of the week, aiming to accrue 150–300 minutes of "moderate" exercise per week.
- Exercise prescription should take into account the woman's current fitness level and factor for the physiological changes of pregnancy as well as the gestation of pregnancy.
- Contact sport or sports with a high risk of trauma should be avoided.

23.8   Summary of weight optimisation in pregnancy

- Excessive weight gain in pregnancy is associated with significant maternal and foetal risks and these risks are further increased in the obese population.
- Excessive weight gain and its attendant risks can be reduced through weight optimisation strategies, which combine professional advice and continued support on healthy eating, lifestyle modification, and exercise.

Further reading

Denison FC, Aedla NR, Keag O, Hor K, et al. Care of women with obesity in pregnancy. Green-top Guideline No. 72. *BJOG.* 2019;126:e62−e106.

IOM (Institute of Medicine). *Nutrition During Pregnancy.* Washington, DC: National Academy Press; 1990.

McAuliffe FM, Killeen SL, Jacob CM, et al. Management of prepregnancy, pregnancy, and postpartum obesity from the FIGO Pregnancy and Non-Communicable Diseases Committee: a FIGO (International Federation of Gynecology and Obstetrics) guideline. *Int J Gynecol Obstet.* 2020.

Thangaratinam S, Rogozińska E, Jolly K, Glinkowski S, Roseboom T, Tomlinson JW, et al. Effects of interventions in pregnancy on maternal weight and obstetric outcomes: *meta-analysis* of randomised evidence. *BMJ.* 2012;344:e2088.

The Royal Australian and New Zealand College of Obstetricans and Gynaecologists. Exercise During Pregnancy. https://ranzcog.edu.au/RANZCOG_SITE/media/RANZCOG-MEDIA/Women%27s%20Health/Statement%20and%20guidelines/Clinical-Obstetrics/Exercise-during-pregnancy-(C-Obs-62)-New-July-2016.pdf?ext = .pdf; 2016.

The Royal Australian and New Zealand College of Obstetricians and Gynaecologists (RANZCOG). Management of obesity in pregnancy. https://ranzcog.edu.au/RANZCOG_SITE/media/RANZCOG-MEDIA/Women%27s%20Health/Statement%20and%20guidelines/Clinical-Obstetrics/Management-of-obesity-(C-Obs-49)-Review-March-2017.pdf?ext = pdf; 2013, updated 2017.

Early pregnancy and obesity

24

Swetha Bhaskar[1] and Ibrahim Alsharaydeh[2]
[1]Department of Obstetrics and Gynaecology, Victoria Hospital, NHS Fife, Scotland,
[2]Department of Obstetrics and Gynaecology, Raigmore Hospital, Inverness, Scotland

Obesity has become a major health problem worldwide and it's an independent risk factor for adverse pregnancy outcome. The World Health Organization classifies obesity into class I BMI (body mass index) 30.0−34.9, class II 35.0−39.9, and class III 40 or greater.

The prevalence of obesity in the general population in the UK has increased markedly since the early 1990s. The prevalence of obesity in pregnancy has also been seen to increase, rising from 9%−10% in the early 1990s to 16%−19% in the 2000s. In England, where the prevalence of obesity in women is among the highest in Europe, one in five women of reproductive age are now obese (BMI $\geq 30 \text{ kg/m}^2$).

Pregnant women who are obese are at greater risk of almost all of early pregnancy-related complications compared with women of normal BMI, including;

- **Maternal Complications:**
  - Infertility: threefold higher in obese than in nonobese due to ovulatory dysfunction and decreased insulin sensitivity.
  - Maternal morbidity and mortality: 30% of maternal death were obese and 22% were overweight (2015 MBRRACE)
  - First trimester miscarriage (OR 1.2)
  - Recurrent miscarriage (OR 3.5)
  - Venous thromboembolism: higher risk of pulmonary embolism (OR 14.9) than DVT (OR 4.4)
  - Preeclampsia (risk doubled with each 5−7 kg/m^2 increase in prepregnancy BMI)
  - Gestational hypertension (OR 2.5−3.5)
  - Gestational diabetes (OR 2.6−4.0)
  - Maternal infections (UTI OR 1.17, genital tract OR 1.24)
  - Anaesthetic complications (25% of cardiac arrest in pregnancy is caused by anaesthesia, of those 75% are obese)
  - Wound infection (OR 1.27)
  - Depression (obese, 33.0%; overweight, 28.6%; normal weight 22.6%)
  - Anxiety and eating disorder (OR 1.4)
  - Serious mental illness
- **Fetal Complications:**
  - Congenital anomalies (NTD OR 1.7 for obese, 3.11 for morbidly obese), other anomalies; hydrocephaly, cardiovascular, and limb reduction abnormalities
  - Stillbirth (OR 3.8)

Handbook of Obesity in Obstetrics and Gynecology. DOI: https://doi.org/10.1016/B978-0-323-89904-8.00030-5

- Prematurity (<32 weeks OR 0.73)
- Macrosomia (>4000 g OR 1.7−1.9) (>4500 g OR 2.0−2.4)
- Neonatal death (OR 3.4)

24.1   Management

24.1.1   Preconception and Early Pregnancy

- **Counselling:**
  - Primary care specialists should counsel all women of reproductive age about the adverse effects of obesity on pregnancy outcome.
- **Interventions:**
  - Clinical and population health practice should focus on interventions to reduce obesity in all women of reproductive age.
  - Women with a BMI 30 kg/m^2 or greater wishing to become pregnant should be advised to take 5 mg folic acid supplementation daily, starting at least 1 month before conception and continuing during the first trimester of pregnancy.
  - Obese women are at high risk of vitamin D deficiency. The evidence on whether routine vitamin D should be given to improve maternal and offspring outcomes remains uncertain.
  - Venous thromboembolism risk assessment should be individually discussed, assessed, and documented at the preconception, first antenatal visit, during pregnancy, intrapartum, and postpartum. One third of pulmonary embolism occurs during the first trimester.
  - Antenatal thromboprophylaxis should be considered based on the risk assessment.

24.1.2   Antenatal

- Where?
  - Care of women with obesity in pregnancy can be integrated into all antenatal clinics, with clear local guidelines and clinical pathways for care available. All pregnant women with a booking BMI 30 kg/m2or greater should be provided with accurate and accessible information about the risks associated with obesity in pregnancy and how they may be minimised. Women should be given the opportunity to discuss this information.
- Interventions:
  - All pregnant women should have their weight and height measured using appropriate equipment, and their BMI calculated at the antenatal booking visit.
  - An appropriate size of cuff should be used for blood pressure measurements taken at the booking visit and all subsequent antenatal consultations.
  - There is a lack of consensus on optimal gestational weight gain. The Institute of Medicine guidelines (USA) recommend different ranges of weight gain for normal weight, overweight, and obese women. (BMI 18.5; recommended weight gain 12.5−18 kg), normal weight (BMI, 18.5−24.9; 11.5−16 kg), overweight (BMI, 25.0−29.9; 7−11.5 kg), and obese (BMI ≥30; 5−9 kg). These guidelines are the most widely used but are not adopted routinely in clinical practice in the UK.
  - Dietetic advice by an appropriately trained professional should be provided early in the pregnancy with focus on a healthy diet rather than prescribed weight gain targets.
  - Antiobesity or weight-loss drugs are not recommended for use in pregnancy.

- Pregnant women with a booking BMI 40 kg/m^2 or greater should be referred to an obstetric anaesthetist for consideration of antenatal assessment like difficulties with venous access, regional and general anaesthesia. Multidisciplinary discussion and planning should occur where significant potential difficulties are identified
- Women with more than one moderate risk factor (BMI of 35 kg/m^2 or greater, first pregnancy, maternal age of more than 40 years, family history of preeclampsia, and multiple pregnancy) may benefit from taking 150 mg aspirin daily from 12 weeks of gestation until birth of the baby.
- Antenatal screening:
  - All women should be offered antenatal screening for chromosomal anomalies; Screening is less effective with a raised BMI.
  - Screening for structural abnormalities between 18 + 0 and 20 + 6 weeks of gestation.
  - Oral glucose tolerance test at 24−28 weeks. Consider early screening at 16 weeks in presence of other risk factors especially previous gestational diabetes.
  - Women with a BMI 30 kg/m^2 or greater should be screened for mental health problems.

Further reading

Available at: http://www.ic.nhs.uk/statistics-and-data-collections/health-andlifestyles/obesity/statistics-on-obesity-physical-activity-and-diet-england-february2009.

Beckett VA, Knight M, Sharpe P. The CAPS Study: incidence, management and outcomes of cardiac arrest in pregnancy in the UK: a prospective, descriptive study. *BJOG*. 2017;124:1374−1381.

Heslehurst N, Lang R, Rankin J, Wilkinson JR, Summerbell CD. Obesity in pregnancy: a study of the impact of maternal obesity on NHS maternity services. *BJOG*. 2007;114:334−342.

on behalf of MBRRACE-UK Knight M, Kenyon S, Brocklehurst P, Neilson J, Shakespeare J, Kurinczuk JJ, eds. *Saving Lives, Improving Mothers' Care − Lessons Learned to Inform Future Maternity Care from the UK and Ireland Confidential Enquiries into Maternal Deaths and Morbidity 2009−12*. Oxford: National Perinatal Epidemiology Unit, University of Oxford; 2014.

Larsen TB, Sørensen HT, Gislum M, Johnsen SP. Maternal smoking, obesity, and risk of venous thromboembolism during pregnancy and the puerperium: a population-based nested casecontrol study. *Thromb Res*. 2007;120:505−509.

Lashen H, Fear K, Sturdee DW. Obesity is associated with increased risk of first trimester and recurrent miscarriage: matched case-control study. *Hum Reprod*. 2004;19:1644−1646.

Molyneaux E, Poston L, Ashurst-Williams S, Howard LM. Obesity and mental disorders during pregnancy and postpartum: a systematic review and *meta*-analysis. *Obstet Gynecol*. 2014;123:857−867.

National Institute of Health and Care Excellence. Antenatal and postnatal mental health: clinical management and service guidance. Clinical guideline 192. Manchester: NICE; 2014.

O'Brien TE, Ray JG, Chan WS. Maternal body mass index and the risk of preeclampsia: a systematic overview. *Epidemiology*. 2003;14:368−374.

Sebire NJ, Jolly M, Harris JP, et al. Maternal obesity and pregnancy outcome: a study of 287,213 pregnancies in London. *Int J Obes Relat Metab Disord*. 2001;25:1175−1182.

Denison FC, Aedla NR, Keag O, et al. on behalf of theRoyal College of Obstetricians and Gynaecologists. Care of Women with Obesity in Pregnancy.Green-top Guideline No. 72. *BJOG*. 2018.

Ultrasound scanning in early pregnancy and foetal abnormality screening in obese women

Smriti Prasad[1] and Asma Khalil[1,2,3]

[1]Vascular Biology Research Centre, Molecular and Clinical Sciences Research Institute, St George's University of London, London, United Kingdom, [2]Twins Trust Centre for Research and Clinical Excellence, St George's University Hospital, London, United Kingdom, [3]Fetal Medicine Unit, Department of Obstetrics and Gynaecology, St. George's University Hospitals NHS Foundation Trust, London, United Kingdom

25.1 Introduction

The prevalence of obesity is on the rise and is a major health concern in pregnancy. It is associated with comorbidities posing risks to the health of both the mother and her unborn foetus. Planned optimum antenatal care commencing from the early weeks of pregnancy can help to identify and mitigate many of these risks, such that adverse outcomes can be avoided.

In this chapter, we will describe ultrasound scanning in early pregnancy and foetal abnormality screening in obese women.

25.2 Ultrasound scanning in early pregnancy

25.2.1 Role of early pregnancy scan in obese pregnant women

- An early pregnancy scan in obese pregnant women can be quite informative for planning antenatal care and is also reassuring because of uncertainty about the last menstrual period.
- Obesity is often associated with increased rates of anovulatory cycles with subsequent irregular cycles and subfertility.
- The increased need for assisted conception leads to an increase in twin and multiple pregnancies.
- A significant proportion of obese pregnant women also have Polycystic Ovarian Syndrome, which has been associated with early pregnancy losses.
- Obesity is also a risk factor for caesarean section in their previous pregnancy.

Approach to an early pregnancy scan

- Assess first by transabdominal examination with full bladder—use linear transducer, anticipate poor visualisation.

Handbook of Obesity in Obstetrics and Gynecology. DOI: https://doi.org/10.1016/B978-0-323-89904-8.00019-6

- Transvaginal examination with high-frequency transducer and with an empty bladder:
  - Confirm intrauterine pregnancy and rule out ectopic pregnancy.
  - Establish viability of pregnancy and document foetal cardiac activity.
  - Measure gestational sac, Crown-rump length (whichever or both if applicable) and yolk sac to correlate with period of gestation.
  - Note the number of foetuses by scanning in both transverse and longitudinal axis of the uterus.
  - Note evidence of any haematoma or bleeding in or around gestational sac.
  - Make a note of the previous caesarean scar (if applicable) and its relation to the gestational sac.
  - Examine both adnexa, pouch of Douglas and uterus to note any pathology like fibroids, ovarian cysts, hyperstimulated ovaries, or Mullerian malformations.
  - Infrequently, major foetal malformations like body stalk anomaly, anencephaly, and conjoined twins can be diagnosed by transvaginal early pregnancy scan.

25.2.2 First Trimester screening (nuchal translucency scan and dual marker)

According to NICE guidelines, all pregnant women should be offered combined first trimester screening to estimate their risks for common aneuploidies. Apart from confirming viability, gestational age, and number of foetuses, an important component of the first trimester ultrasound is to carry out the nuchal translucency (NT) measurement in order to screen for chromosomal abnormalities, and it may even detect major foetal structural anomalies.

Dual marker consists of maternal serum beta HCG and PAPP-A measurements using specific assays. Both NT and dual marker measurements are used to derive an individualised numerical risk for common aneuploidies (Trisomy 21, 18, and 13) against the woman's background risk, which is determined primarily by her age.

The early foetal anatomic assessment may prove most satisfactory when carried out by transvaginal ultrasound, rather than transabdominal ultrasound in the obese population.

Transvaginal scan is an important tool to detect major congenital abnormalities like anencephaly, holoprosencephaly, anterior abdominal wall defects, etc. Numerous studies have shown that visualisation and hence detection of major structural defects, including cardiac abnormalities can be as high as 90%with transvaginal examination. The importance and utility of a transvaginal sonographic assessment in obese women cannot be overemphasised.

Considerations specific to screening in pregnant women with raised BMI

- NT measurements in obese women can be difficult and often require longer scanning time and repeated measurements. Additionally, imaging of foetal nasal bone may be suboptimal due to maternal habitus.
- The levels of biochemical markers like b-HCG and PAPP-A are variably influenced by maternal weight and the values of these analytes are lower in obese women presumably due to increased plasma volume and dilutional effects. Inaccurate information can lead to significant alterations in the estimated risk and result in screening failure. Therefore, it is important to provide laboratories with correct information about maternal weight for accurate risk estimation.

- Noninvasive prenatal testing (NIPT) by cell-free foetal DNA requires a foetal fraction above 3%−4% to ensure the reliability of results. Increased maternal BMI is associated with a decrease in the foetal fraction and a higher rate of "no call" NIPT results, hence the need for repeated samples or further invasive testing with their inherent risks. Women should be counselled about known limitations of NIPT in the setting of maternal obesity.

  Caution regarding use of Dopplers in the first trimester: Embryonic period (from conception till 9 + 6 weeks) is characterised by rapid cell division and development of the embryonic organs.

  According to updated ISUOG safety guidance 2021, spectral Doppler, colour flow imaging, power imaging, and other Doppler ultrasound modalities should not be used routinely in the embryonic period. If use of Doppler is clinically indicated, then the exposure time should be kept to a minimum.

  In the foetal period (CRL > 45 mm), Doppler ultrasound modalities may be used routinely for certain clinical indications, such as screening for trisomy and cardiac anomalies. When performing Doppler ultrasound, the displayed thermal index should be ≤1.0 and exposure time should be kept as short as possible (usually no longer than 5−10 minutes).

25.2.3   Fetal abnormality screening in obese women

Foetuses of obese mothers are more likely to have congenital malformations compared to foetuses of mothers with normal prepregnancy weight, as depicted in Fig. 25.1.

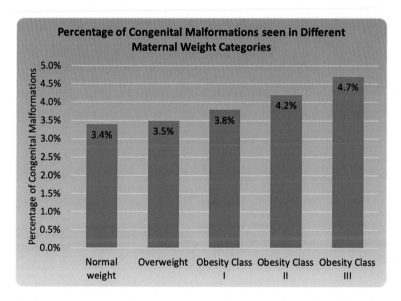

Figure 25.1 Incidence of congenital malformations seen in different maternal weight categories.
Source: Adapted from Persson et al. Risk of major congenital malformations in relation to maternal overweight and obesity severity: cohort study of 1.2 million singletons. BMJ 2017;357(j2563).

In a meta-analysis by Stothard et al., associations between maternal obesity and cardiovascular anomalies (OR, 1.30; 95% CI, 1.12−1.51), cleft palate (OR, 1.23; 95% CI, 1.03−1.47), cleft lip and palate (OR, 1.20; 95% CI, 1.03−1.40), anorectal atresia (OR, 1.48; 95% CI, 1.12−1.97), and limb reduction (OR, 1.34; 95% CI, 1.03−1.73) have been described. Foetuses of obese pregnant women are also at significantly increased risk of neural tube defect (OR, 1.87; 95% CI, 1.62−2.15). The pathophysiology of malformations in obese mothers is multifactorial, with interactions between genetic and environmental factors.

Adiposity poses a great challenge to optimal visualisation of foetal anatomy. The fat issue is echogenic and leads to signal attenuation as the sound waves traverse through it, compounded by factors like artefacts, reverberation, and surrounding noise signals.

The image quality is inversely correlated to the thickness of maternal abdomen; therefore maternal obesity has a negative impact on the detection of foetal anatomic defects. In a secondary analysis of the data from the First and Second Trimester Evaluation of Risk (FaSTER) trial ($N = 8555$ women), it was reported that maternal BMI >30 kg/m^2 significantly decreased the likelihood of the detection of any congenital anomalies ($P = .001$).

Unfortunately, role of transvaginal approach, which is of immense benefit in first trimester anatomy assessment, is limited in the second trimester.

Some technical tips to improve imaging of foetal anatomy in obese women:

1. The visualisation of foetal anatomy in obese women has been reported to be better at later gestational ages. Anomaly scans in mothers may be scheduled at 21−22 weeks compared to the more common window of 18−20 weeks.
2. One of the main organ systems that are poorly visualised in mothers with high BMI is foetal cardiovascular system. Referral for foetal echo with a cardiologist is desirable in such women.
3. Preexamination counselling of the obese woman must point out the likely impact of adiposity on image clarity, and possible nonvisualisation of foetal anatomy should be clearly communicated to the patient so that the expectations are realistic.
4. Allowing more time for scan and expertise of the sonographer is directly related to successful completion of anatomy assessment.
5. Thinner areas of maternal abdomen may be employed as better acoustic windows such as retraction of panniculus upwards, the subcostal approach, or the periumbilical approach.
6. Use of novel pre- and postprocessing filters and techniques lead to substantial improvement in image quality, such as tissue harmonic imaging, speckle reduction technology.
7. MRI may be employed in selected cases to facilitate diagnosis in suspected cases as it is independent of maternal BMI, foetal position, etc.
8. Examination with full maternal bladder (discussed in next chapter) leads to better visualisation.
9. Examination in Sims position, as proposed by Benacerraf, may be tried. Sims position results in shorter skin to amniotic cavity insonation distance with potential for improved image quality.
10. Last but not least, scanning when foetus is quiescent in supine position would result in better visualisation of foetal anatomy.

25.3   Conclusion

Maternal obesity limits the performance of the available screening modalities in both first and second trimester, which is concerning as obesity itself is associated with increased risks. Effective communication with women about these limitations, systematic approach to examination, and use of new imaging techniques would be key to improved antenatal care in these women. Moreover, the emphasis should also be on preconception counselling and women should be motivated to lose weight before planning pregnancy.

Conflict of interests

None declared.

Funding source

None.

Further reading

Benacerraf BR. A technical tip on scanning obese gravidae. *Ultrasound Obstet Gynecol.* 2010;35(5):615−616. Available from: https://doi.org/10.1002/uog.7550. PMID: 20052658.

Denison FC, Aedla NR, Keag O, et al. Royal college of obstetricians and gynaecologists. care of women with obesity in pregnancy: green-top guideline no. 72. *BJOG.* 2019;126 (3):e62−e106. Available from: https://doi.org/10.1111/1471-0528.15386. Epub 2018 Nov 21. PMID: 30465332.

Mumford V, Khalil A. Ultrasound scanning in early pregnancy and fetal abnormality screening in obese women. In: Mahmood TA, Arulkumaran S, Chervenak FA, eds. *Obesity and Obstetrics.* 2nd ed. Elsevier; 2020:61−68. ISBN 9780128179215. Available from: https://doi.org/10.1016/B978-0-12-817921-5.00007-2.

Paladini D. Sonography in obese and overweight pregnant women: clinical, medicolegal and technical issues. *Ultrasound Obstet Gynecol.* 2009;33(6):720−729. Available from: https://doi.org/10.1002/uog.6393. PMID: 19479683.

Antenatal care for obese women

26

Gamal Sayed[1] and Tahir Mahmood[2]
[1]Women's Wellness and Research Centre and Clinical Department, College of Medicine, Qatar University, Doha, Qatar, [2]Department of Obstetrics & Gynaecology, Victoria Hospital, Kirkcaldy, United Kingdom

26.1 Introduction

- Obesity poses an increased risk for women before, during, and after pregnancy.
- Traditionally obesity is calculated using weight and height and is classified by the WHO (World Health Organization) into class I BMI (Body Mass Index) 30.0–34.9, class II 35.0–39.9, and class III 40 or greater
- Obesity is a worldwide pandemic but the prevalence is not uniform across different regions and countries.

26.2 According to the WHO (2016) report

- In the United States, approximately 70% of women are overweight or obese.
- The prevalence was 67% in United Kingdom;
- 59% in Sweden;
- 29% in Japan; and.
- Only 20% in India.
- Although these figures may help target the approaches against obesity associated comorbidities, individualised approaches are required in different regions.
- Nevertheless, the principles of management remain the same and will be addressed in this chapter.

26.3 Antepartum care of obese women

Obesity in pregnancy is associated with increased risks of the following:

- Miscarriage
- Congenital abnormalities
- Gestational diabetes
- Hypertensive disorders of pregnancy
- Venous thromboembolism (VTE).

Handbook of Obesity in Obstetrics and Gynecology. DOI: https://doi.org/10.1016/B978-0-323-89904-8.00014-7

- Foetal risks, including macrosomia which increases the risk of shoulder dystocia with its sequalae as brachial plexus injury, still birth, and prematurity.
- Intrapartum changes and risks as difficulties in placing venous access, in palpating the presenting parts or foetal size, and in recording external foetal heart rate tracing.
- Dysfunctional and prolonged labour, increased caesarean section, less rates of successful vaginal birth after caesarean section
- Anaesthetic complications: obese women are at higher risk of anaesthesia-related complications including higher initial failure rate of epidural insertion, increased resite rates and failed intubation.
- Obesity is regarded as a significant risk factor to maternal mortality from anaesthetic complications.
- Postpartum risks including postpartum haemorrhage, wound infections, and wound dehiscence.
- VTE.
- Lower rates of initiation and maintenance of breast feeding.
- Postpartum weight retention for the mother.
- COVID-19: obesity increases the morbidity and mortality from Coronavirus-19 infection and this seems independent of other risk factors.

26.4 Prepregnancy counselling

- An excellent opportunity to initiate and counsel regarding healthy lifestyle, eating, and exercise as well as optimising health and weight loss before embarking on the pregnancy.
- If the woman has undergone bariatric surgery, she should be advised to wait for 12−18 months before getting pregnant.
- During this time these women are usually actively losing most weight to optimise the weight loss. Additionally, it gives time to address nutritional deficiencies.

26.5 Antenatal care

26.5.1 Location, facilities, and personnel

The care for obese women in pregnancy can be integrated in all the antenatal clinics. There should however be clear local pathways and guidelines in place with regard to their management.

There should be availability of specific equipment, which includes:

- Large blood pressure cuffs
- Larger sized compression stockings and pneumatic compression devices
- Appropriate weighing scales including sit on types
- Large wheelchairs
- Larger chairs without arms
- Appropriately sized ultrasound scan couches, examination beds, delivery beds, ward beds, with appropriately sized mattresses, operating theatre tables, and transfer equipment

26.6   The first antenatal visit

Obese women should be counselled appropriately about the risk of increased maternal, foetal, and neonatal complications associated with obesity.

Although ideally this should have been done in the prepregnancy counselling visit, it is important to remember that around 45% of women can have unplanned pregnancies.

Counselling should ideally include:

- Maternal, foetal, and neonatal risks
- Plan of care during pregnancy
- Referrals including dietician and anaesthetist and others when needed
- Lifestyle modifications including exercise and diet in pregnancy
- Avoiding excessive weight gain during pregnancy
- Warning signs, early identification, management of medical and obstetric complications
- Timing, mode, and location of delivery

26.7   Measuring weight and height for risk stratification

- The initial measuring of weight and height is usually performed at the first antenatal visit.
- This enables stratification of risk as regards thromboembolism, gestational diabetes, and preeclampsia.
- It also enables planning care during pregnancy and identifying resources during the pregnancy.
- Many local protocols stratify the risk according to this initial BMI. This is important in order to streamline the services as well as to offer services for those who really need it and avoid overstraining the health services.
- For example, a formal anaesthetic review in clinics is warranted for BMI >40 (although ideally if resources allow, it can be offered for all obese women).
- Ideally BMI should be reassessed during the early phase of the third trimester, to restratify the risks for foetal well-being and planning for labour and delivery.

26.8   Diet and lifestyle modifications

- Counselling with regard to various proven approaches to diet and lifestyle changes early in pregnancy, if they have not been already underway in the prepregnancy counselling.
- It is important to involve a dietician and health education reaffirmed around various types of diet and calories.
- Women in uncomplicated pregnancy should be advised that moderate physical activity is very important in mental and overall physical health.

26.9 Weight gain in pregnancy

- Physiological weight gain related to pregnancy amounts to approximately 11 kg.
- There are however differing views and lack of consensus on the optimal gestational weight gain (GWG).
- The RCOG guidelines focus on a healthy diet rather than prescribed weight gains.
- There is more emphasis on initial BMI rather than GWH. With that in mind the guidance is not to weigh again in pregnancy (or maybe do so once again in third trimester).
- Measuring the weight frequently in pregnancy may lead to unnecessary anxiety with little added benefit.
- The North American guidelines however base their recommendations on the Institute of Medicine (IOM, USA), now National Academies of Sciences, Engineering, and Medicine.
- This stratifies the weight gain per each BMI category as follows for singleton pregnancies. The more the BMI the less the recommended GWG:
- BMI <18.5 kg/m^2 (underweight)—weight gain 12.5 to 18.0 kg
- BMI $18.5-24.9$ kg/m^2 (normal weight)—weight gain 11.5 to 16.0 kg
- BMI $25.0-29.9$ kg/m^2 (overweight)—weight gain 7.0 to 11.5 kg
- BMI ≥ 30.0 kg/m^2 (obese)—weight gain 5 to 9.0 kg
- Although the authors in IOM state that the risk of adverse maternal and foetal outcomes did vary with the GWG, there is insufficient evidence for the IOM to include preeclampsia and gestational diabetes commonly encountered in obese women in pregnancy in their guidelines.
- Furthermore, a key finding of that analysis was that the prepregnancy BMI was more associated with adverse maternal and foetal outcomes than GWG.
- This challenges the usefulness of the ranges in clinical practice, and various other guidelines focus more on prepregnancy BMI as the focus of intervention rather than GWG in pregnancy.
- As a general principle, obese women should be encouraged to optimise their weight before pregnancy, avoid excessive weight gain in pregnancy and postpartum weight retention, and avoid trying to lose weight during the pregnancy.
- Until further evidence is available, a healthy diet, physical activity, and adhering to the above general principles can be more productive and achievable in lowering the risks of adverse outcomes than tightly prescribed GWG targets.

Thromboembolism risk assessment

- Obesity increases the risk of thromboembolic events in pregnancy.
- According to the RCOG guidelines for VTE in pregnancy (2015), a booking (or early pregnancy) BMI of >30 scores a 1 and those with a BMI >40 score a 2 in risk assessment. Other preexisting comorbidities are also given a score.
- Women with a score of 3 or more require antenatal thromboprophylaxis with low-molecular-weight heparin (LMWH) from 28 weeks,
- And those with a score of 4 or more require antenatal thromboprophylaxis with LMWH from the first trimester and this is to continue for 6 weeks postpartum
- It is important to give the appropriate doses of LMWH which are usually based on early pregnancy weight.
- It is important to use the current weight when calculating the dose for LMWH rather than relying upon first trimester weight.

- This underpins the importance of having local protocols as this may vary between different hospitals.

Hypertensive disorders in pregnancy

- Obesity is a risk factor to develop hypertensive disorders in pregnancy (HDP) including preeclampsia.
- It is recommended to use appropriately sized large cuffs in obese women as failure to use large cuffs can lead to overestimation of blood pressure, leading to unnecessary investigations and interventions.
- Women should be prescribed with low-dose aspirin from 12 weeks of pregnancy until delivery as it can reduce the incidence of HDP in obese women, especially those with other risk factors.
- Other associated moderate risk factors include age >40 years, family history of preeclampsia, first pregnancy, and multiple pregnancy.
- Recent evidence suggests that for obese women of BMI >35, using 150 mg aspirin per day rather than 75–100 mg may be more beneficial.
- The benefits of aspirin maybe enhanced by taking it in the evenings/night rather than during the day.

Gestational diabetes

- Obesity increases the risk of developing gestational diabetes by threefold as compared to women with normal weight.
- Furthermore, pregnancy may unfold preexisting undiagnosed diabetes, especially in those women who have not been seen regularly with healthcare services before the pregnancy.
- During booking, estimation of fasting/random blood sugar and HBA1C be considered to diagnose preexisting diabetes and, if detected, treated appropriately according to local protocols.
- If these results are normal all obese women should have an oral glucose tolerance test (OGTT) at 24–28 weeks.
- Following delivery those who have been diagnosed with GDM should have a follow-up at 8 weeks to ensure that GDM has resolved.
- This can usually be done by testing fasting blood glucose or HBA1C.
- A 75-g 2-hour OGTT is not usually performed routinely and reserved if clinically indicated.
- Women should be informed of the risk of developing diabetes and tested annually using a fasting blood glucose or HBA1C.
- During a future pregnancy an OGTT should be performed in these women at 16 weeks.

Medications and supplements

- Stop medications that were used for management of weight loss.
- Continue with folic acid 5 mg which is ideally started 1–3 months before pregnancy, during the first trimester.
- Offer vitamin D 10 μg daily as obese women are at a higher risk of vitamin D deficiency.
- Aspirin 150 mg per day from 12 weeks until delivery in women with BMI >35 and an additional moderate risk factor for developing preeclampsia (see below).
- Women who had bariatric surgery will need special multivitamin preparations including vitamin B12, folic acid, iron, and fat-soluble vitamins.

26.10   Overview of antenatal tests appointments and scans

- Offer the routine antenatal tests as Rubella, HIV, Hepatitis B, Hepatitis C (high risk groups only), urine culture, antibody screen, complete blood count, syphilis and chlamydia (in North America).
- Refer to dietician early in pregnancy.
- Thyroid function tests.
- Oral glucose tolerance test at 24−28 weeks.
- Consider early screening at 16 weeks in the presence of other risk factors especially previous gestational diabetes.

Ultrasound scanning:

- Early pregnancy scans for accurate estimation of gestational age.
- Anomalies scan at 18−20 weeks and may need another assessment at 24 weeks if any concerns about the anatomy of heart.
- Foetal growth scans at 28, 32, and 36 weeks as it is difficult to assess foetal growth clinically.
- More scans are needed if complications arise,

26.11   Mode and timing of delivery

- The decision regarding the mode and timing of delivery should be individualised considering other risk factors, comorbidities, and the woman's wishes.
- A detailed obstetric and anaesthetist plan should be included in the mother's notes.
- Women with high BMI have an increased risk of a caesarean section (CS).
- Women should be counselled as regards an increased risk of significant morbidity and mortality.
- Obesity on its own is not an indication for a CS.
- Elective induction of labour at term for obese women may reduce the chance of a CS without increasing the risk for adverse outcomes.
- Induction of labour for foetal macrosomia has shown a reduction of risk of shoulder dystocia and foetal fractures. This, however, is irrespective of maternal BMI.
- Women with an estimated foetal weight >5000 grams (those with gestational diabetes) and EFW >4500 grams should be offered an elective CS, irrespective of BMI.

26.12   Planning for delivery

- When in labour, the duty anaesthetist should be informed.
- Epidural analgesia during labour can be technically difficult due to body habitus and loss of landmarks, so it is better place to place regional anaesthesia early on, rather than as an emergency
- Obesity can pose a technical difficulty in accessing veins and placing a canula, hence a canula should be inserted during early labour.
- Issues related to induction and management of labour have been discussed separately.

26.13   Post delivery

- Women should be counselled as regards the benefits of breast feeding during antenatal classes.
- There are lower initiation and maintenance rates of breast feeding among obese women.
- Appropriate support should be given to encourage breast feeding via breast feeding counsellors.

Further reading

WHO. Fact sheet—obesity and overweight; Updated February 2018.

Yaktine AL, Rasmussen KM, eds. *Weight Gain During Pregnancy: Reexamining the Guidelines*. Washington, DC: National Academies Press; 2009.

Care of women with obesity in pregnancy RCOG; 2018.

Obesity in pregnancy ACOG; 2015 (re-affirmed 2020).

Obesity and pregnancy. Royal College of Physicians of Ireland; 2013.

Obesity in pregnancy. Canada: SOGC; 2010.

Reducing the risk of venous thromboembolism during pregnancy and the puerperium RCOG; 2015.

Management of pregnancy in elderly obese women

27

Tahir Mahmood[1] and Gamal Sayed[2]
[1]Department of Obstetrics & Gynaecology, Victoria Hospital, Kirkcaldy, United Kingdom,
[2]Women's Wellness and Research Centre and Clinical Department, College of Medicine, Qatar University, Doha, Qatar

Pregnancy in elderly obese women poses increased maternal and foetal risks;
The World Health Organization classifies obesity into:

- class I BMI (Body mass index) 30.0–34.9
- class II 35.0–39.9
- class III 40 or greater

There has been an increase in prevalence of obesity worldwide.

Prevalence of obesity is around 30% in the United States, increasing to near 60% when both obese and overweight (BMI 25.0–29.9) are combined.

- This chapter is focused on the care of women being >40, as many of the higher risks of pregnancy complications are related to delayed childbearing, development and access to assisted conception technology (ART), multiple pregnancies, and of high parity. Presence of obesity in this cohort of women has an additive effect for complications.
- It is recognised that whether or not the maternal age is considered as being advanced is affected by sociologic, ethnic, and cultural considerations as well.
- It is difficult to define the effect a specific age threshold for pregnancy outcome, as the effects of increasing age on pregnancy outcomes seems to occur more as a continuum.

27.1 Prevalence

The mean age at childbearing in Europe rose to 29.4 in Europe in 2015. According to the CDC report, there was a significant increase in the number of births to women aged 45–49 years between 1990 and 2010, rising from 0.39/1000 deliveries in 1990 to 1.79/1000 in 2010. According to another US report from 2006–2007 to 2014–2015, there was there was an increase in number of births of 8% for women aged 40–44 years, and 26% for women aged 45–54 years.

Handbook of Obesity in Obstetrics and Gynecology. DOI: https://doi.org/10.1016/B978-0-323-89904-8.00008-1

27.2 Pregnancy outcomes in relation to advanced maternal age

Fertility gradually declines as a woman matures, starting from 32 years of age, due to a decline in oocyte quality and quantity, and fecundability subsequently declines at around the age of 37 years.

Therefore there is an increasing demand for ART services.

Newer developments in ART have helped women with pathological conditions —poor oocyte quality diminished ovarian reserve or primary ovarian insufficiency —who can benefit from oocyte donation (OD).

OD programmes have also enabled perimenopausal and even postmenopausal women to conceive.

Conception issues:

Decreased fertility OR(odds ratio) 2
Increased demand for ART OR:1.5−2

First trimester complications:

- A four- to eightfold increased risk of ectopic pregnancy due to accumulation of risk factors over time (previous pelvic infections, prolonged smoking, tubal pathology, decreased tubal function, and a delay of oocyte transportation).
- A higher incidence of multiple pregnancies.
- Increased risk of spontaneous miscarriages, associated with increased risk of trisomy and other euploidy, especially Down syndrome.
- The increased incidence of nonchromosomal congenital malformations and birth defects, as neonates born to women aged >40 years had a twofold increased risk for cardiac defects, oesophageal atresia, and craniosynostosis even after adjusting for BMI and ART.

27.3 Late pregnancy complications

Chronic hypertension and type 2 diabetes are more common among older pregnant women, and women >35 years of age are at a two to fourfold higher risk of having hypertension compared to women <35.

There is higher risk of developing preeclampsia in women aged over 40 years, The incidence of both pregestational and gestational diabetes mellitus is three to sixfold in gravid women over the age of 40 years, approaching to almost 30% in gravid women aged >50.

There is increased incidence of placental abruption, as well as placenta praevia. Nulliparous women aged >40 years have a 10-fold increased risk for placenta praevia compared to nulliparous women aged 20−29 years (absolute risk of 0.25% vs. 0.03% respectively).

There is a higher incidence of low birth weight and preterm births and the relative risk was 1−9 in women >35 years of age, compared to women aged 20−24 years old.

Among ART pregnancies, especially those achieved via OD, there was a higher risk for preeclampsia, preterm birth, and small for dates, compared to those who conceived naturally,

According to CDC data, the mortality rate among pregnant women <35 years of age is 10.8 per 100,000 births, compared to 38 deaths/100,000 births in women aged >35.

27.4 Intrapartum and postpartum complications

Obesity poses a higher incidence of elective caesarean sections and second stage labour dystocia and operative vaginal delivery. A significantly increase risk of postpartum haemorrhage, need for blood transfusion, prolonged hospitalisation and admission to intensive care unit. Additionally, increased maternal age is associated with an increased risk of still birth.

Risks related to obesity which will have cumulative effect on pregnancy outcome:

- Miscarriage: the higher the BMI, the higher the risk of spontaneous Miscarriages
- Increased congenital abnormalities as neural tube defects, hydrocephaly, cardiovascular and limb reduction abnormalities
- Hypertensive disorders of pregnancy
- Gestational diabetes
- Venous thromboembolism (VTE)
- Sleep apnoea
- Foetal macrosomia (increases shoulder dystocia with its sequalae as brachial plexus injury)
- Stillbirth
- Increased caesarean delivery
- Higher risk of dysfunctional and pronged labours
- Increased incidence of induction of labours (IOL), and more failed IOL
- Low success rates at vaginal birth after caesarean section
- Anaesthetic complications especially with citing epidural in labour and general anaesthesia
- Endometritis, wound infection and dehiscence, and surgical site infections
- For those who had bariatric surgery, there is increased risk of nutritional deficiencies, advise them to wait for 12−18 months at least after surgery to address nutritional deficiency before embarking on pregnancy.
- Among those with prior bariatric surgery, there is a higher incidence of intrauterine growth restriction, thus requiring close monitoring of foetal development.
- Postpartum depression

27.5 Management

27.5.1 Prepregnancy

- Advanced maternal age at childbearing is associated with higher rates of coexisting medical conditions that may affect the course of pregnancy and labour.
- Chronic hypertension, thyroid disorders, or haematological disorders are much more common in women >45 than women of 30−35 years old.

- Advanced age is a risk factor for all cancers especially ovarian cancer, cervical cancer, breast cancer, and endometrial cancer among obese weight group.
- There is also a higher incidence of uterine fibroids. Fibroids are independently associated with preterm labour, malpresentation, placental abruption, dysfunctional labour, and caesarean delivery.
- Being obese and the age >40 would have a compounding effect on various parameters in pregnancy.
- All women should be seen at a prepregnancy clinic for total health assessment, lifestyle assessment, counselling, risk assessment during pregnancy, review of medication and advice with regard to remedial actions deemed necessary.
- Consider preconception genetic screening for couples and women who are known carriers of a genetic disease, who may consider the option of preimplantation genetic diagnosis.
- Consider screening for HIV, Hepatitis B, Hepatitis C.
- Depending on the medical history and existing comorbidities, the following may be considered for women >50:
  - o Mammogram
  - o Pap smear
  - o Pelvic/abdominal ultrasound scan
  - o Assessment of renal, liver, thyroid function tests, lipid profile
  - o Blood pressure profile
  - o Oral glucose tolerance test
  - o Assessment of cardiac function (resting electrocardiogram/echocardiography)
- Weight loss advice: (surgical and nonsurgical interventions improve chance of conceiving, and reduce miscarriage rate, gestational diabetes, preeclampsia, foetal macrosomia, and VTE.
- Stop medications for weight loss before pregnancy as many are not safe.
- Review current medication to identify medications with teratogenic potential are replace them with safe alternative ones.
- Check rubella immunity and offer vaccination for measles, mumps, and rubella at least a month before pregnancy.
- Start folic acid 5 mg (1−3 months before pregnancy) and continue at least during first trimester.
- Advise additional calcium and vitamin D3.
- Universal recommendations: smoking cessation, avoiding recreational drugs, reducing caffeine intake, healthy diet, and exercise.
- Assessment for possible mental health issues with the use of a validated questionnaire.

27.6   Antenatal

27.6.1   First trimester

- Offer individualised care at a high-risk clinic.
- Arrange ultrasound during first trimester for correct assessment of gestational age, number of sacs, and nuchal translucency thickness.
- Maternal serum biochemical screening testing between 11 + 0 and 13 + 6 weeks and NIPT (cell free DNA).
- Chorionic villous sampling and inherent risks related to foetal loss.
- Explain risks of obesity and age.

- Record BMI at first visit, initiate discussion on weight management in pregnancy, give exercise advice, and provide information leaflet.
- At first antennal visit, check rubella status, and undertake other routine pregnancy tests as HIV, Hepatitis B, Hepatitis C (high risk groups only), urine culture, antibody screen, complete blood count, syphilis and chlamydia (in North America)
- Folic acid 5 mg daily to continue until 13 weeks of gestation
- Vitamin D 10 μg for bone and teeth health and development
- Aspirin 75−150 mg from 12 weeks until delivery and calcium 1 gm per day, to reduce the risk of preeclampsia.
- Thyroid function tests if not already done.
- Assess for risk of thromboembolism. Daily prophylactic enoxaparin as per local hospital protocol scoring.

27.6.2   Second trimester

Offer amniocentesis as required based on foetal risk.

Detailed foetal abnormality scan at 19−20 weeks and may require a repeat detailed assessment around 23−24 weeks for foetal cardiac imaging.

- Oral glucose tolerance test at 24−28 weeks. Consider early screening at 16 weeks in the presence of other risk factors especially previous gestational diabetes.
- Serial growth scans 28, 32, 36 weeks (as it is difficult to assess foetal growth clinically).
- Additional scans may be indicated needed if complications arise.
- Close monitoring of foetal and maternal well-being at the day assessment unit.

27.6.3   Third trimester

- Close monitoring of foetal and maternal well-being at the day assessment unit.
- Serial growth scans 28, 32, 36 weeks (as it is difficult to assess foetal growth clinically).
- Anaesthesia review around 36 weeks gestation for pain relief and choice of anaesthesia if operative intervention is anticipated.
- Timing of delivery and mode of delivery should be individualised.
- It has been estimated that the gestational age associated with the lowest cumulative risk of perinatal death is 38 weeks.
- Data from the United States has reported that the cumulative risks of stillbirth for women <35 years, 35−39 years, and older than 40 years old are 6.2, 7.9, and 12.8 per 1000 pregnancies. Nulliparous women had a higher risk of stillbirth.
- A planned IOL at 39−40 weeks is recommended with careful close monitoring of mother and the foetus. Some women may opt for elective caesarean birth based upon their risk assessment.

27.6.4   Intrapartum care

- Facilities and equipment considerations for large weights, larger cuffs for blood pressure, birthing beds, operating room tables.
- Anaesthetic team should be involved in the care at the outset when in labour. Early epidural analgesia during labour (can be technically difficult due to body habitus and loss of landmarks, so better place early than in emergency).
- If BMI ≥ 40 consider a second cannula.

- Foetal scalp electrode for foetal monitoring once forewater amniotomy has been performed.
- Progress of labour should be carefully monitored and an early resort to caesarean section should be considered. The duration of the second stage of labour does not appear to be affected by increasing BMI.
- For caesarean sections, a 2 g prophylactic cefazolin dose is used if the woman is more than 80 kg. This dose increases to 3 g if she weighs more than 120 kg.
- Closure of subcutaneous tissue (of more than 2 cm) at caesarean section should be done to reduce incidence of wound infection and dehiscence. Drains. however, should not be routine.
- Active management of third stage (risk of atonic postpartum haemorrhage) should be instituted

27.6.5 Postnatal

- Appropriate support to initiate breast feeding (lower initiation and maintenance rates).
- Risk scoring for thromboembolism as per local hospital protocol. Pneumatic compression devices (mechanical) or pharmacological (low molecular weight heparin) or both. Adjust the prophylactic dose to BMI or weight.
- Avoid oral contraceptive pills (increased risk of thromboembolism).
- Can use barrier methods as condom (but higher failure rate), progesterone-only pills, progesterone implants, or injections and copper or progesterone-based coils (inserted following birth).
- Arrange monitoring of all vital signs at home until 6 weeks postnatally.
- Arrange follow-up testing for those who had gestational diabetes.
- Referral to weight management services.
- Arrange a postdelivery briefing meeting couple of months later.

Further reading

American College of Obstetricians and Gynaecologists. Obesity in pregnancy ACOG; 2015 (re-affirmed 2020).

Mathews TJ. Mean age of mothers is on the rise: United States, 2000−2014. *NCHS Data Brief.* 2016;(232)8.

Royal College of Obstetricians and Gynaecologists. Care of Women with Obesity in Pregnancy RCOG; 2018.

Royal College of Obstetricians and Gynaecologists. Induction of Labour at Term in Older Mothers, RCOG; 2013.

Royal College of Obstetricians and Gynaecologists. Reducing the Risk of Venous Thromboembolism during Pregnancy and the Puerperium RCOG; 2015.

Sauer MV. Reproduction at an advanced age and maternal health. *Fertil Steril.* 2015;103 (5):1136−1143.

Storgaard M, Loft A, Bergh C, et al. Obsteric and neonatal complications in pregnancies conceived after oocyte donation: a systematic review and *meta*-analysis. *BJOG An Int J Obstet Gynaecol.* 2017;124(4):571−572.

Yogev Y, Melamed N, Bardin R, et al. Pregnancy outcome at extremely advanced maternal age. *Am J Obstet Gynecol.* 2010;203(6):558. e1−7.

Novel viruses and pregnancy

28

Nirmala Mary[1] and Nithiya Palaniappan[2]
[1]Royal Infirmary Edinburgh, Edinburgh, United Kingdom, [2]Department of Obstetrics and Gynaecology, Victoria Hospital, Kirkcaldy, Scotland

28.1   Introduction

Throughout human civilisation pandemics have occurred regularly in history.

There is a fertile ground for spread of viruses due to the increased contact with vectors and conducive environmental factors, including climate change, with the ever-expanding humankind: increasing population, sprawling cites, spreading human settlements, and fast and easy travel.

Globalisation has unfortunately increased the risk of having epidemics and pandemics from endemic infections.

Lately pandemics have been related to viral infections instead of bacterial infections. Human immunodeficiency virus, HIN1influenza A virus, severe acute respiratory syndrome coronavirus (SARS-CoV) and Middle East respiratory syndrome coronavirus (MERS-CoV) have already afflicted humanity. The world is currently ravaged by a new coronavirus—the COVID-19 pandemic.

28.2   Virus and pathogenesis

A virus is an infective agent that typically consists of a nucleic acid molecule (RNA/DNA) surrounded in a protein coat (capsid). Some virus have a phospholipid membrane with spike projections surrounding the capsid derived from the host cell membrane. A virus is too small to be seen by light microscopy and is able to multiply only within the living cells of a host. Viruses are abundant and diverse pathogens present in almost every ecosystem on earth. They have evolved over millions of years to produce precise mechanisms to target and infect host cells.

Viruses are essentially parasites which must invade host cells in order to replicate. They cannot synthesise proteins because they lack ribosomes for the translation of viral messenger RNA. The viral protein attaches to specific receptors on the host cell surface inducing a conformational change in the viral capsid protein leading to a host of mechanisms causing fusion of viral and cell membranes and release of viral genomic material, replication by transcription or translation using host

Handbook of Obesity in Obstetrics and Gynecology. DOI: https://doi.org/10.1016/B978-0-323-89904-8.00035-4

genetic material, maturation and virion release by lysis or budding. The new virions infect and destroy further host cells.

Viral infections could be asymptomatic or lead to severe disease.

Transmission:

- Ingestion
- Inhalation
- Sexual transmission
- Transmission through bodily fluids and blood products
- Vectors—bites of insects (mosquitoes, some flies, ticks, etc.), birds or animals (bats, monkeys, dogs, etc.)

Types of viral infections:

- Respiratory infections: Coronaviruses, HIN1, MERS
- Gastrointestinal tract: Norovirus, Rotaviruses
- Liver: Hepatitis
- Nervous system: Rabies virus, West Nile Virus, Polio
- Skin: Human papilloma virus, Chicken pox
- Placenta and foetus: Rubella, Cytomegaloviruses, Zika virus.

Pregnancy is a state of immunological suppression in order to promote foetal implantation. The hormonal and immune system changes generated in pregnancy may explain woman's vulnerability to infections. Pregnancy-related immunosuppression, physiological changes of pregnancy, and foetal demands interfere with maternal response to infections in some women, though the exact mechanisms are unclear. There are increased rates of complications and higher mortality associated with some viral infections in pregnant women compared to the general population.

Other risk factors:

Obesity and metabolic syndrome:

Obesity is associated with a significantly increased risk of development of asthma and diabetes. Obesity and metabolic syndrome has also been associated with increased risks of infection. This is evident in the recent pandemics. There is a significantly increased risk of hospitalisation, need for mechanical ventilation, and mortality (60−80 times). Obesity causes a chronic state of metainflammation affecting immunity. There is a delayed or blunted response to influenza virus infection leading to higher viral loads, increased and prolonged viral shedding, and poorer recovery.

Hypertension:

Hypertension is associated with a dysregulated inflammatory immune response. Hypertension affects the central nervous system increasing the sympathetic nerve activity and dysregulating the parasympathetic and antiinflammatory pathways, thereby contributing to the end-organ damage.

Pregnant women with other risk factors, especially obesity, are at an increased risk of community acquired respiratory infections, increased severity of infections possibly due to altered immune functionality, decreased lung functionality, and physiological changes of pregnancy. Obesity and pregnancy independently lead to an immunocompromised state and their combination could have a synergistic effect.

Diagnostic tests:

- antibody-capture enzyme-linked immunosorbent assay
- antigen-capture detection tests
- serum neutralisation test
- reverse transcriptase polymerase chain reaction (RT-PCR) assay
- electron microscopy
- virus isolation by cell culture.

Careful consideration should be given to the selection of diagnostic tests, which take into account technical specifications, disease incidence and prevalence, and social and medical implications of test results. It is strongly recommended that diagnostic tests that have undergone an independent and international evaluation be considered for use.

Prevention and control—general principles:

- Good outbreak control relies on applying a package of interventions including case management, surveillance, contact tracing, a good laboratory service, safe burials, and social mobilisation.
- A centralised approach with clear evidence-based protocols of management ensuring availability of staff who are trained to provide safe and effective care is essential in managing novel virus infections.
- Community engagement is key to successfully controlling outbreaks.

Avoidable factors noted from recent pandemics:

- Inequality and socioeconomic access barriers to care.
- Delayed admission to hospital and/or consideration of the likelihood of a viral infection in the differential diagnosis. Delayed confirmation of diagnosis; delay in administering antiviral medication.
- Lack of clear clinical leadership in overall case management.
- Unequal access to research/experimental therapies in pregnancy possibly due to medical anxiety/lack of inclusion at planning stages leads to inequalities of care and increased risk for pregnant women.

28.3a   Ebola virus (EBV) (Filoviridae: 1976–2021)

It is a severe and frequently lethal disease leading to multiorgan failure.
Average case fatality: 50% (25%–90%); Case fatality in pregnant women—72% (46%–100%).

Endemic areas:

- Sub-Saharan Africa—Sudan, Guinea, Sierra Leone, Liberia, Democratic Republic of Congo.

Transmission:

- No natural host has been definitively identified yet. There is speculation secondary to retrospective epidemiological investigation that EBV is a zoonosis, possibly due to contact with bodily fluids, blood, secretions, organs of wild animals/carcasses (fruit bats, chimpanzees, gorillas, antelopes).
- Human to human transmission—contact with blood/bodily fluids from person who is sick or has died from EBV.
- Vertical transmission—mother to foetus.

Incubation period:

- 2–21 days; continue to be infectious until virus present in blood.

Signs and symptoms:

EBV infection typically starts as a nonspecific febrile illness followed by severe gastrointestinal symptoms. High viremia patients develop dysregulated immune response leading to severe multiorgan dysfunction syndrome and fatality.

- Nonspecific viral prodromal symptoms, that is, fever, fatigue, muscle pain, headache, sore throat
- Gastrointestinal symptoms—nausea, vomiting, diarrhoea, and abdominal pain
- Hypotension from dehydration and hypovolemia
- Rash and sometimes haemorrhagic manifestations—haemoptysis, haematemesis, and melena
- Multiorgan failure:
    - o Renal failure—acute tubular necrosis secondary to viral infection, myoglobin pigment injury related to rhabdomyolysis or cytokine-mediated nephrotoxicity
    - o Respiratory failure due to acute lung injury secondary to systemic inflammatory response syndrome, respiratory muscle fatigue, pulmonary oedema, and viral pneumonia
    - o liver failure, neurological manifestations like headaches, seizures, memory loss, insomnia, and encephalopathy
    - o cardiac dysfunction myocarditis and pericarditis.

Diagnosis:

- Reverse transcriptase PCR test
- Rapid antigen detection tests with appropriate training for front-line healthcare workers in remote settings where PCR tests are not readily available.

These tests are recommended for screening and as part of surveillance activities.
Patient with viral load greater than 10 million genome copies/mL were noted to have poorer prognosis. Other risk factors: age >45 year, pyrexia >38°C.

Differential diagnosis:

Other infectious diseases like malaria, typhoid fever, other viral infections like Lassa fever, meningitis.

Treatment:

- Supportive measures
  - o Early aggressive supportive care with good management of fluid and electrolyte losses with rehydration using oral fluids/IV crystalloids depending on clinical condition.
  - o Correction of electrolyte imbalance and maintaining acid base balance to prevent lethal arrhythmias and fluid shifts and other symptomatic treatment.
- Critical care: patients with critical illness including multiorgan dysfunction may require advanced support in critical care units.
- Monoclonal antibodies (Inmazeb and Ebanga) are approved.
- A range of potential treatments including blood products, immune therapies, and drug therapies are currently being evaluated.

There are no robust data on ideal treatment of pregnant women with EBV disease but the WHO advises that pregnant and breastfeeding should be offered early supportive care. Vaccination and experimental treatment should be offered under same conditions as for nonpregnant women.

Clinical management of babies born to infected mothers, parental presence with its associated risks and benefits also presented difficult challenges for healthcare staff.

As the typical setting of EVD outbreaks are in areas with limited access to trained medical staff, facilities including infrastructure, alongside the presence of armed conflicts, the logistics of containment and care are significantly difficult to attain.

A coordinated approach led by national and international health agencies like the WHO along with input from international humanitarian agencies, financial aid, and nongovernmental organisations are crucial for effective management and care.

Clinical trials/experimental therapies need to be ethically approved and may not be acceptable to population who have genuine suspicions and concerns about safety of the therapies offered and intentions in a postcolonial setting.

Outcomes:

- Mortality for pregnant women with EBV disease is significant though not higher than general population.
- Nearly all women had adverse pregnancy outcome of foetal loss (miscarriage, stillbirth), neonatal death, and obstetric haemorrhage. EBV has been noted to be in amniotic fluid, breast milk, and foetal tissue up to 30 days after recovery from viremia. Amongst the known cases the disease was found mainly in the third trimester (56%) and a high rate of vertical transmission was noted. There is one report of neonatal survival with no relapse; it is rare but known to happen.
- Long-term sequelae: Ebola survivors can have long-term health problems including generalised nonspecific symptoms, uveitis, loss of vision, diplopia, hearing loss, tinnitus, cough, shortness of breath, paroxysmal nocturnal dyspnoea, palpitations, arrhythmias, neurological sequelae like headache, dizziness, memory loss, insomnia, seizures, change in mental health status with conditions like anxiety, depression, or posttraumatic stress disorder.

Prevention:

- The strategy to prevent spread of on going outbreak is interruption of community and nosocomial transmission. Creation of clear protocols is of high importance to maintain standards of infection control.
- Screening all women regarding symptoms and enquire regarding recent travel.
- Community engagement, Raising awareness
- Contact tracing and followup for 21 days
- Reducing the risk of wildlife-to-human transmission
- Reducing human-to-human transmission
  - o Isolation
  - o Healthcare workers who are pregnant should not care for patients with EVD.
  - o Treatment in specialist holding centres with treatment beds in areas to isolate patients with suspected or confirmed EVD
  - o Strict infection control measures of handwashing
  - o Safe injection practices
  - o Dedicated or disposable patient care equipment and provision
  - o Use of appropriate personal protection equipment, that is, close contact (within 1 m)—face shield or medical mask and goggles, a clean nonsterile long sleeved gown and gloves.
  - o Strategies for donning and doffing PPE
  - o Attention to environmental infection control
  - o Outbreak containment measures like separating sick from the healthy and safe an dignified burial
  - o Reducing the risk of sexual transmission (safe sex until 12 months from onset of symptoms or until semen tests negative twice for EBV)
- Reducing the risk of transmission from pregnancy-related fluids and tissue
  - o Community support to survivors for safe antenatal care and meet their need for sexual and reproductive care and respect their choices.
  - o Healthcare staff should be informed of risk of virus persisting in blood and bodily fluids or contaminated surfaces or materials like beddings or clothing for their own safety and safety of the women they care for.
  - o Obstetric management of pregnant women with EVD, particularly decisions about mode of delivery for women in labour, needs to consider risks to the woman, risks of exposure for healthcare providers, and potential benefits to the neonate. EBV was found in amniotic fluid after women was noted to be PCR negative for EBV.
  - o If breastfeeding breast milk should be tested for virus before starting.
- Visitors should be screened before entering the patient area and they should not have any direct contact with the patient. Visitors should be trained to safely put on the same type of PPE recommended for healthcare workers. Visitors also should be observed and assisted with PPE to prevent or reduce the risk of infection.

Vaccines:

- Ervebo vaccine (Live attenuated recombinant vaccine) (18 years or older, not for pregnant or breastfeeding women)
- Zabdeno (Monovalent vaccine containing modified adenovirus) and Mvabea (Modified virus) two-component vaccine given 8 weeks apart for 1 year or older

28.3b   Zika virus (Flavivirus: 2015—current)

Zika virus in pregnancy causes congenital Zika virus syndrome which results in congenital brain abnormalities including microcephaly.

Epidemiology

- First discovered in 1947 in Rhesus monkeys in Uganda.
- First outbreak in 2007 in African and Asian countries and further spread quickly and widely to central and Southern American countries and the Caribbean.
- Epidemic subsided in 2017, although infection is considered as an endemic.

Transmission

- Transmission predominantly is by the bite of a female mosquito of the *Aedes aegyptii* species mostly, and occasionally of the other *Aedes* species.
- Sexual and blood transfusion from infected host is also reported.
- Maternal—foetal transmission has been reported.

Incubation period:

- 3—12 days

Signs and symptoms:

- Asymptomatic or mild self-limiting flu like illness in about one in five infected lasting about 2—7 days
- Macular or maculopapular rash
- Fever
- Myalgia/arthralgia/arthritis
- Headache/vomiting
- Conjunctivitis

Diagnosis:

- RT-PCR within 1 week of symptoms (virus can clear by 5—7 days for symptoms) within 2 weeks of travel from endemic area or sexual partner travelling from endemic area within 2 weeks.
- Zika virus is a notifiable disease.
- If Zika antibodies are not detected in serum for 4 or more weeks after exposure infection can be excluded.

Zika infection in pregnancy

- Pregnancy is not a risk factor and does not make a woman more susceptible for infection.
- Pregnancy with infection at an earlier gestation is more likely to be affected.
- 5%—15% of pregnancies with Zika virus infection can be affected.
- Zika virus can also trigger Guillain-Barré syndrome.
- Refer to Table 28.1 for congenital malformations in affected pregnancies.

Table 28.1 Congenital malformations in Zika virus affected pregnancy.

Cranial abnormalities	Extra-cranial abnormalities
• Microcephaly • Ventriculomegaly • Periventricular cysts • Callosal abnormalities • Microphthalmia • Cerebellar atrophy (transverse diameter <5th percentile) • Vermian agenesis • Blake's cyst • Mega cisterna magna (> 95th percentile) • Choroid plexus cyst • Brain atrophy leading to microcephaly (abnormally small brain) • Cortical and white matter abnormalities (e.g., agyria)	• Foetal growth restriction • Talipes

Management of pregnancies exposed to Zika virus infection

- Supportive treatment if symptoms prevail
- If there is a positive history of exposure:
  - o Offer baseline ultrasound at 18−20 weeks, followed by growth scan at 28−30 weeks.
  - o If the head circumference falls less than 2 standard deviations, then refer to foetal medicine services for further evaluation.
  - o Microcephaly is usually diagnosed when the head circumference falls well below the 2 standard deviations.
  - o If microcephaly observed along with other changes in the brain, then consideration needs to be given for amniocentesis for an RT-PCR test after cautious counselling.
  - o Amniocentesis carries a risk of miscarriage, however the sensitivity of the test and the extent to which a foetus is affected despite being infected is not known.
  - o Foetal MRI can be considered if changes are seen on ultrasound.
  - o If diagnosis of microcephaly is confirmed, termination of pregnancy can be offered in any stage of pregnancy, due to the poor outcome.

Prevention:

- Avoid travel to endemic areas and contraception while in endemic areas.
- Pregnant women must postpone nonessential travel to endemic areas.
- If recent travel should seek advice from healthcare team regarding appropriate testing and monitoring.
- If partner travelled to endemic area abstain from intercourse or use condoms for the duration of pregnancy.
- Personal protection from malaria (*Aedes* mosquitoes are active during the day)
- Clothing to cover all parts of body, mosquito nets or repellents can be considered

- Avoid conception within 2 months of travel if female partner alone has travelled; 3 months if male partner or both have travelled by effective contraception and barrier methods.

28.3c MERS-CoV (2012–current)

MERS-CoV is a zoonotic coronavirus infection which ranges from mild respiratory to SARS and death.

Case fatality: Estimated to be around 35%−40%.

This may be an overestimate as its only from lab-confirmed cases. Mild illness may have been unrecognised. *Severe cases were noted in 77% of known infected pregnant women.*

Transmission:

- Dromedary camels—close contact with nasal/bodily secretions. Dromedary camels are a known animal reservoir. The actual origins of MERS-CoV are still unconfirmed but thought to be possibly transmitted from bats to camels in the past.
- Human to human transmission occurs through close household contacts, crowded places (pilgrimages-like Hajj and Umrah), and in healthcare settings, likely through respiratory droplets/aerosol generating procedures and environmental surfaces.

Endemic areas:

- Middle East—majority of cases reported in Saudi Arabia
- Africa
- South Asia (Republic of Korea)

Incubation period:

- 2–14 days

Signs and symptoms:

- Fever, cough, shortness of breath
- Gastrointestinal symptoms including diarrhoea
- Pneumonia (71%)
- Severe illness leading to respiratory failure and multiorgan failure leading to death
    Illness noted to cause more severe disease in older population, people with weakened immune systems and chronic illness like renal disease, cancer, chronic lung disease, and diabetes.

Diagnosis:

- RT-PCR
- Antibody testing
- Seroconversion can be detected by antibody test 14 days apart
    (Detected more frequently in lower respiratory samples than upper respiratory samples, blood in unwell patients and stool samples.)

Treatment:

- Supportive treatment based on clinical condition.
- No specific antiviral or immunomodulator like glucocorticoid have been shown to be beneficial.
- No vaccine currently available (trials on going).

Prevention:

- Avoid contact with camels, drinking raw camel milk/camel urine/meat that hasn't been cooked properly. Pasteurisation or cooking should be used to render animal products safe for consumption.
- Good hand hygiene at farms, markets, barns, or other areas where there is contact especially with sick animals.
- Healthcare settings—infection prevention and control should be implemented with good hand hygiene, and personal protection is important though it is difficult to identify people with mild illness/nonspecific symptoms.
- Enhanced surveillance of severe respiratory illness.
- Maintain high vigilance of travellers from endemic areas.

28.3d HIN1 virus (Ortho-myxovirus: 2009−2010)

The virus contains a mixture of genetic material from human, pig, and bird flu viruses.

Mostly people had upper respiratory tract infection, but approximately 49% had viral pneumonia/ARDs and 20% had bacterial infections like *Streptococcus pneumonia/pyogenes, Staphylococcus aureus, Haemophilus influenza*.

Case fatality: Mostly mild illness with 50% recovering within 7 days and a further 25% within 10 days.

Risk groups:

- Socially deprived
- Underlying medical problems like diabetes, asthma, immunosuppression cardiac disease/renal disease
- Morbid obesity
- Pregnant women because of altered immune and physiological adaptations, especially in second and third trimesters (7%−9% of ICU admissions)
- Black and ethnic minority women.

Transmission:

- Respiratory droplets/aerosol generating procedure
- Touching infected surfaces

Incubation period: 2−10 days. Highly contagious during the first 5 days of illness.

Signs and symptoms:

- Headaches, aching muscles, exhaustion, or fatigue
- Sore throat, coryza
- High temperature >38°C with chills and shivers
- Gastrointestinal side effects like diarrhoea, vomiting, or stomach upset.
- Sometimes myocarditis associated with marked tachycardia—usually good prognosis for recovery
- Neurological symptoms like seizures, encephalitis with altered mental state.

Poor prognosis: Dyspnoea, respiratory rate >30/minute, supplementary oxygen requirement, significant tachycardia, pneumonia, altered levels of consciousness.

Diagnosis:

Nasopharyngeal viral swab

1. RT-PCR
2. Rapid antigen detection tests as a point of care test (less sensitivity 10%−70%).

Treatment:

- General supportive measures like rest, paracetamol/oxygen.
- Suspected patients should be managed in isolation and barrier nursing.
- Antiviral treatment (neuraminidase inhibitors—act by preventing the virus from budding and escaping from the host cells) ideally commenced within 48 hours of symptoms. Treatment within 48 hours was associated with 84% reduction in odds of ICU admission.
  o Oseltamivir (Tamiflu)—Oral capsule
  o Zanamir (Relenza)—Diskhaler (first line as acts in respiratory system with no absorption into bloodstream).
- Consider appropriate antimicrobial treatment for suspected bacterial infection.
- Early multidisciplinary input and care from obstetrician, obstetric anaesthetist, respiratory physician, microbiologist, and haematologist are required.
- Consideration to antenatal steroids for preterm labour/birth

(exclude preeclampsia/venous thromboembolism/pulmonary embolism. Watch for disseminated intravascular coagulopathy).

- ECMO (extracorporeal membrane oxygenation) if not responding to mechanical ventilation.
  Vaginal birth will be tolerated by most women with good analgesia and hydration.
  Critically unwell women may require delivery by caesarean section after stabilisation and correction of any complications like DIC in maternal interest to help with recovery and mechanical ventilation.
- Breastfeeding should be promoted—Oseltamavir preferred antiviral medication.
  A threefold increased risk of preterm delivery and fivefold increase in perinatal mortality were noted.

Differential diagnosis:

- Chorioamnionitis
- Severe urinary infection

4 of Obesity in Obstetrics and Gynecology

- Group A/B streptococcal infection
- Malaria

Prevention:

- Strict hygiene and infection control measures.
- Good respiratory hygiene covering mouth and nose when coughing and sneezing.
- Good hand hygiene.
- Patient isolation and healthcare professionals having appropriate personal protection like face masks and plastic aprons.
- FFP3 masks for aerosol generating procedures.
- Vaccination: HIN1 vaccine contains inactivated virus. This is similar to seasonal flu vaccine.

All healthcare workers and vulnerable population should be invited for vaccination
The safety aspect of the vaccine is drawn from its similarity to seasonal flu vaccine. A good antibody response has been noted in pregnant women similar to general population.

28.3e   Coronavirus—SARS-CoV-2 (COVID-19)

SARS-CoV-2 is a coronavirus strain causing the latest pandemic engulfing the world. The new variants may cause increased transmissibility and more severe disease. One in seven hospital admissions with delta variant needed intensive care unit admission.

Case fatality: Majority of pregnant women (73%–86%) are asymptomatic and most symptomatic pregnant women have mild or moderate cold/flu like symptoms. Pregnant women are more likely to need intensive care unit admission with higher needs for ventilation and extracorporeal membrane oxygenation particularly in the third trimester. Risk of death remains low 2.4/100,000 maternities in the United Kingdom. The mortality rates were noted to be 22 times higher in institutions from less developed regions. The PregCOV-19 Living Systematic Review Consortium analysis showed 0.02% of pregnant women with confirmed with COVID-19 died of any cause and 0.2% of pregnant women with COVID-19 required ECMO. Women with prepregnancy comorbidities, especially obesity, were noted to have higher adverse outcomes.

Symptomatic maternal COVID is associated with iatrogenic preterm birth due to maternal compromise. It may be associated with small for gestational age babies. Stillbirth remains rare but COVID-19 is associated with an increased risk of stillbirth.

Transmission:

- Respiratory droplets or secretions
- Aerosol generating procedures carry high risk of spread.
- Fomites
- Faeces

Spread is mainly through close contact with infected person and less commonly from contaminated surfaces.

Pregnant women are not more likely to acquire infection than general population.

Vertical transmission is rare. There is evidence of trans-placental passage of COVID-19 antibodies, suggesting passage of passive immunity.

Risk factors for hospital admission:

- Black, Asian, or other minority ethnic background
- BMI $>30 \text{ kg/m}^2$
- Maternal age >35
- Socioeconomic deprivation
- Prepregnancy comorbidities like preexisting diabetes and hypertension

Incubation period:

- $1-14$ days; average time $5-6$ days

Signs and symptoms:

- Mild cold/flu like illness with fever, sore throat, myalgia, dry cough, loss of sense of taste or smell, or diarrhoea.
- Shortness of breath, pneumonia.
- Severe disease—multiorgan failure, septic shock, and venous thromboembolism
  Fever and shortness of breath were noted to be associated with worse morbidity and mortality.

Diagnosis:

- RT-PCR
- Rapid antigen tests
  Other investigations: full blood count, coagulation screen, arterial blood gas renal function, C reactive protein, CT chest showing ground glass appearance/consolidation.

Treatment:

- Home isolation if mild symptoms with advice and counsel about signs and symptoms of deterioration and advice to seek help if needing urgent care, that is, difficulty breathing, chest pain, etc.
- Advice about isolation at home
  o Improve air circulation by keeping doors and windows open
  o Psychosocial support
  o Manage other medical conditions closely (like diabetes, asthma, etc.) and seek help early if worsening symptoms or poor control.
- Symptomatic treatment
  o Antipyretics/analgesia/antitussives as required
  o Mobilisation, hydration, and good diet.
- Assess and admit women with moderate to severe COVID.
- Oxygen to maintain saturations above 94%, escalating with e.g., nasal prongs, masks, CPAP, IPPV, ECMO.

- Hourly fluid input/output chart to monitor fluid balance.
- No antibiotics unless additional bacterial infection suspected.
- LMWH for VTE prophylaxis or higher dose thromboprophylaxis in severe COVID after multidisciplinary discussion.
- Steroids if oxygen is needed (e.g., oral prednisolone 40 mg once daily or IV hydrocortisone 80 mg twice daily, with intramuscular dexamethasone 6 mg twice daily for four doses followed by oral prednisolone as below if foetal lung maturity is also required).
- MDT review—consider appropriate escalation of care, location of care, timing of delivery.
- Strongly consider tocilizumab (400 mg/600 mg/800 mg single IV infusion depending on weight) if C-reactive protein at or above 75 mg/L or in ICU.
- Strongly consider REGEN-COV monoclonal antibodies (8 g single IV infusion) in those with no SARS-CoV-2 antibodies.
- Critically ill antenatal women who have recovered should have a discharge plan with Consultant Obstetrician including foetal biometry in 2 weeks.

Azithromycin, hydroxychloroquine, and lopinavir/ritonavir have been shown to be ineffective and should not be offered.

Prevention:

- Joint international and national collaboration, guidance, and interventions in pandemic.
- Societal education, awareness, and social distancing, avoiding crowded poorly ventilated places whenever possible following local guidance.
- Personal protection (face protection) if inside/close contact, good hand hygiene according to guidance.
- Protect family member especially vulnerable adults especially if unvaccinated—pregnant, multiple morbidities, and elderly.
- Early detection and isolation.
- Special care in healthcare setting with appropriate PPE, isolation to prevent spread to and through healthcare staff.
- Visitors having similar personal protection as healthcare staff for visiting.

Vaccination:

1. Pfizer—mRNA vaccine
2. Moderna—mRNA vaccine
3. Astrazeneca—recombinant, replication-deficient chimpanzee adenovirus vector encoding the SARS-CoV-2 spike glycoprotein. Produced in genetically modified human embryonic kidney 293 cells.

Strongly recommended in pregnancy at any time in pregnancy and while breastfeeding. All vaccines have been tested and no animal reproductive toxicity was noted

The vaccines generate robust humoral immunity in pregnant and lactating women. Some studies have shown that the immune responses were greater than the response to natural infection. Some preliminary studies have shown that most vaccinated mothers pass passive immunity to their newborns.

Preliminary data from a couple of studies have shown no increased risks of miscarriage. Reassuring data is available regarding safety of mRNA vaccine in later pregnancy with no increase in stillbirth, preterm delivery, small for gestation, or neonatal death.

Minor common adverse events similar to nonpregnant people have been reported. A very rare vaccine-induced thrombosis and thrombocytopenia have been reported with use of viral vector (1 in 250,000 people) vaccine (AstraZeneca/ Jansen). It is thought to be an idiosyncratic autoimmune response where platelet factor 4 develops within 3 weeks of administration of vaccine.

Further reading

Coronavirus infection in pregnancy. Information for healthcare professionals. RCOG guideline. Published Aug 2021.

Erik AKarlsson, Glendie Marcelin, et al. Review on the impact of pregnancy and obesity on influenza virus infection. *Influenza Other Respir Viruses*. 2012;6(6):449−460. Available from: https://doi.org/10.1111/j.1750-2659.2012.00342.x.

Jacob ST, Crozier I, Fischer WA, et al. Ebola virus disease. *Nat Rev Dis Primers*. 2020;6 (13). Available from: https://doi.org/10.1038/s41572-020-0147-3.

Jocelyne P, Boivin G. Review article pandemics throughout history. *Front. Microbiol.* 2021;15. Available from: https://doi.org/10.3389/fmicb.2020.631736. January.

John A, Elena S, et al. Clinical manifestations, risk factors, and maternal and perinatal outcomes of coronavirus disease 2019 in pregnancy: living systematic review and *meta*-analysisfor PregCOV-19 living systematic review consortium. *BMJ*. 2020;370:m3320.

Lim Boon H, Mahmood Tahir A. Influenza A H1N1 2009 (swine flu) and pregnancy. *J Obstetr Gynecol India*. 2011;61(4):386−393. Available from: https://doi.org/10.1007/ s13224-011-0055-2. July−August.

Megan EF, Carolina C, et al. Pregnancy and breastfeeding in the context of Ebola:a systematic review. *Lancet Infect Dis*. 2020;20:e149−e158.

RCOG/RCM/PHE/HPS Clinical guidelines—updated 27/02/19.

Rebekah H, Stacey S-C. Impact of obesity on influenza a virus pathogenesis, immune response, and evolution. *Front. Immunol.* 2019. Available from: https://doi.org/10.3389/ fimmu.2019.01071. 10 May.

Ricardo Wesley Alberca, Nátalli Zanete Pereira, et al. Pregnancy, viral infection, and COVID-19. *Front. Immunol.* 2020. Available from: https://doi.org/10.3389/fimmu.2020.01672. 07 July.

Singh Madhu V, et al. The immune system and hypertension. *Immunol Res*. 2014;59(1-3): 243−253. Available from: https://doi.org/10.1007/s12026-014-8548-6.

WHO. Available from: https://www.who.int/news-room/fact-sheets/detail/ebola-virus-disease.

Yaseen MArabi, Hanan HBalkhy, et al. Middle East Respiratory Syndrome. *N Engl J Med*. 2017;376:584−594. Available from: https://doi.org/10.1056/NEJMsr1408795.

Zika Virus Infection and Pregnancy, Information for Healthcare Professionals; RCOG; Feb 2019.

Infections during pregnancy

Tahir Mahmood
Department of Obstetrics & Gynaecology, Victoria Hospital, Kirkcaldy, United Kingdom

- Many infections in pregnancy have little effect on maternal health, and may result in minimal or no maternal symptoms unless the mother is immunocompromised.
- Acute maternal infection can have significant consequences on the mother right through pregnancy because of immunological adjustments.
- However certain infections are known to be teratogenic to the developing foetus, and this damage is determined by the gestational age at exposure to infection.

29.1 Immunology of pregnancy

Pregnancy alters the maternal immune system to allow the genetically foreign foetus to develop without causing rejection in a variety of ways.

29.2 In nonpregnant state

- T helper cells produce cytokines in response to infection.
- T helper cells can be further subdivided as:
  - o type 1 T helper cells produce interferon, interleukin, and tumour necrosis factor, which promote an antibody response;
  - o whilst type 2 helper cells produce differing ILs which stimulate a cell-mediated innate immune response.
- Cytotoxic T cells kill the pathogens directly.

29.3 In pregnancy

- There is a shift of T helper cells from type 1 to type 2.
- Therefore, the antibody-mediated immune response is suppressed, and this is compensated by an increased activation of the innate immune system.
- The innate system is less efficient in clearing viruses and bacteria than the specific antibody response.
- Hence the pregnant woman is at an increased risk of certain pathogens: some intracellular viruses, bacteria, and parasites.

Handbook of Obesity in Obstetrics and Gynecology. DOI: https://doi.org/10.1016/B978-0-323-89904-8.00045-7

There is an increased susceptibility of infection to the following pathogens:

- Malaria
- Measles
- Toxoplasmosis
- Listeriosis
- Leprosy
- *Pneumocystis carinii*

Some pathogens result in an increased severity of disease in pregnancy:

- Influenza
- Varicella
- Psittacosis
- Viral haemorrhagic fever (Ebola and Lassa)

29.4   Effect of obesity on the immunology

Adipose tissue is known to have endocrine, steroidogenesis, and immunological function. Although the full extent of dysregulation of the immune system is not fully understood, and nor is the compounding effect of being pregnant, the following mechanisms have been postulated:

- Suppression of functionality of both CD4 T cells and CD8 T cells
- T cell diversity
- Impaired NK cells
- Decreased cytokine production
- Lower levels of CD8 and NK cells may account for the increased susceptibility of pregnant obese women to severe illness or mortality from the H1N1 influenza epidemic in 2009.
- Obesity is a state of chronic systemic inflammation associated with oxidative stress and with higher baseline levels of inflammatory cytokines and monocytes.
- It has been postulated that this chronic inflammatory state may desensitise the response of the immune system and may subsequently diminish the ability to mount an acute cytokine response to an infection.
- Oxidative stress in obesity results in an increased proliferation of B cells at the expense of T cells.
- Increased B cells levels should be protective against previously encountered pathogens, but obesity has been shown to change the memory T cell response causing individuals to be at a risk of repeat infections.

29.5   How to interpret the screening tests when there is a suspicion of infection?

- IgG and IgM negative = absence of infection or extremely recent acute infection.
- IgG positive, IgM negative = indicative of previous infection.
- IgM (+/− IgG) positive = indicative of current infection (though not in all cases).

- IgM and IgG positive = indicative of recent infection or a false-positive test result.
- It is recommended that the test may be repeated 2−3 weeks later, and a fourfold rise in IgG antibody titre indicates recent infection.
- IgG antibodies can cross the placenta.

29.6 Urinary tract infection

- NICE guidance screening and treatment for asymptomatic bacteriuria as the risk of ascending infection and consequent risk of pyelonephritis is significant (up to 40%).
- 17%−20% women report urinary tract infection (UTI) symptoms with symptomatic bacteriuria in pregnancy.
- 2%−9% of pregnant women have bacteriuria in the first trimester and 10%−30% go on to develop ascending infection in the second or third trimester.
- E. coli is thought to be the most common organism, as a bowel commensal.UTIs should be treated to reduce maternal morbidity.
- UTIs treatment also reduces risks, including prematurity, as infection can be a trigger for the initiation of labour at any gestation.
- Appropriate antibiotics course of 7 days is generally recommended depending upon result of urine culture sensitivity and local antibiotic protocol.
- A test of cure repeat urine culture should be carried out 7 days after completion of treatment, especially in women with symptomatic bacteriuria.
- It is advisable to avoid Nitrofurantoin in the third trimester due to the risk of haemolytic anaemia,
- Trimethoprim should be avoided in the first trimester due to its antifolate effect.

29.7 Group B Streptococcus (*Streptococcus agalactiae*)

- Group B streptococcus (GBS) is the commonest cause of severe early onset infection in the newborn.
- Intrapartum antibiotics reduce the incidence of early onset GBS.
- Antibiotics treatment have no impact on late onset (after 7 days) which is not associated with maternal GBS.
- There is no universal agreement whether routine antenatal screening for GBS should be offered or not, and policy differ globally.

29.8 Management of positive group B streptococcus result

- GBS positive bacteriuria should be treated antenatally
- *Intrapartum antibiotics should be given to women in the following groups:*
- GBS UTI is identified in pregnancy;
- GBS identified on low vaginal and/or anorectal swab during current pregnancy;
- had a previous baby affected by GBS.

29.9   Managing women in labour with positive group B streptococcus screening

- Delivery should be expedited for woman with ruptured membranes at term.
- Delivery should be expedited for women with the preterm rupture of membranes beyond 34 weeks.
- For pregnancy <34 weeks, management depends on gestational age, maternal and foetal condition, case by case.

29.10   Genital herpes

- Genital herpes is a sexually transmitted infection caused by either Herpes Simplex Virus 1 or 2 (HSV1 or HSV2),
- It is important to ascertain whether HSV infection is likely to be primary or secondary.
- If it is confirmed to be primary, then the timing of the episode in relation to gestational age is important to ascertain potential risks to the baby.

29.11   Classical features of primary herpes simplex virus infection are

- Multiple, small vulval vesicles, presenting as ulcers
- Significant pain
- Dysuria and urinary retention
- Painful inguinal lymph nodes may be present
- Flu like symptoms
- Among immunocompromised, significantly serious symptoms may be seen (encephalitis, hepatitis and transverse myelitis)
- Women with secondary HSV may have few fewer symptoms
- HSV can increase risk of miscarriage or preterm labour
- No evidence of increasing congenital defects
- Intrauterine (congenital) infection is rare
- *Neonatal infection risk is 1−2 per 100,000 and can present as:*
- Localised Herpes—localised to skin, eyes and mouth
- CNS Herpes (e.g., encephalitis)
- Disseminated Herpes—widespread involvement of multiple organs, with significant mortality up to 70% and survivors at risk of long-term sequelae (mental retardation).

29.12   Treatment is influenced by

- the gestational age as risks are higher for baby with primary infection in third trimester.
- within 6 weeks of delivery as the baby will not have had the chance to gain protection from trans- placental antibodies; and

- and if there is ongoing viral shedding, exposure increases.
- A care pathway for managing women with HSV is shown in Table 29.1.
- In general, antiviral treatment with acyclovir (usually oral but IV for disseminated Herpes) is recommended at onset.
- Antiviral treatment helps to reduce the duration and severity of symptoms.
- Antiviral treatment also decreases the duration of viral shedding.
- IgG antibodies to HSV-1 and HSV-2 testing should be undertaken and compared to HSV isolated from the viral PCR swab as the presence of matching antibodies in both samples suggest recurrence
  - HIV positive women with active HSV lesions, are more likely to transmit HIV infection and should be offered daily oral acyclovir from 32 weeks until delivery.

Table 29.1 Management of women with HSV during pregnancy.

	Management	Intrapartum care	Neonatal risk
Primary Herpes Onset up to 28 weeks	Refer to Genito-Urinary Medicine (GUM) clinic Viral PCR swab to confirm diagnosis (DX) Commence Acyclovir at onset for 5 days Consider Acyclovir from 36 weeks	Aim for vaginal delivery unless otherwise indicated	Low: development of trans-placental antibodies
Primary Herpes Onset beyond 28 weeks	Refer to GUM Viral PCR swab, HSV 1 and 2 IgG assessment, Commence Acyclovir and continue until delivery	Delivery by caesarean section (if infection within 6 weeks of delivery) If wishes vaginal delivery (Acyclovir in labour), Avoid invasive procedures[a]	High risk of transmission: not enough interval for trans-placental antibodies
Secondary Herpes	Liaise with GUM Viral PCR swab, HSV IgG 1 and 2 Consider starting Acyclovir at 36 weeks especially if frequent flare-ups, If HIV positive, start at 32 weeks.	Aim for vaginal delivery Can use invasive procedures[a]	Low risk: localised disease if affected

[a]Invasive procedures during labour include attaching foetal scalp electrode, foetal blood sampling, artificial rupture of membranes, instrumental delivery.

29.13   Chlamydia trachomatis

- Any woman with symptoms of infection and/or has a positive chlamydia result should be treated.
- Azithromycin is recommended as first line treatment.
- There are potential associations with preterm rupture of membranes and premature delivery.
- The baby is at increased risk of neonatal conjunctivitis and neonatal pneumonia, therefore a test of cure is advisable at 6 weeks.
- Patient should be referred for counselling, partner notification, full sexual health screen, and treatment.

29.14   Chicken pox (varicella zoster virus)

- At first antenatal visit, history should enquire whether she had chicken pox or shingles in the past.
- A majority of women are immune and incidence of primary infection in pregnancy is very low, around 3 in 1000.
- Pregnant women should be advised to avoid exposure to any child with suspected chicken pox, as even a short exposure of up to 15 minutes could be quite significant.
- The incubation period is 14−21 days.
- The period of infectivity begins in the prodromal phase, 1−2 days prior to rash appearing and continues until all lesions have crusted over.
- VZV immunity should be checked if the woman had no previous history of infection [IgG suggests immunity and IgM suggests acute infection requiring Varicella-zoster immuno-globulin (VZIG)] which can be given up to 10 days after contact.
- These women should be treated as potentially infectious from 8−28 days after exposure if they receive VZIG and 8−21 if they do not.
- Occasionally, a second dose may be needed if subsequent exposure occurs and more than 3 weeks has elapsed since the last dose.

29.15   Exposure to chicken pox in adulthood

- Young adults and pregnant women are more susceptible, and can result in potentially life-threatening hepatitis, pneumonitis, and encephalitis.
- Treatment for positive cases should be initiated with Oral Acyclovir within 24 hours of rash appearing to reduce risk of pneumonitis, and intravenous route should be considered for those who are unwell.
- Women should be treated in isolation while in the hospital
- The risk of foetal varicella syndrome is greatest in the first 16 weeks and occurs in about 2% of infected pregnancies.
- Herpes zoster reactivation in utero may be the most likely cause.

29.16   The features of foetal varicella syndrome

- Skin scarring in dermatomal distribution
- Eye lesions (microphthalmia, chorioretinitis, or cataracts)
- Hypoplasia of the ipsilateral limb as well as
- Neurological abnormalities
- Ultrasound features include limb deformity, microcephaly, hydrocephalus, soft tissue calcification and foetal growth restriction.

29.17   Labour management and neonatal varicella

- There is around 20% risk of neonatal varicella if infection is acquired in the last 4 weeks of pregnancy.
- Ideally delivery should be delayed for >7 days after the onset of rash to ensure some passive transfer of antibodies.
- VZIG should be considered for newborns at risk of VZV infection.
- The neonatal team should decide what examinations and follow-ups are required.

29.18   Rubella

- Rubella is an RNA virus, spread by respiratory droplets, have an incubation of 14−21 days.
- The person is infectious in the prodromal phase for 7 days before and after the appearance of the rash, and it is recommended that such persons should be screened for both rubella and parvovirus.
- Paired samples should be used, taken during the acute and convalescent phase to differentiate between acute infection and past immunity.
- Infection occurring before 13 weeks can lead to microcephaly and mental retardation, eye lesions (microphthalmia and congenital cataracts), cardiac, and sensorineural deafness.
- Around a third of foetuses exposed to infection between 13−16 weeks are affected by deafness.
- Infection after 16 weeks may rarely lead to foetal infection.

29.19   Toxoplasmosis

- Toxoplasmosis affects around 1 in 500 pregnancies.
- It is caused by *Toxoplasma gondii* (protozoan parasite).
- Usually spreads by eating food contaminated with cat faeces or undercooked meat.
- The incubation period is < 2 days.
- Approximately 20% of women are immune.
- Most affected individuals are asymptomatic.
- A few experience vague symptoms (fever, myalgia, lymphadenopathy, and, rarely, visual disturbance).
- The immunocompromised can develop encephalitis, myocarditis, pneumonitis, and hepatitis.

- The diagnosis requires serological testing with paired samples 28 days apart.
- It can cause congenital toxoplasmosis (severe neurological or ocular disease as well as cerebral and cardiac abnormalities).
- There is no routine screening for toxoplasmosis in the United Kingdom.
- Screening should be considered in high risk group or if ultrasound shows suspicious features (hydrocephalus, intracranial calcifications, microcephaly/ventriculomegaly, IUGR, ascites, hepatosplenomegaly).
- Most affected foetuses will have a normal scan and may be unaffected at birth.
- Infection during first trimester likely to result in more severe infection than in later gestations.
- Infection later in pregnancy is less severe but there is an increased risk of vertical transmission.
- If foetal infection is either suspected or confirmed by positive PCR of amniotic fluid, then treatment is given to the mother to try to prevent mother-to-child transfer.
- Pregnant women should be advised about washing hands before handling food, eating thoroughly cooked meat, washing fruit and vegetables, and avoiding contact with cat faeces.

29.20 Parvovirus B19

- Parvovirus is spread by respiratory droplets.
- An incubation period of 4–20 days.
- The mother is often relatively asymptomatic but may have a malar rash, "slapped cheek."
- Both rubella and parvovirus B19 should be suspected when woman presents with a malar rash, fever and arthralgia.
- The parvovirus is infectious prior to the appearance of rash and not so once the rash has appeared.
- Immunocompromised women are at increased risk of haemolysis and aplastic anaemia, with swollen legs, hypertension, proteinuria, and anaemia.
- Intrauterine foetal infection can occur in around 25%–30% of cases, leading to suppression of erythropoiesis, thrombocytopenia, resultant anaemia, and cardiotoxicity. These changes can lead to cardiac failure and hydrops.
- The diagnosis is confirmed by checking paired serological samples >10 days apart.
- Amniocentesis may be indicated to confirm infection.
- Ultrasound may show evidence of pleural and pericardial effusions, ascites, skin oedema, and cardiomegaly with infection.
- Serial ultrasound scanning at 1–2 weekly interval, including measuring middle cerebral artery peak systolic velocity (to assess need for intrauterine foetal blood transfusion).
- Timing and mode of delivery is determined by gestational age and degree of foetal compromise.

29.21 Cytomegalovirus infection

- Incidence of primary infection in pregnancy is around 1 in 100.
- women are generally asymptomatic.
- May also present with nonspecific symptoms such as fever, malaise, and lymphadenopathy.

- Primary infection is usually diagnosed with two tiered maternal blood sampling 4−6 weeks apart: there is seroconversion of negative to positive IgM or a fourfold increase in IgG antibody titre.
- Vertical transmission risk of 30%−40% in primary infection.
- Around 10%−20% will be symptomatic and have evidence of infection at birth
- Ultrasound features: intracranial calcifications, echogenic bowel, hepatosplenomegaly, IUGR, microcephaly, and evidence of hydrops may be suggestive of cytomegalovirus (CMV) infection.
- Most of these ultrasound features are nonspecific to CMV and only around 15% CMV-infected foetus will have ultrasound changes.
- Amniocentesis may be advised to confirm infection after 20 weeks as false negative results are possible (the virus is slow growing and foetal kidney may not excrete the virus in high quantities).
- A third of those who become infected die and two thirds will have long-term impairments.
- Newborn exposed to infection may have growth restriction, microcephaly, hepatospleno-megaly, chorioretinitis, psychomotor retardation, and sensorineural hearing loss.
- Infection should be confirmed by isolating CMV in foetal urine and/or saliva postnatally.

29.22 Malaria

- Malaria results from the bite of a female anopheles' mosquito.
- The majority of infections are caused by *Plasmodium falciparum* (75%), mainly in Africa; it sequestrates in the placenta and this facilitates evasion of splenic processing and filtration.
- Other species are *Plasmodium vivax*, *Plasmodium malariae*, and *Plasmodium ovale*, mainly in Asia, with unique ability of vivax and ovale sporozoites to persist as dormant in the liver, causing relapses months or years later.
- Malarial infection can also result from blood transfusion or vertical transmission.
- Decreased immunity and the physiological change of increased skin blood flow during pregnancy also increases infection risk.
- Prevention includes adopting antimosquito measures: using skin repellents such as 50% DEET, knockdown room sprays containing permethrin and pyrethroid, insecticide-treated bed nets (pyrethroid impregnated) and clothing that covers as much of the body as possible.
- Chemoprophylaxis in pregnancy does not completely remove the risk of malaria.
- Chemoprophylaxis:
- it can be either causal (directed against liver schizont stage and medication would need to continue for 7 days after leaving a malaria endemic area);
- or suppressive (directed against erythrocyte stage and needs to continue for 4 weeks after leaving a malaria area). This should be discussed with the experts.

29.23 Presentation

- Symptoms include fevers/sweats, headache, myalgia, cough, general malaise, and GI upset including abdominal pain, nausea, vomiting, and diarrhoea.
- Signs include pyrexia (fever pattern varies for each species), jaundice, sweating, pallor, splenomegaly, tender hepatomegaly and respiratory distress.

- Markers of severe disease include end organ damage: impaired consciousness, hypoglycaemia, respiratory distress, convulsions, pulmonary oedema, circulatory collapse, abnormal bleeding and DIC, jaundice, haemoglobinuria, and acute renal failure.
- The gold standard for diagnosis is a blood film:
- A thick film allows diagnosis and thin film identifies the species.
- Serial samples may be needed to confirm a diagnosis.

Management: Discuss with infectious disease experts and needs multidisciplinary team input:

- Supportive measures (hydration treating anaemia, hypoglycaemia, sepsis, hypotension)
- Antimalaria treatment (according to the species)
- Treatment of complications
- Admission for monitoring (both mother and baby)
- ITU admission in severe malaria with end organ damage.

29.24   Risks during pregnancy

- Maternal and foetal mortality
- Miscarriage
- Stillbirth
- Preterm labour
- Foetal growth restriction
- Low birth weight
- Maternal and foetal anaemia
- Susceptibility of infant to malaria

All babies of affected women should be screened with thick and thin blood film at birth and weekly until 28 days.

There is some evidence that Intrauterine exposure to infection can result in reduced response to vaccination in infancy.

Phase 1 clinical trials with VAR2CSA (falciparum-derived protein) vaccine to prevent malaria in pregnancy and its adverse consequences for offspring have started.

29.25   Summary

1. The timing of infection in relation to gestation has significant impact on the timing of mode of delivery and thus the potential impact and consequences of the infection on the foetus.
2. A multidisciplinary input is important for developing a patient focused individualised care plan.

Further reading

Antenatal care for uncomplicated pregnancies. NICE clinical guideline 62. Guidance issued March (2008); last modified December 2014.

BASHH and Royal College of Obstetricians and Gynaecologists (RCOG). Management of Genital Herpes in pregnancy. London; 2014.

Carlson A, Norwitz ER, Stiller RJ. Cytomegalovirus infection in pregnancy: should all women be screened? *Rev Obstet Gynecol.* 2010;3(4):172−179.

NICE. Clinical knowledge summaries—parvovirus b19 infection scenario: suspected parvovirus B19 or possible exposure—pregnant women <http://cks.nice.org.uk/parvovirus-b19-infection#!scenario:2> Accessed 15.09.15.

Paquet C, Yudin MH. Toxoplasmosis in pregnancy: prevention, screening, and treatment. *SOGC Clinical Practice Guideline.* 2013. Available from: http://sogc.org/wp-content/uploads/2013/02/gui285CPG1301E-Toxoplasmosis.pdf.

Puder KS, Treadwell MC, Gonik B. Ultrasound characteristics of in utero infection. *Infect Dis Obstet Gynecol.* 1997;5:262−270.

Royal College of Obstetricians and Gynaecologists (RCOG). The diagnosis and treatment of malaria in pregnancy, Green-top guideline 54a and 54b. London: Royal College of Obstetricians and Gynaecologists (RCOG).

Royal. College of Obstetricians and Gynaecologists (RCOG). Chicken pox in pregnancy-Green-top guideline number 13; 2015, London: Royal College of Obstetricians and Gynaecologists (RCOG).

Stegman BJ, Carey JC. TORCH Infections. toxoplasmosis, other (syphilis, varicella-zoster, parvovirus B19), rubella, cytomegalovirus (CMV), and herpes infections. *Curr Womens Health Rep.* 2002;2(4):253−283.

Weight management during pregnancy

Dominique Baker and Tahir Mahmood
Department of Obstetrics & Gynaecology, Victoria Hospital, Kirkcaldy, United Kingdom

30.1 Introduction

- The antenatal period can be a difficult time for pregnant obese patients to manage their weight.
- There is often the misconception that a pregnant person needs to "eat for two" and a patient may need educating that a healthy, balanced diet should be maintained.
- The weight gained while pregnant can be attributed to the foetus, placenta, amniotic fluid, and maternal adipose deposits, along with plasma expansion.
- Healthcare professional should raise the issue of raised body mass index (BMI) with the patients as an important health issue.
- However, many healthcare professionals may lack the training to discuss issues regarding obesity in pregnancy and fear that such discussion could adversely affect their relationship with their patients.
- All patients should have their BMI assessed at the first antenatal booking appointment as this will guide clinicians about their weight management for the remainder of their pregnancy.
- It is important to note that prepregnancy weight has more influence on the health of the mother and baby, rather than actual weight gain during pregnancy itself.
- This is why preconception and postnatal advice regarding maintaining a healthy lifestyle is of vital importance.

30.2 Weight gain guidelines

- The American College of Obstetricians and Gynaecologists (ACOG) and the Society of Obstetricians and Gynaecologists of Canada (SOGC) advise that patients should be informed that the recommended weight gain during pregnancy for obese pregnant people should be between 5 and 9.1 kg for singletons and 11.3−19.1 kg for twin pregnancies.
- These recommendations were produced by the Institute of Medicine (IOM) in the United States, and are derived from observational studies that these weight gains reduce the risk of foetal and maternal adverse outcomes.
- It is important to note that these recommendations are based on patients living in the United States, and may not be applicable throughout the world.

Handbook of Obesity in Obstetrics and Gynecology. DOI: https://doi.org/10.1016/B978-0-323-89904-8.00004-4

- In addition, these weight gain targets are not suitable for adolescent pregnancies, as the weight gain requirements for these patients are likely to be higher than the recommendations made by the IOM.
- A more recent guideline from the Royal College of Obstetricians and Gynaecologists (RCOG) advises that there is disagreement on what constitutes a safe and healthy weight gain in pregnancy.
- Therefore, the RCOG advise that rather than a weight gain goal, women should concentrate on a achieving a healthy diet until more evidence is exists.

30.3   Risks of inappropriate weight gain

- Maternal risks of excessive weight gain include an increased chance of caesarean section, higher postpartum weight retention, preterm birth, preeclampsia, and gestational diabetes.
- Obesity has the additional long-term associations of cardiovascular disease, metabolic syndrome, type II diabetes and early mortality.
- Therefore, gestational weight gain has the potential to have wide-ranging public health implications and should be a priority for policy makers.
- In addition, an Australian study has demonstrated that patients with weight gains above the IOM recommendations have a hospital stay that is 20% longer than those patients with an appropriate weight gain. This has important financial associations for healthcare and the wider society.

The risks for the child when there is excessive gestational weight gain include:

- being born large for gestational age;
- childhood obesity;
- asthma; and
- diabetes mellitus Type 1.

30.3.1   Information gap

- When questioned patients appear to have a broad understanding of the associated maternal risks but less knowledge of the risks for their unborn child.
- This highlights a potential area for patient education, as well as a motivational tool for patients to maintain a healthy lifestyle while pregnant.

30.3.2   Inadequate weight gain is another risk factor

- Obese patients are also at risk of inadequate weight gain while pregnant due to poor nutrition and fears over weight gain.

This in itself has associated *risks for both mothers and babies including*:

- small for gestational age babies;
- preterm birth;
- neural tube defects; and
- reduction in breast feeding rates.

However, when obese patients gain less than the specified weight gain, they have a reduced requirement of epidural anaesthesia during labour.

30.4 Should women be weighed during antenatal period?

30.4.1 Practice in the United Kingdom

- The National Institute of Clinical Excellence (NICE) advises that BMI should be recorded at booking but patients shouldn't be reweighed unless it will change their management.
- However, if there are concerns about nutrition, for example in hyperemesis gravidarum or in eating disorders, it may be appropriate to reweigh a pregnant patient.

Practice in the United States:

- Guidelines from the United States advise that obese patients should weigh themselves throughout pregnancy and healthcare professionals should discuss weight gain at each appointment.
- The rationale behind this is that reweighing patients at regular intervals throughout their pregnancy may reduce the chance of excessive gestational weight gain.
- However, it is important to note that patients may find this emotionally distressing and care should be taken to ensure this is discussed in a sensitive manner.

30.5 Exercise

- Pregnant women should be reassured that it is safe to take moderate level physical activity during pregnancy.
- Patients should be advised to incorporate physical activity into their daily routines as it can improve their likelihood of maintaining a healthy weight.
- Ideally a pregnant woman should take 30 minutes of moderate level physical activity each day.
- A pregnant woman can continue their previous physical activity level while pregnant as long as their pregnancy is low risk.
- Supervised exercise programmes appear to reduce excessive weight gain by 0.9 kg for obese patients and seem to be more successful if initiated before 12 weeks gestation.
- However, the exact nature of how much and how often physical activity is needed to reproduce these results requires further research.

30.6 Diet

- Pregnant women should be signposted towards dietician advice early on in their pregnancy regarding healthy eating routines.
- Woman should be advised how to maintain a healthy weight during and after pregnancy.
- Woman should be advised to limit portion sizes and centre meals around wholegrain starches with plenty of fibre, fruits and vegetables.

- It is recommended that there is a requisite for an additional 200 kcal per day in the third trimester of pregnancy only.
- It is important to inform woman that dieting with the aim of weight loss while pregnant isn't advised, as it may lead to poor foetal outcomes.
- Reviews of records from times of famine show that a calorie intake less than 1500 a day in the third trimester is associated with small for gestational age babies.
- This has important public health implications as small for gestational weight infants have higher infant mortality and morbidity rates, as well as increased rates of chronic disease in adulthood, such as cardiovascular disease.
- Furthermore, restricted calorie intake in the first trimester has been linked to the development of neural tube defects in the foetus due to reduced folate intake during this period.

30.7 Achieving healthy lifestyles

- Pregnant women find the weight gain targets difficult to keep to, as reported in the United States where 48% of patients gained over the recommended amount of weight.
- This may be due to the fact that patients are unaware of these guidelines, as 69.4% of New Zealand patients overestimated how much weight they should gain while pregnant.
- Interestingly, it was the obese patients who were consistently more likely to overestimate how much weight they should gain while pregnant.
- Healthcare professionals should be aware of the fact that certain groups are more at risk of having excessive gestational weight gain.
- The dietary and weight gain advise should be more focused on the obese population and those who have had previous pregnancies with excessive weight gain.
- Educational information should be tailored to the woman's level of understanding and education, as it has been shown that patients with lower levels of education are at higher risk of excess gestational weight gain.
- An additional risk factor for excessive weight gain is patients who underestimate their prepregnancy weight.
- A patient's perception of their weight should be assessed, as it could allow more focused and intensive interventions to be targeted towards those at higher risk of excessive weight gain.
- Women from lower socioeconomic groups may be more prone to excessive weight gain and in the United Kingdom, they are supported by a special scheme which aims to provide fresh or frozen vegetables and fruits for those eligible patients.
- Obese patients appear to have less excessive gestational weight gain when dietary information is presented to them in regular, face-to-face sessions concentrating on a patient specific diet that regulates and monitors food consumption.
- The mean reduction in weight gain ranges from 6.5 to 7.8 kg for obese patients who receive dietetic assistance.
- Stand-alone sessions or leaflets appear less effective at achieving a healthy pregnancy weight gain.

Supervised programs have been found to suffer from poor attendance and low numbers of patient retention for the following reasons:

- work commitments;
- access;

- expensive parking;
- pregnancy complications;
- too tired;
- bad timing; or
- diet not culturally relevant.

The physical discomfort experienced during pregnancy prevents some patients from taking physical exercise, along with pregnancy nausea and vomiting.

30.8 Postpartum

- The postnatal check should be used as an opportunity to provide personalised guidance on weight loss and exercise with signposting to services equipped to help with this goal.
- If the woman has had an uncomplicated vaginal delivery they can resume a low-impact physical exercise regime straight after birth.
- However, for any caesarean deliveries or complicated vaginal births it is advisable to wait 6−8 weeks before recommencing prepregnancy exercise routines.
- Excessive weight gain above the ACOG recommended levels is associated with long-term obesity as women find it difficult to lose the weight they have gained.
- The weight that a woman carries forward into any subsequent pregnancy can affect her risk of caesarean section, large for gestational age infants and preeclampsia.
- Healthcare professionals should provide advice and assistance to mothers wishing to breast feed, as well as education around its benefits.
- Patients can be reassured that healthy diet and exercise will not affect their ability to breast feed.
- The evidence regarding whether breast feeding assists in weight loss postpartum is currently conflicting and further research is required in this area.

30.9 Conclusion

- Current national guidelines regarding the specifics of how much weight the obese pregnant person should gain are conflicting.
- However, there is agreement that a healthy lifestyle before, during and after pregnancy has long-term impacts on maternal, foetal and national health.
- Healthcare professional and policy makers should support the development of schemes which aim to provide focused education and targeted exercise schemes for these women.

Recommendations for further reading

Weight Management Before, During and After Pregnancy. National Institute for Health and Care Excellence. July 2010.

Weight Gain During Pregnancy. Committee Opinion, the American College of Obstetricians and Gynecologists. Number 548; January 2013 (Reaffirmed 2020).

Care of Women with Obesity in Pregnancy. Royal College of Obstetricians & Gynaecologists, Green-top Guideline No. 72. November 2018.

Stein Z, Susser M. The Dutch Famine, 1944—1945, and the reproductive process. I. Effects on six indices at birth. *Pediat Res.* 1975;9:70—76.

Williamson CS. 31 *Nutrition in Pregnancy*. London, UK: British Nutrition Foundation; 2006:28—59.

Centers for Disease Control and Prevention. https://www.cdc.gov/reproductivehealth/maternalinfanthealth/pregnancy-weight-gain.htm. Accessed January 2021.

Guidance for Healthy Weight Gain in Pregnancy. 2014. New Zealand Ministry of Health.

Hector D, Hebden L. *Prevention of Excessive Gestational Weight Gain: An Evidence Review to Inform Policy and Practice*. Sydney: Physical Activity Nutrition Obesity Research Group; 2013.

Management of pregnancy in women with history of weight loss surgery

Omar Thanoon[1], Asma Gharaibeh[2] and Tahir Mahmood[1]
[1]Department of Obstetrics & Gynaecology, Victoria Hospital, Kirkcaldy, United Kingdom,
[2]Royal Infirmary of Edinburgh, Edinburgh, Scotland

31.1   Introduction

- The proportion of the female population either overweight or obese has increased significantly in recent years and pregnancy in this group of women is associated with significant health problems.
- Pregnancy in obese women is associated with increased risk of metabolic syndrome, including impaired glucose tolerance and raised blood pressure.
- Of late, more and more young women are seeking bariatric surgery (BS) where nonsurgical methods have failed to achieve or maintain significant weight loss.
- A Cochrane review has concluded that that BS is more effective and cost-effective than nonsurgical measures in patient's follow-up for 2 years with measurable long-term health benefits.
- Women with a history of BS are a challenge for an obstetrician in antenatal clinic due to their unfamiliarity with the issues surrounding the surgery.
- This unfamiliarity creates the risk that women may be suboptimally managed and that serious complications may go unrecognized resulting in morbidity.

31.2   Types of bariatric surgery

- Restrictive procedures:
  - o these surgical procedures reduce oral intake by minimizing the available space in their stomach: laparoscopic adjustable gastric banding (LAGB) and laparoscopic sleeve gastrectomy.
  - o In LAGB, an inflatable band is positioned around the fundus of a patient's stomach. A variable amount of sterile water is injected into the band via tubing and it reduces stomach volume to a small pouch above the band.
  - o In laparoscopic sleeve gastrectomy, a narrow gastric sleeve is created by removing the greater curvature of the stomach.
- Malabsorptive procedures:
  - o These procedures are rarely performed especially in young women with reproductive potential.

Handbook of Obesity in Obstetrics and Gynecology. DOI: https://doi.org/10.1016/B978-0-323-89904-8.00005-6

- Combination of restriction and malabsorption:
  - o Roux-en-Y gastric bypass is an example of a combined procedure. This operation creates not only anatomical restriction but also has an element of malabsorption.
  - o In this operation, the stomach is restricted into a smaller gastric pouch with its contents directed via a food channel into the jejunum, bypassing the main body of the stomach and the duodenum, resulting in malabsorption.
  - o This type of surgery does result in better weight loss compared to purely restrictive procedures.
- BS is associated with deficiency of essential nutrients (folate, vitamin D, vitamin B12, calcium, iron, and copper).
  - o Folic acid absorption is reduced because of reduced production of gastric acid secretion and reduced surface area of small intestine, thus increasing risk of macrocytic anaemia in mother and neural tube defect in the baby.
  - o Vitamin B12 deficiency is quite high after BS. This is secondary to reduced secretion of intrinsic factor, reduced acidity, and lack of absorption in duodenum (where bypassed), with possible risk of macrocytic anaemia.
  - o There is increased risk of iron malabsorption due to an increase in gastric pH and restriction of iron absorption in duodenum and ileum. This would increase risk of iron deficiency anaemia and neonatal anaemia. If unable to tolerate oral iron supplements, then intravenous administration should be considered.
  - o Vitamin A is fat soluble and there is a strong association of retinol isoforms with foetal malformations. The deficiency also increase risk of nocturnal blindness.
  - o Vitamin B1 deficiency can be exacerbated by nausea and vomiting of pregnancy. There is a case report of Wernicke's encephalopathy in a pregnancy following BS.
  - o Calcium and vitamin D are mainly absorbed in small intestine and will require close monitoring and supplementation. Women who have undergone a procedure involving partial gastrectomy or gastric bypass require large doses of vitamin D supplementation.
  - o Dietary recommendations for post BS are shown in Table 31.1.

31.2.1   Complications of bariatric surgery

- Serious complication rate in BS is around 5%.
- Most common complication is reflux.

Table 31.1 Recommended dose of supplements for pregnant women after bariatric surgery.

Folic acid	5 mg daily preconception until the end of the first trimester
Thiamine	12−50 mg daily, extra 200−300 mg if vomiting
Vitamin B12	1000 µg daily or 1000 µg IM injection 4−12 weekly
Iron	45−60 mg daily, consider IV if persistently low despite oral replacement
Calcium	1500 mg daily
Vitamin D	Routine replacement not recommended unless testing identifies a deficiency
Vitamin A	1000−5000 IU daily in the form of beta-carotene
Vitamin E	15 mg daily
Vitamin K	90−300 µg daily

31.2.2 Complications specific to laparoscopic adjustable gastric banding

- Band slippage causing vomiting and epigastric pain
- Band erosion or migration
- Late port infection
- Obstruction due to adhesions
- Dumping syndrome (sleeve gastrectomy)

31.2.3 Complications specific to Roux-en-Y gastric bypass

- Small bowel obstruction
- Cholelithiasis
- Internal hernias with bowel obstruction during pregnancy
- Risk of ischaemia and perforation during pregnancy
- Dumping syndrome (early and late)
  - o *Early dumping syndrome:*
  - o Occurs following rapid transit of food into small intestine, and symptoms include abdominal pain, nausea, diarrhoea, flushing, palpitations, tachycardia, and hypertension.
  - o *Late dumping syndrome:*
  - o Occurs between 1 and 3 hours after eating carbohydrate-rich food, causing a rapid hyperinsulinaemia and a reciprocal hyperglycaemia. The symptoms include tremors, poor concentration, palpitations, and rarely even loss of consciousness.

31.2.4 Reproductive health after bariatric surgery

- BS results in improvement in comorbidities such as hypertension and diabetes, and this has benefits for pregnancy as it also reduces the incidence of pregnancy-related hypertension, preeclampsia, and gestational diabetes.
- Weight loss will improve fertility in women who have previously been unable to conceive due to obesity-related conditions such as polycystic ovaries.
- This improvement in fecundity is multifactorial. Loss of a significant amount leads to reduced insulin resistance, lower levels of insulin and leptin, and improvement in the release of pulsatile gonadotrophin secretion.

31.2.5 Preconception counselling

- Women are advised to delay pregnancy until at least a year after surgery, with 2 years being recommended if they wish to achieve optimal weight loss and stabilise their nutritional status before pregnancy.
- Good contraceptive advice should be offered to all women following BS.
- Women should be advised that the effectiveness of oral contraceptives (OC) could be reduced by BS and that OC should be avoided in favour of nonoral methods of contraception.
- Long acting reversible methods (intrauterine device, depot, or progesterone implant) are safe and not associated with risk of venous thromboembolism.
- Should be advised to take higher dose of folic acid 5 mg daily.

31.2.6   Effects of pregnancy on surgery

- All patients are recommended to take a number of dietary supplements for replacement.
- Nutritional deficiencies can be even more challenging in pregnancy.
- Hyperemesis gravidarum, gastric reflux, bloating, and abdominal discomfort from enlarging uterus will make maintaining healthy diets and tablet consumption difficult.
- Maintaining a balanced diet is more difficult in pregnancy as nausea and vomiting affect up to 80% of pregnancies.
- Pregnancy increases the chances of an acute intraabdominal complication related to adhesions from the bariatric procedure, as the gravid uterus will displace other intraabdominal organs.

31.2.7   Effects of surgery on pregnancy

- Women who had undergone surgery had significant lower risks of gestational diabetes and large for date infants compared to women with increased BMIs.
- Some studies suggest there is an increased risk of foetal growth restriction and an increase in perinatal mortality in pregnancy after BS.
- Pregnancy in women who have had BS is associated with an increased risk (RR of 1.20) of heart and musculoskeletal birth defects in their infants (probably due to folic acid deficiency).
- BS has no implication on the mode of delivery, and caesarean section is only recommended for the usual obstetric indications.

31.2.8   Antenatal management

- Women who become pregnant post BS require regular blood tests to screen for nutritional deficiencies in every trimester.
- A number of nutritional supplements should be recommended to women throughout the pregnancy (Table 31.1).
- Pregnancy-related nausea and vomiting should be treated with early recourse to antiemetic medication and dietary advice.
- Women who develop severe protracted vomiting need review by their surgical team to exclude the possibility that the vomiting is not due to a complication of their surgery before it is labelled as pregnancy-related hyperemesis.
- Patients with a gastric band have the benefit of being able to reverse the restriction. This can be achieved by deflating the balloon; this allows better control of weight loss and it also reduces nutritional deficiencies.
- Women should be informed that pregnancies in the restriction maintained group had reduced rate of hypertension and small for gestational age babies, whereas those in the deflation group had increased gestational weight gain and larger babies.
- Antenatal care pathway should include the following:
  o Early pregnancy dating scan, followed by detailed foetal anomaly scanning during second trimester.
  o Regular antenatal care for assessment of weight gain, glucose metabolism, and blood pressure measurements.

o Women should be offered weight monitoring in each trimester using medical scales with bariatric capability, especially if BS has been carried out in the last 12 months or for those who have not yet achieved their target weight.

o Serial foetal growth assessment by ultrasound scanning during third trimester with the recognition that increased BMI is significantly associated with decreased clinical and sonographic estimated foetal weight accuracy.

31.2.9 Recommended investigations in pregnancy after bariatric surgery

- Women post BS will be at increased risk of developing gestational diabetes (unless they have achieved a normal BMI postsurgery) and should have screening for gestational diabetes.
- An oral glucose tolerance test is not appropriate because the rate of absorption of a glucose load is unpredictable (dumping syndrome).
- Home monitoring of pre and postprandial blood capillary glucose levels from 24 to 28 weeks is a better alternative to oral glucose tolerance test.
- In view of the possible risk of foetal growth restriction, serial growth scans to monitor foetal well-being should be recommended.
- Women will require measurement of ferritin, B12, folate, calcium, and vitamin D at their first antenatal appointment and then at intervals through the pregnancy.
- Regular appointments with a midwife or obstetrician should be arranged to allow careful measurement of blood pressure and weight gain.

31.2.10 Management of acute attendance during pregnancy

- The possibility of intraabdominal complications must be considered in any woman with a history of previous BS and who presents with gastrointestinal symptoms such as severe vomiting or abdominal pain.
- The usual obstetric causes such a urinary tract infection and uterine causes should be considered but the possibility of internal hernia and bowel obstruction should also be part of a differential diagnosis.
- A full assessment requires clinical examination by a senior obstetrician.
- Blood tests should include inflammatory markers (C-reactive protein), blood count, midstream urine, renal and liver profile, and lactate measurement.
- In pregnancy the desire to avoid radiation exposure to the foetus must not prevent the appropriate investigations such as CT scan with a low exposure technique.
- As MRI scanning is associated with a much lower radiation dose than CT scanning, it should be considered as an alternative in pregnancy.

31.2.11 Postpartum management

- Women who have had BS are more at risk of gastric ulceration and therefore nonsteroidal antiinflammatory drugs should be avoided. Alternatives such as paracetamol and codeine should be considered.
- Assessment of a woman's risk for venous thromboembolism is essential in all postnatal women; BS does not independently increase the risk but the risk from residual elevation

of BMI should always be considered with prophylactic anticoagulation recommended as appropriate.
- It is good practice to reinforce the need for followup with the woman's bariatric surgical team.
- Contraceptive options should be discussed, especially if there have been complications in the pregnancy related to the BS or where pregnancy occurred early prior to weight optimization.

Further reading

ACOG. Bariatric surgery and pregnancy—ACOG practice bulletin no. 105. *Obstet Gyncol.* 2009;113:1405–1413.

Brown M, Penna L. Management of pregnancy in women with history of weight loss surgery. In: Mahmood T, Arulkumaran S, Chervenak FA, eds. *Obesity and Obstetrics.* 2nd ed. London: Elsevier; 2000:53–60. Available from: http://doi.org/10.1016/B978-0-12-5.00006-0.

Health survey for England 2017. <http://www.digital.nhs.uk>.

Mengesha BM, Carter JT, Dehlendorf CE, et al. Perioperative pregnancy interval, contraceptive counselling, and contraceptive use in women undergoing BS. *AJOG.* 2018;1(81): e1–e9.

Milone M, De Placido G, Musella M, et al. Incidence of successful pregnancy after weight loss interventions in infertile women: a systematic review and *meta*-analysis of the literature. *Obes Surg.* 2016;26:443–451.

NICE. Obesity: clinical assessment and management. <http://www.nice.org.uk>; 2016.

Welbourn R, Hollyman M, Kinsman R, et al. Bariatric surgery worldwide: baseline demographic description and one-year outcomes from the fourth IFSO global registry report 2018. *Obes Surg.* 2019;29(3):782–795.

Challenges of third trimester scanning in obese women

32

Smriti Prasad[1] and Asma Khalil[1,2,3]

[1]Fetal Medicine Unit, Department of Obstetrics and Gynaecology, St. George's University Hospitals NHS Foundation Trust, London, United Kingdom, [2]Vascular Biology Research Centre, Molecular and Clinical Sciences Research Institute, St George's University of London, London, United Kingdom, [3]Twins Trust Centre for Research and Clinical Excellence, St George's University Hospital, London, United Kingdom

32.1 Introduction

- Obesity has been recognised as an important public health problem, and worldwide the prevalence of obese pregnant women continues to rise.
- In the United States, approximately 61%−64.5% of the total population is classified as either overweight or obese; in the United Kingdom, 33% of women are overweight and 23% are obese, giving a total of 56% of women over the recommended BMI.
- Recent reports of the UK Confidential Enquiry into Maternal Deaths have reported obesity to be a factor associated with direct maternal deaths. Over a quarter (29%) of the women who died in this triennium were obese and a further 26% were overweight.

32.2 Why do obese pregnant women need a third trimester scan?

- Obese pregnant women have increased chances of comorbidities like type 2 diabetes mellitus (DM), chronic hypertension, and chronic kidney disease, compared to pregnant women with normal BMI. All these diseases are known to be associated with placental insufficiency, and hence have implications for impaired foetal growth. Another spectrum of disordered growth is occurrence of macrosomia/large for gestational age foetuses in diabetic mothers. The prevalence of macrosomia is reported to be higher in diabetic mothers who are also obese.
- Obesity itself is associated with development of various antenatal complications, such as gestational diabetes mellitus (GDM), preeclampsia, and venous thromboembolism, which necessitate third trimester scans for foetal growth and well-being. Meta-analysis has shown that the estimated risk of stillbirth may be twice as high in obese pregnant women as that in normal weight pregnant women.
- In the maternity units in the UK, it is not a routine practice to offer a third trimester scan to low-risk pregnant women, rather the need for a third trimester scan is guided by clinical assessment of foetal growth by symphysio fundal height (SFH) assessment. Obesity precludes correct assessment of SFH, which is difficult to perform in such women due to

excessive maternal adipose tissues. The RCOG recommends that women with BMI >35 kg/m^2 should be referred for serial ultrasound assessment of foetal growth as obesity contributes to the limited predictive accuracy of SFH.
- Knowledge of foetal lie and presentation is essential to help pregnant women formulate an appropriate birth plan. Unfortunately, maternal obesity often results in inconclusive clinical assessment of foetal presentation, hence the need for an ultrasound scan.

With advances in technological advances, ultrasound now is considered as a safe modality for foetal assessment, especially after the embryonic stages of development, provided the ALARA principle is adhered to.

32.3   Challenges in third trimester screening

- Several observational studies have reported that maternal obesity is associated with major limitations in the ability to adequately examine the foetal anatomic structures in the second trimester.
- In the third trimester, the primary objective of an ultrasound scan is to assess foetal growth. In this section, we will describe the challenges in third trimester scanning in obese pregnant women and how to overcome these limitations.

32.3.1   Poor visualisation of anatomic structures due to impaired acoustic window

There are three main reasons to evaluate foetal anatomy in the third trimester:

- **firstly**, features of some abnormalities may be apparent only in the third trimester, such as microcephaly, craniosynostosis, nonlethal skeletal dysplasia like achondroplasia, oesophageal or bowel atresias, and renal problems such as hydronephrosis and lower urinary tract obstruction;
- **secondly**, to use this opportunity to pick up abnormalities which might have been missed on previous scans; and
- **thirdly**, to diagnose abnormalities which develop only in the third trimester like foetal ovarian cysts or acquired causes of ventriculomegaly like haemorrhage or infection. A recent study by Ficara et al. in 2019 reported that around 0.5% of foetal abnormalities are first diagnosed in the third trimester.

The first acoustic limiting factor in these cases is the depth of insonation required to produce good visualisation, which is directly proportional to the thickness of maternal pannus.

The second limiting factor is the absorption of ultrasound energy by the maternal tissues such that the final signal to background noise ratio is significantly reduced, leading to poor imaging.

Over the years, a number of imaging pre and postprocessing techniques have been developed to overcome these challenges. Some of the commonly used techniques are tissue harmonic imaging, speckle reduction filters, and compound imaging.

Studies have shown that using these techniques in obese pregnant women resulted in significant improvement of image quality, contrast, and better edge recognition.

The expertise of the ultrasound personnel and their understanding and ability to modify these techniques for a particular case is very likely to have a bearing on the final quality of the image.

Therefore, if it is logistically feasible, ultrasound scans for obese pregnant women may be undertaken by the more experienced personnel in a particular unit. As prevalence of obesity continues to reach epidemic proportions, it may not be financially viable to refer all pregnant women to tertiary foetal medicine centres.

32.3.2   Coassociated factors

Obesity is also associated with increased chances of subfertility and infertility, and hence conceptions by Assisted Reproductive Technology, which results in increased multiple pregnancy rates.

Ultrasound scans for multiple pregnancies are technically difficult independent of maternal BMI.

Obese pregnant women are also more likely to have had a previous caesarean birth, further limiting the acoustic window. These additional factors contribute to technical challenges that sonographers face while scanning obese pregnant women.

32.3.3   Challenges in estimated foetal weight calculation

Foetal macrosomia is reported to be a risk factor for adverse intrapartum and neonatal outcomes. Although the occurrence of intrapartum complications like shoulder dystocia are largely unpredictable, it is important to recognise the risk factors to ensure optimal outcomes. Statements from professional bodies recommend offering induction of labour at term or even discussing elective surgical delivery, when a diagnosis of macrosomia is made; therefore, correct assessment of EFW is desirable as it has a major impact on antenatal and labour management.

Regarding the accuracy of EFW calculation in obese mothers, the evidence is conflicting. Few studies report that increasing maternal BMI is associated with greater errors in EFW calculation, however the newer reports conclude that accuracy of ultrasonographic estimation of foetal weight is independent of maternal obesity. Nonetheless, obstetricians, sonographers, and women should be aware of the potential limitations in estimating foetal weight in obese pregnant women.

32.3.4   Ergonomics

Obesity is reported as the most significant barrier to practicing ergonomically sound scanning technique as maximum transducer pressure needs to be applied to obtain good quality images in obese subjects.

As the proportion of obese pregnant women continues to rise, the incidence of work-related musculoskeletal injury is expected to increase with potential workforce shortages and financial implications.

32.3.5   Medicolegal implications

Medicolegal issues may arise where a foetal anomaly may be missed owing to poor ultrasound visualisation of a foetus in an obese mother.

Many women requiring sonography do not fully understand the limitations of the scanning with regard to foetal anatomy assessment.

How to overcome these challenges?

Challenges	Tips to overcome these challenges
Poor visualisation of foetal anatomy	1. Position the obese woman in an oblique or decubitus position. This avoids the risk of aortocaval compression which can be due to a combination of uterine enlargement and a heavy abdominal pannus. This also allows the pannus to fall away and provide a better window at the lateral flank.
	2. To deal with the problem of adipose tissue absorbing the ultrasound energy:
	a. Use all possible preprocessing and postprocessing techniques and filters to increase the signal-to-background noise ratio but only within the recommended output range of energy levels by the Food and Drug Administration in the United States or by the corresponding European Community healthcare bodies. This can be done with each of the following, or all in combination:
	i. Tissue harmonic imaging, to improve image quality in obese women.
	ii. Speckle reduction filters, to improve image quality and contrast in obese women.
	iii. Compound imaging, to reduce speckle artifacts and improve contrast resolution.
	3. To use the areas with relatively less accumulation of abdominal adipose tissue as a scanning window—the periumbilical area, the iliac fossa and the infra pannicular area.
	4. Ultrasound scan with full bladder as the uterus is displaced towards the umbilicus which is relatively devoid of adiposity.
	5. Wait for the foetus to be in an optimal position on the ultrasound scan, such as with a posterior spine.
Incorrect assessment of EFW and difficulty in recording Dopplers	1. Use above tips to obtain images with sharp edge definition.
	2. Sonographers with expertise in scanning obese pregnant women.
	3. To take multiple biometric measurements if they indicate significant alterations in growth velocities between scans, to reduce margins of error.

(Continued)

(Continued)

Challenges	Tips to overcome these challenges
	4. To employ processing filters like reducing pulse repetition frequency, increasing colour gain, and usage of power Doppler settings within recommended energy levels. Use of previously defined preset settings for obese patients may reduce the time taken to assess foetal Dopplers.
	5. Wait for the foetus to be quiescent.
Medicolegal implications	1. Documented discussion with the patient regarding the technical limitations of ultrasound and to ensure realistic expectations.
	2. Give the parents an information leaflet which explains that obesity, caesarean section scars, twin/multiple pregnancies and fibroids reduce ultrasound foetal visualisation and thus reduces detection rate of congenital anomalies.
Ergonomic barriers	1. Ergonomic training of sonographers.
	2. Access to ergonomically beneficial equipment.
	3. To allow proper scheduling of scan with rest time.

Ultrasound scans in obese pregnant women are limited by technical challenges, most of which can be overcome by image optimisation tricks. Effective communication with women about limitations and proper documentation is the key to avoid disappointment and litigation.

Further reading

Denison FC, Aedla NR, Keag O, et al. Royal College of Obstetricians and Gynaecologists. Care of women with obesity in pregnancy: green-top guideline no. 72. *BJOG*. 2019;126 (3):e62−e106. Available from: https://doi.org/10.1111/1471-0528.15386. Epub 2018 Nov 21. PMID: 30465332.

Paladini D. Sonography in obese and overweight pregnant women: clinical, medicolegal and technical issues. *Ultrasound Obstet Gynecol*. 2009;33(6):720−729. Available from: https://doi.org/10.1002/uog.6393.

Victoria Mumford AK. Ultrasound scanning in early pregnancy and fetal abnormality screening in obese women. In: Mahmood TA, Arulkumaran S, Chervenak FA, eds. *Obesity and Obstetrics*. 2nd ed. Elsevier; 2020:61−68. ISBN 9780128179215. Available from: https://doi.org/10.1016/B978-0-12-817921-5.00007-2.

Screening for gestational diabetes mellitus

Tahir Mahmood

Department of Obstetrics & Gynaecology, Victoria Hospital, Kirkcaldy, United Kingdom

- It has been estimated that diabetes affects an estimated 415 million people globally and is projected to increase to 642 million by 2040.
- There is also an equally high burden of prediabetes, approximately 318 million people and likely to increase to 481 million by 2040.
- These findings are accompanied by high levels of overweight and obese adults. The WHO data show that more than half of the world's adult population was overweight (39%) or obese (13%), including an estimated 42 million pregnant women.
- Hyperglycaemia in pregnancy (HIP) is a broad term that encompasses various forms of glucose dysregulation seen during pregnancy [DM T1, DMT2, gestational diabetes mellitus (GDM)]
- GDM is a condition with abnormal glucose intolerance diagnosed for the first time during pregnancy at routine testing.
- The international Diabetes Federation has estimated that 21 million live births—one in six (16.8%)—occur in women with some form of HIP, of which 2.5% may be due to overt diabetes in pregnancy. The remaining 14.3% (one in seven pregnancies) is due to GDM.
- More than one third of people with diabetes and a majority of people with prediabetes remain undiagnosed and unaware of the condition.
- The diagnosis is most often made in the second and third trimester by an oral glucose tolerance test.
- GDM affects up to almost 20% of the pregnancies in Europe.
- Besides glucosuria, women with GDM most often do not have any signs of the condition, unless there is excessive foetal growth that influences the well-being of the women.
- Women diagnosed with GDM are at increased risk of excessive foetal growth and birth complications.
- Women developing GDM may be more insulin resistant even before pregnancy or may not be capable of increasing Insulin secretion sufficiently to maintain a normal glucose tolerance throughout pregnancy.

33.1 Prepregnancy risk factors include

- Obesity/overweight before conception
- A family history of GDM
- History of impaired glucose tolerance
- Ethnicity with high diabetes prevalence

Handbook of Obesity in Obstetrics and Gynecology. DOI: https://doi.org/10.1016/B978-0-323-89904-8.00028-7

- A previous birth of a large for gestational age infant
- A family history of diabetes (both first and second line relatives)
- Prolonged glucocorticoid exposure
- PCOS
- Subfertility or conception using assisted reproductive technologies
- Recent research suggests that the microbiota could play a role as well
  During pregnancy there may be cumulative risk factors as placental hormones also play a role in increasing insulin resistance:
- Multiple pregnancies
- Excessive weight gain
- Excessive foetal growth
- Polyhydramnios

33.2 Effect of gestational diabetes mellitus on both mother and baby

33.2.1 Mother

- Hypertensive disorders (preeclampsia, gestational hypertension)
- Excessive foetal growth (macrosomia)
- Hydramnios
- Preterm labour
- Venous thromboembolism
- Risk of operative birth (Caesarean section, assisted instrumental birth)
- Birth Trauma (shoulder dystocia)
- Postoperative/postpartum infection
- Postoperative/postpartum haemorrhage

33.2.2 Postnatal

- Failure to initiate and/or maintain breast feeding

33.2.3 Foetal/neonatal

- Respiratory distress disease
- Cardiomyopathy
- Neonatal hypoglycaemia
- Neonatal polycythemia
- Neonatal hyperbilirubinaemia
- Neonatal hypocalcaemia
- Stillbirth
- Neonatal death
- Nonchromosomal congenital malformations

33.3   Long-term impact

33.3.1   Mother

- Weight retention
- GDM in subsequent pregnancy
- Future overt diabetes
- Future cardiovascular disease X twofold increased risk

33.3.2   Newborn

- Programming and imprinting foetal origins of disease
- Increased risk of early onset obesity, diabetes, hypertension

33.3.3   Screening for gestational diabetes mellitus

- The first set of criteria for diagnosing GDM was published in 1979 by the National Diabetes Data Group (NDDG) and the criteria were set to detect women at markedly increased risk of later type 2 DM.
- The most recent set of criteria for diagnosing GDM was published by IADPSG (Tables 33.1 and 33.2) and was based on pregnancy and neonatal outcome.
- The IADPSG criteria have now been accepted by several authorities across the world, including WHO and ADA, and implemented in more or less modified versions all over the world.
- GDM screening is usually done after 24 weeks of gestation.
- Some countries screen all pregnant women for GDM (universal screening).

Table 33.1 Selection of available criteria for GDM diagnosis.

Criteria	No of abnormal values	Oral glucose load	Glucose cut-offs Mmol/L (mg/dL)
2013 WHO/ 2010 IADPSG	>/ − 1	75 g	Fasting >/ − 5.1 (92) 1 h >/ − 10.0/(180) 2 h >/ − 8.8/(153)
1999 WHO	>/ − 1	75 g	Fasting >/ − 7.0 (126) 2 h >/ − 7.8/(140)
Carpenter and Coustan	>/ − 2	100 g	Fasting >/ − 5.3 (95) 1 h >/ − 10.0/(180) 2 h >/ − 8.6/(155) 3 h >/ − 7.8 (140)
NICE	>/ − 1	75 g	Fasting >/ − 5.6 (100) 2 h >/ − 7.8/(140)
EASD	>/ − 1	75 g	Fasting >/ − 6.0 (108) 2 h >/ − 9.0/(162)

Screening may be universal (with or without a glucose challenge test) or risk factor based.
EASD, European Association for the Study of Diabetes; *IADPSG*, International Association of Diabetes in Pregnancy Study Group; *NICE*, National Institute for Health and Care Excellence; *WHO*, World Health Organization.

Table 33.2 Time to screen for GDM and diagnostic criteria.

Authority	Population to screen	Timing	Test	No. of abnormal values	Fasting glucose mg/dL	1 h post loading (PL)	2 h (PL)	3 h (PL)
ACOG 2013	Selective	First visit	2 step 3 h 100 g	>/– 2	95	180	155	140
ACOG 2013 NDDG	Selective	First visit	2 step 3 h 100 g	>/– 2	105	190	165	145
ADA 2017	Universal	24–28 weeks	1 step 2 h 75 g	>/– 2	95	180	155	NA
ADA 2017	Universal	First visit	2 step 3 h 100 g	>/– 2	95	180	155	140
FIGO 2013	Universal	24–28 weeks	1 step 2 h 75 g	>/– 1	92	180	153	NA
IADPSG 2010	Universal	24–28 weeks	1 step 2 h 75 g	>/– 1	92	180	153	NA
WHO 2013	Universal	24–28 weeks	1 step 2 h 75 g	>/– 1	92	180	153	NA

ACOG, American College of Obstetricians and Gynecologists; *ADA*, American Diabetes Association; *FIGO*, International Federation of Obstetrics and Gynaecology; *IADPSG*, International Association of Diabetes in Pregnancy Study Group; *NDDG*, National Diabetes Data Group; *WHO*, World Health Organization.

- In most of low—middle-income countries, screening is based on risk factors (selective screening).
- Most countries now follow one-step screening strategies but some are using a two-step approach.
- Different countries follow different screening methodology and criteria as shown in Tables 33.1 and 33.2.
- FIGO has now recommended one-step universal screening for all pregnant women globally may not be capable of increasing Insulin secretion sufficiently to maintain a normal glucose tolerance globally.

33.4 Screening methods

33.4.1 Oral glucose tolerance test

- This test is performed in the morning after an overnight fast, with an oral glucose load of either 75 or 100 g.
- Glucose levels are measured either by capillary or venous blood samples at different time intervals according to the methodology being followed (fasting/1-hour/2-hour/3-hour).
- The cut-off differs between tests, and so does the number of values that should be exceeded to fulfil the criteria for having GDM (Tables 33.1 and 33.2)
- If one or more values exceeds the cut-off of 5.1, 10.00, and 8.5 mmol/L at fasting, 1-hour and 2-hours, the woman is deemed to have hyperglycaemia of pregnancy.

33.4.2 Glucose challenge test

- A glucose challenge test (GCT) is performed for screening a large population to select the ones with higher glucose levels that should be examined further with an Oral glucose tolerance test (OGTT).
- This test can be performed at any time of the day with no restriction to meal time.
- At a GCT, the glucose load is 50 g and blood glucose levels are measured after 1 hour.
- In case of a positive result, the woman is referred to an OGTT.

33.4.3 Glycated haemoglobin

- Glycated haemoglobin (HbA1c) reflects the mean blood glucose level for the last 4 weeks and is used for evaluation of treatment effect during pregnancy.
- HbA1c is a screening tool for type 2 diabetes outside of pregnancy.
- A very high HbA1c in the first trimester may indicate pregestational diabetes.
- If HbA1c is 48 mmol/L or higher, or plasma glucose is 7 mmol/L or higher, then the woman is classified as having preexisting diabetes.
- HbA1C is not a reliable tool for diagnosing GDM.

33.5   Should all pregnant women be tested for hyperglycaemia in pregnancy?

- Globally, 130 million births occur annually, of which 85% occur in low- and low—middle-income countries with limited resources.
- There are no published data on the level of adherence to recommended policies for HIP screening at individual country level.
- Multiple studies have shown that the sensitivity to detect GDM using risk-based factors is poor.
- Concerns have been expressed that universal testing of all pregnant women (and increased diagnosis) would consequently place additional logistic and economic challenges on healthcare systems.
- Health economic studies favour universal testing, particularly when accompanied by long-term postpartum preventive care for women with GDM and their offspring.
- The FIGO has issued a pragmatic guideline on HIP which is accommodative to the operational, logistical, and resource restraints in different parts of the world and advocates a single-step approach.

33.6   Should all women be tested by a 75 g oral glucose tolerance test?

- Some professional bodies recommend a two-step approach: a 50 g GCT, followed by a 75 g OGTT in women who have tested at initial screening.
- This approach reduces the number of OGTT and ensures that only women with significant glucose intolerance are diagnosed.
- It has been reported that the GCT misses around 25% of cases who may have significant OGTT abnormalities, especially those with only hyperglycaemia.
- A meta-analysis of four RCTs, comparing the two-step with the one-step approach, showed more women with GDM were diagnosed by using the one-step approach. There were better neonatal and maternal outcomes in the latter group.
- Diagnostic cut-off values also vary among different guidelines. Mean glucose values for fasting, 1 and 2 hour for 75 g OGTT are based on an acceptable odds ratio of 1.75 for markers of diabetic foetopathy (large for gestational age, excess foetal adiposity, and foetal hyperinsulinaemia) as in HAPO study. These values have been accepted by the WHO.
- It has been proposed that lower 2-hour values should be used in South Asia as women are relatively small.

33.7   Postpartum follow-up after gestational diabetes

- After delivery of the placenta, glucose tolerance will normalise within hours or days.
- The offspring should be tested around 2 hours after birth for hypoglycaemia, and if low, early feeding may be necessary.
- Breast feeding should be encouraged as lactation has been shown to reduce risk for progression to overt diabetes.

- The women should be advised that breast feeding for >10 months has also been reported to decrease the risk of diabetes mellitus at 2 years after delivery by 57% in women with a history of GDM.
- After birth, the women should be advised to attend for glucose tolerance testing about 2 month's postpartum to exclude overt diabetes.
- Women should be advised about the effect of lifestyle interventions, diet, and exercise and weight management.
- Women with previous history of GDM have a higher risk of developing GDM during future pregnancies.
- Some published data suggest that about 40% women with history of GDM will develop type 2 DM within 10 years of index pregnancy, and they are at twice the risk of cardiovascular event.
- Regular glucose tolerance testing with 1−3 years interval is therefore recommended.
- Women should be informed that offspring of mothers with HIP are also at increased risk of early onset of obesity, diabetes T2, and cardiometabolic disorders as a consequence of intrauterine developmental programming, hence early lifestyle interventions are advised.

33.8 Interventions for prevention of gestational diabetes mellitus

33.8.1 Interventions which work

- Dietary and physical activity interventions aimed at women who are overweight or obese.
- Lose weight prior to pregnancy and/or limit weight gain in the early part of pregnancy.

33.8.2 Interventions where there is insufficient evidence

- Metformin use does not appear to reduce the risk of GDM development in pregnant women with obesity or PCO.
- General lifestyle interventions.
- Supplementation with vitamin D.
- Myo-insitol.
- Probiotics.

Further reading

Committee on Practice Bulletins-Obstetrics. Practice bulletin no 137: gestational diabetes mellitus. Obstet Gynecol 2013;122:406.

Hod M, Kapur A, Sacks DA, Hadar E, et al. The international federation of Gynaecology and Obstetrics (FIGO) initiative on gestational diabetes mellitus: a pragmatic guide for diagnosis, management, and care. *Int J Gynaecol Obstet*. 2015;131(Suppl 3):S173−S211.

Hod M., Kapur A., Lapolla A., Mahmood T. Hyperglycaemia in pregnancy—a global challenge and an opportunity for improving future health (special issue), diabetes research and clinical practice: 145; November 2018 (ISSN 0168-8227), Elsevier, Amsterdam.

Kapur A, Mahmood T. Universal testing for hyperglycaemia in pregnancy. In: Mahmood T, Arulkumaran A, Chervenak F, eds. *Obesity and Obstetrics*. 2nd ed. Elsevier; 2020:165−171. Available from: https//doi.org/10.1016/B978-0-12-817921-5.00017-5.

National Institute of Clinical Excellence and Healthcare (NICE). Diabetes in pregnancy: management from preconception to post natal care (NG3). <https://www.nice.org.uk/search?q = Gestational + diabetes>; 2020.

World Health Organisation. Diagnostic criteria and classification of hyperglycaemia first detected in pregnancy: a World Health Organisation guideline <http://www.who.int>; 2013.

Management of obese pregnant women with pre-diabetes and type 1 and 2 diabetes mellitus

A.M. Egan[1], C. Newman[2] and F.P. Dunne[2]
[1]Division of Endocrinology, Diabetes and Metabolism, Mayo Clinic, Rochester, MN, United States, [2]College of Medicine Nursing and Health Sciences, National University of Ireland Galway, Galway, Ireland

34.1   Background

Diabetes is the most common metabolic complication of pregnancy and is independently associated with adverse pregnancy outcomes. Some of these adverse outcomes are in common with those occurring in women with obesity alone (e.g. caesarean delivery), while others are unique to diabetes (e.g. ketoacidosis).

Pregestational diabetes (including type 1 and type 2 diabetes) affects approximately 1% pregnancies and is associated with a two to fivefold increased risk of major foetal complications including congenital anomaly (related to early pregnancy hyperglycaemia) and stillbirth. Women with pregestational diabetes are at risk of diabetic ketoacidosis (type 1 > type 2), and have higher rates of caesarean delivery and hypertensive disorders of pregnancy.[1]

Gestational diabetes mellitus (GDM) prevalence is most often reported as 2%–6% pregnancies, but can vary widely depending on the screening approach and population under evaluation.[2,3] It results in milder degrees of hyperglycaemia compared to pregestational diabetes, but is associated with a 30% increased risk of caesarean delivery, a 50% risk of gestational hypertension, and is a strong risk factor for future type 2 diabetes.[3] Infants of women with GDM are 30% more likely to be large for gestational age, and are at higher risk of childhood obesity.[4]

The majority of women with GDM are obese. However, recent data from the United Kingdom (UK) highlights that 23% women with type 1 diabetes and 65% women with type 2 diabetes also have a body mass index (BMI) ≥ 30 kg/m^2 based on the first recorded weight in pregnancy.[1] Contemporary, multicentre data from Ireland reveal similar results with 17% women with type 1 diabetes and 52% women with type 2 diabetes having a BMI ≥ 30 kg/m^2 at the initial antenatal visit.[5]

The large body of research on this topic may be summarised by stating that the impact of maternal obesity and maternal hyperglycaemia on pregnancy outcomes would appear to be independent and additive.[4,6] This underpins the clinical

Handbook of Obesity in Obstetrics and Gynecology. DOI: https://doi.org/10.1016/B978-0-323-89904-8.00010-X

importance of considering both conditions when treating affected women during pregnancy and postpartum.

Ideally all pregnant women with diabetes will be offered immediate contact and ongoing care in a joint diabetes/antenatal clinic. This clinic should provide access to a multidisciplinary team who can provide expert care tailored to their individual needs.

34.2 Treatment of gestational diabetes mellitus

For all women with GDM, if the fasting plasma glucose concentration is <7 mmol/L at diagnosis, a trial of diet and exercise modification (Box 34.1) with home glucose monitoring (Table 34.1) is recommended.[7]

If glucose targets are not met within 1−2 weeks (Table 34.2), treatment should be intensified (necessary in approximately 20% women with GDM).[7]

In the UK, the National Institute for Health and Care Excellence (NICE) recommend metformin as first-line therapy for women with GDM.[6] It is typically introduced at a dose of 500 mg once daily and increased as tolerated to 1000 mg twice daily. Broadly speaking, this is considered a good option for women with obesity as it is weight neutral (see Chapter 4).

Box 34.1 Diet and exercise recommendations for women with diabetes and obesity during pregnancy.

- Dietary reference intake for pregnancy (minimum): 175-g carbohydrate, 71-g protein, 28-g fibre
- Typical energy requirement: 1800−2500 kCal/day
- Ideal meal plan: 3 small-to-moderate−sized meals and 3 snacks
- Positive food choices: whole grains, fresh vegetables, some fruits, low glycaemic index foods. Avoid added sugars and eliminate sugar-sweetened beverages.
- Moderate exercise (e.g. walking for 30 minutes after a meal) is recommended throughout pregnancy unless there is a specific medical contraindication.

Table 34.1 Approach to antenatal glucose monitoring.

Type of diabetes	Daily testing
GDM or Type 2 diabetes controlled with lifestyle +/− oral agents	Fasting and 1-hour post meal
GDM or Type 2 diabetes requiring insulin	Fasting, premeal, 1-hour post meal and bedtime glucose daily
Type 1 diabetes	Fasting, premeal, 1-hour post meal and bedtime glucose daily[a]

[a]Continuous glucose monitoring (CGM) should be offered to women with type 1 diabetes as it improves neonatal outcomes. Intermittently scanned CGM may be used if CGM is not possible.[7]

Table 34.2 Glycaemic goals for women with diabetes during pregnancy.

Timepoint	Glucose target (aim for 85% values within goal)
Fasting	<5.3 mmol/L
1-hour postprandial	<7.8 mmol/L
2-hour postprandial	<6.4mmo/L
Women using continuous glucose monitoring	Target range is 2.5−7.8 mmol/L. Aim for time in range of >70% and time above range <25%

If taking insulin, glucose should be maintained >4.0 mmol/L at all times. The need to avoid hypoglycaemia may require individualisation of the glycaemic goals.

If this approach does not achieve success within 1−2 weeks, or if the initial fasting plasma glucose is ≥7.0 mmol/L, insulin should be added (with or without metformin) (Table 34.3).

However, the pharmacological approach to GDM is not standardised. For example, the American Diabetes Association recommends insulin as first-line therapy for women with GDM, and either metformin or glyburide/glibenclamide as alternative options. The latter option is a long-acting sulfonylurea, typically initiated at 2.5 mg/day and increased to maximum of 20 mg per day.[7]

The Institute of Medicine guidelines (Table 34.4) on gestational weight gain should be followed.[8] This is especially important for those with both obesity and diabetes, as excessive gestational weight gain will further fuel abnormal foetal growth[9]

Weight loss is not recommended during pregnancy, but increasing physical activity and dietary modifications may be of significant benefit in controlling gestational weight gain and improve the pregnancy outcome.[6]

34.3 Treatment of preexisting diabetes during pregnancy

HbA1c in early pregnancy is a useful marker of pregnancy risk and ideally will be ≤6.5%. While the greatest focus will be on safely reaching the glycaemic goals outlined in Table 34.2 using daily self-monitoring (Table 34.1), further HbA1c testing can be completed in the second and third trimesters of pregnancy to provide further detail on level of risk.[7]

Women with type 2 diabetes previously managed with lifestyle modifications will typically require additional therapy as pregnancy progresses. Consideration should be given to the use of metformin as it reduces excessive maternal weight gain, insulin dosage, and improves glycaemic control in women with type 2 diabetes. However, women should be counselled on the increase in small-for-gestational age infants among those exposed to metformin.[10]

Additional therapeutic options for hyperglycaemia in women with type 2 diabetes during pregnancy include glyburide/glibenclamide and insulin. Insulin therapy

Table 34.3 Insulin therapy during pregnancy.

GDM	Type 2 diabetes	Type 1 diabetes
Aspart and lispro are safe and effective rapid-acting insulin options and are preferable to regular (soluble) insulin. Isophane insulin/neutral protamine Hagedorn (NPH) insulin is considered first choice for basal insulin during pregnancy. Many clinicians use the long-acting insulin analogues detemir or glargine. Insulin should be titrated every 2–3 days depending on glycaemic control. For example, if morning fasting blood glucose is elevated, NPH can be initiated at 0.2 units/kg of body weight. NPH is frequently intensified to a TID regimen, particularly in the presence of premeal hyperglycaemia. Postprandial hyperglycaemia can be managed with premeal rapid-acting insulin analogues (e.g. 1 unit for 10 g carbohydrate).	Women established on mixed insulin prior to pregnancy should be transitioned to a multiple daily injection regimen. Women requiring insulin therapy de novo during pregnancy should have therapy introduced in a similar fashion to those with GDM, understanding that treatment will likely start earlier in pregnancy and insulin requirements may be higher.	A multiple daily injection regimen or insulin pump therapy is recommended. Aspart and lispro are the rapid-acting insulins of choice. Many clinicians continue the long-acting insulin analogues detemir or glargine, rather than switching to isophane insulin. Insulin pump therapy can be continued during pregnancy. However, hybrid-closed loop functions might need to be deactivated as their fasting goals are typically above that recommended for pregnancy. Initiation of insulin pump therapy during pregnancy should be reserved for those with persistently unstable glucose concentrations or recurrent severe hypoglycaemia.

GDM, Gestational diabetes mellitus.
Note: The results of the EVOLVE (https://clinicaltrials.gov/ct2/show/NCT01892319) and EXPECT (https://clinicaltrials.gov/ct2/show/NCT03377699) studies are expected to be published in the coming months and will provide additional information on basal insulin analogues during pregnancy.

should be introduced in a similar fashion to those with GDM (Table 34.3). All alternative glucose-lowering agents are not considered safe in pregnancy.

Treatment of hyperglycaemia in women with type 1 diabetes can be challenging and women need regular review and support to adjust their insulin doses in response to varying degrees of insulin sensitivity and maternal weight gain (see

Table 34.4 Institute of Medicine guidelines for gestational weight gain.

Pregestational BMI category	BMI (kg/m^2)	Recommended total weight gain (kg)	Recommended mean weight gain: trimester 2 and 3 (kg/week)
Underweight	< 18.5	12.5−18.0	0.51 (0.44−0.58)
Normal weight	18.5−24.9	11.5−16.0	0.42 (0.35−0.50)
Overweight	25.0−29.9	7.0−11.5	0.28 (0.23−0.33)
Obese	≥ 30.0	5.0−9.0	0.22 (0.17−0.27)

BMI, Body mass index.

Chapter 4). The input of an experienced diabetes educator and dietician can be invaluable. Options for insulin therapy are outlined in Table 34.3. Women should use either a multiple daily injection regimen or continuous subcutaneous insulin infusion (CSII/insulin pump therapy), with no definitive evidence to recommend one regimen over the other. A systematic review and meta-analysis of 47 studies (predominantly nonrandomised studies), reported that in pregnancies complicated by type 1 diabetes, CSII resulted in better first trimester glycaemic control compared to multiple daily injection insulin therapy, and lower risk of small-for-gestational age; but was associated with higher gestational weight gain and rates of large for gestational age.[11] Additional potential advantages of pump therapy include greater flexibility and a reduction in severe hypoglycaemia.

On the other, continuous glucose monitoring is associated with improvements in maternal glucose and neonatal health outcomes, is cost-effective, and recommended by NICE for all pregnant women with type 1 diabetes.[12]

34.4   Additional tips for caring for women with pregestational diabetes during pregnancy

- Prescribe folic acid 5 mg per day until 12 weeks gestation to reduce the risk of a neural tube defect in the baby (higher risk with obesity and diabetes).
- Women with type 1 or 2 diabetes should be prescribed low dose aspirin (75−150 mg daily) starting at 12−16 weeks gestation to lower the risk of preeclampsia.[9]
- Review medications list at the first antenatal visit and replace/remove any medications with teratogenic potential (Chapter 4).
- Diabetic ketoacidosis is a medical emergency and is associated with foetal demise during pregnancy. Women with type 1 diabetes should have the ability to test for ketones and all pregnant women with diabetes should be urgently tested for ketonaemia if hyperglycaemic or unwell. Diabetic ketoacidosis in pregnancy can occur with near-normal blood glucose readings. This diagnosis requires treatment in a critical care setting.
- Women with pregestational diabetes should have retinal evaluation on at least two occasions during the pregnancy (for example, in the first trimester and again in the late second trimester). Additional evaluation is needed for those with known retinopathy.[8]

- Renal assessment should occur in the first trimester of pregnancy with nephrology referral if the serum creatinine is ≥ 120 micromol/L, the urinary albumin:creatinine ratio is >30 mg/mmol or the total protein excretion is >0.5 g/day. Estimated glomerular filtration rate is not reliable in pregnancy.[8]
- The combination of diabetes and obesity places a woman at high risk of developing hypertensive disorders of pregnancy. See Chapter 4 for additional information on pharmacological options.
- Provide education on hypoglycaemia diagnosis and management where applicable. Women with type 1 diabetes should receive glucagon to be used in the setting of severe hypoglycaemia.

34.5 Treatment of newly diagnosed, 'overt' diabetes during pregnancy

As outlined in the chapter discussing the management of diabetes in pregnancy, some women are diagnosed with overt diabetes for the first time during pregnancy. The diagnosis is typically made in the setting of a fasting plasma glucose ≥ 7 mmol/L, a 2-hour postprandial (or random) glucose $\geq$ 11.1 mmol/L, or HbA1c $\geq$ 6.5% (48 mmol/mol).

Unfortunately, the best approach to screening (particularly in early pregnancy) is unclear and national and international guidelines vary. Nonetheless, obese women are at increased risk of undiagnosed type 2 diabetes. If diagnosed for the first time during pregnancy women should receive intensive education and treatment similar to those with preexisting diabetes. Close postpartum follow-up is vital to ensure their diabetes is appropriately classified and treated.

34.6 Monitoring foetal well-being

Early pregnancy scanning will confirm foetal viability, and a detailed foetal anatomy scan should occur at 18- to 20-weeks' gestation. Selected women may proceed to foetal echocardiography, particularly if there is suboptimal glycaemic control.[7]

Diabetes and obesity both increase the risk of congenital anomalies, but unfortunately excess subcutaneous fat in obese women can limit the sensitivity of ultrasound in detecting foetal anomalies.

Pregnant women with diabetes should receive ultrasound monitoring of foetal growth and amniotic fluid volume every 4 weeks from 28−36 weeks[9]

NICE does not recommend routine monitoring of foetal well-being (e.g. biophysical profile) earlier than 38 weeks unless there is risk of foetal growth restriction— for example, in the setting of maternal macro- or microvascular disease. However, the American College of Obstetricians and Gynecologists state that starting at 32 weeks' gestation, it is reasonable to initiate once- or twice-weekly monitoring such as the nonstress test, biophysical profile, or modified biophysical profile in women with pregestational diabetes or those requiring pharmacological intervention with GDM.

34.7   Timing and mode of delivery

Women with pregestational diabetes should be delivered between 38 weeks and six days of gestation.[8]

Women with GDM should be delivered before 40 + 6 weeks gestation.[7]

Earlier delivery may be required in the presence of pregnancy complications.

Diabetes should not be considered a contraindication to vaginal birth after a previous caesarean, but obesity will reduce the risk of a successful vaginal birth.

It is reasonable to consider elective caesarean delivery if the estimated foetal weight is >4500 g, bearing in mind the limitations of ultrasound in obese women.

Obese women will benefit from anaesthesiology consultation in late pregnancy to review the birthing plan and analgesic options.

34.8   Intrapartum control of glucose

In women with diabetes, capillary blood glucose should be monitored every hour during active labour.

Tight glycaemic control (4−7 mmol/L) is recommended intrapartum, and this may reduce the risk of foetal distress and neonatal hypoglycaemia.

Women with diabetes who required pharmacologic intervention during pregnancy, or those whose intrapartum glucose is not within goal are typically managed with an intravenous (IV) insulin and IV glucose infusion and hourly glucose testing.[7] The IV insulin should be adjusted hourly according to the capillary blood glucose, and each centre should have a protocol readily available for this purpose. The protocol may need to be modified based on individual patient needs. One example of an IV protocol is outlined in Box 34.2.

Electrolytes should be measured at least every 4−6 hours while IV insulin is ongoing.

Typically basal insulin and oral hypoglycaemic agents can be continued throughout labour.

Following delivery, the IV insulin rate can be reduced to 50%. When necessary, subcutaneous insulin should be reintroduced once the woman can eat and drink, and there should be at least a 30-minute overlap before stopping the IV insulin.

CSII can be continued during labour if appropriate expertise is available. However, regular adjustment to the basal rates will be necessary, and IV insulin should be initiated if not at goal.

Women with diabetes undergoing elective caesarean delivery will benefit from being scheduled in the early morning to avoid prolonged fasting.

Box 34.2 Intravenous insulin and glucose during labour and delivery.

5% dextrose is infused intravenously at 100 mL/hour throughout.

   To make insulin infusion: add 50 units of actrapid to 49.5 mL of NaCl 0.9% to make up 50 mL of fluid in a 50 mL syringe (1 unit of insulin = 1 mL). Start the infusion as directed by the blood glucose estimation as outlined below.

Blood glucose (mmol/L)	Insulin infusion rate (mL/hour)
≤ 4.0	0.5a
Observe for hypoglycaemia	
4.1−6.0	1.0
6.1−8.0	2.0
8.1−10.0	3.0
If >9 for 2 consecutive hours consider adjusting insulin rates	
10.1−12.0	4.0
> 12.1	6.0
Check for ketones and inform medical team	

[a]Infusion rate of 0 mL/hour for glucose ≤ 4.0 mmol/L may be used in gestational or type 2 diabetes, but for women with type 1 diabetes, start infusion at 0.5 mL/hour.

   Note: Infusion rates need to be reviewed regularly and adjusted if plasma/capillary blood glucose is not within goal.

Adapted from Guidelines for the Management of Diabetes in Pregnancy (atlanticdipireland.com).

34.9   Postpartum care

34.9.1   Mother

Breastfeeding should be encouraged and supported where desired. It assists with postpartum weight loss and is possibly associated with reduced future risk of obesity and diabetes in the infants.[13] In women with GDM, breastfeeding is associated with a reduced risk of future type 2 diabetes.

   Insulin sensitivity dramatically increases postpartum, and women on insulin therapy are at risk of hypoglycaemia.

   Women with GDM should discontinue all glucose-lowering therapy but should have glucose testing prior to discharge from hospital to exclude persistent diabetes. Formal assessment for type 2 diabetes is recommended at 6−12 weeks postpartum, and at regular intervals lifelong.

   Women with pregestational diabetes should revert to their prepregnancy insulin doses following delivery, with even further dose reductions if breastfeeding of about 25%. For those with type 2 diabetes, metformin is considered safe while breastfeeding but other noninsulin, glucose-lowering agents are contraindicated.

Weight loss goals should be reviewed at the postpartum visit and ongoing support provided where available.

Pregnancy spacing and postpartum contraception should be also discussed at the postpartum visit.

34.9.2 Infant

A neonatologist should review the infant(s) following delivery and delivery units should have the ability to provide enhanced neonatal care.

Neonatal blood glucose testing should occur 2−4 hours after delivery, with earlier evaluation if clinical signs such as irritability or seizure-like activity are present.

Intravenous dextrose or tube feeding are generally reserved for situations when the glucose is <2 mmol/L despite feeding.[7]

References

1. Murphy HR, Howgate C, O'Keefe J, et al. Characteristics and outcomes of pregnant women with type 1 or type 2 diabetes: a 5-year national population-based cohort study. *Lancet Diabetes Endocrinol.* 2021;9(3):153−164. Available from: https://doi.org/10.1016/S2213-8587(20)30406-X, Epub 2021 Jan 28. PMID. Available from: 33516295.

2. Buckley BS, Harreiter J, Damm P, et al. Gestational diabetes mellitus in Europe: prevalence, current screening practice and barriers to screening. A review. *Diabet Med.* 2012;29(7):844−854. Available from: https://doi.org/10.1111/j.1464-5491.2011.03541.x, PMID. Available from: 22150506.

3. O'Sullivan EP, Avalos G, O'Reilly M, et al. Atlantic diabetes in pregnancy (DIP): the prevalence and outcomes of gestational diabetes mellitus using new diagnostic criteria. *Diabetologia.* 2011;54(7):1670−1675. Available from: https://doi.org/10.1007/s00125-011-2150-4, Epub 2011 Apr 15. Erratum in: Diabetologia. 2016 Apr;59(4):873. PMID. Available from: 21494772.

4. Simmons D. Diabetes and obesity in pregnancy. *Best Pract Res Clin Obstet Gynaecol.* 2011;25(1):25−36. Available from: https://doi.org/10.1016/j.bpobgyn.2010.10.006, Epub 2011 Jan 17. PMID. Available from: 21247811.

5. Egan AM, Dow ML, Vella A. A review of the pathophysiology and management of diabetes in pregnancy. *Mayo Clin Proc.* 2020;95(12):2734−2746. Available from: https://doi.org/10.1016/j.mayocp.2020.02.019, Epub 2020 Jul 28. PMID. Available from: 32736942.

6. NICE. Diabetes in pregnancy: management from preconception to the postnatal period. <http://www.nice.org.uk/guidance/NG3>; 2020.

7. American Diabetes Association. 14. Management of diabetes in pregnancy: standards of medical care in diabetes—2021. *Diabetes Care.* 2021;44(suppl 1):S200−S201.

8. Institute of Medicine (US) and National Research Council (US) Committee to Reexamine IOM Pregnancy Weight Guidelines. In: Rasmussen KM, Yaktine AL, eds. *Weight Gain During Pregnancy: Reexamining the Guidelines.* Washington, DC: National Acadamies Press (US); 2009.

9. ACOG practice bulletin no. 201 summary: pregestational diabetes mellitus. *Obstet Gynecol*. 2018;132(6):1514−1516.
10. Brand K. Metformin in pregnancy and risk of adverse long-term outcomes: a register-based cohort study. *BMJ Open Diabetes Res Care*. 2022. Available from: https://doi.org/10.1136/bmjdrc-2021-002363.
11. Rys P. Continuous subcutaneous insulin infusion vs multiple daily injections in pregnant women with type 1 diabetes mellitus: a systematic review and meta-analysis of randomised controlled trials and observational studies. *Eur J Endocrinol*. 2018. Available from: https://doi.org/10.1530/EJE-17-0804.
12. Feig DS. Continuous glucose monitoring in pregnant women with type 1 diabetes (CONCEPTT): a multicentre international randomised controlled trial. *The Lancet*. 2017. Available from: https://doi.org/10.1016/S0140-6736(17)32400-5.
13. O'Reilly MW, Avalos G, Dennedy MC, O'Sullivan EP, Dunne F. Atlantic DIP: high prevalence of abnormal glucose tolerance post partum is reduced by breast-feeding in women with prior gestational diabetes mellitus. *Eur J Endocrinol*. 2011;165 (6):953−959. Available from: https://doi.org/10.1530/EJE-11-0663, Epub 2011 Sep 21. PMID. Available from: 21937504.

Drug interactions for women with diabetes and obesity during pregnancy

C. Newman[1], F.P. Dunne[1] and A.M. Egan[2]
[1]College of Medicine Nursing and Health Sciences, National University of Ireland Galway, Galway, Ireland, [2]Division of Endocrinology, Diabetes and Metabolism, Mayo Clinic, Rochester, MN, United States

35.1 Diabetes during pregnancy

While early pregnancy is a time of relative insulin sensitivity, insulin resistance increases throughout the second and early third trimesters of pregnancy (Table 35.1).

This resistance occurs to preserve carbohydrate for the developing foetus and results from a combination of placental hormones, increased free-fatty acids and changes in the phosphorylation of the insulin receptor. These changes occur in all pregnancies; however, they only become clinically apparent in those with diabetes (either gestational or pregestational).

These physiological alterations can impact the use/dosage of medications used to treat both diabetes itself and its complications.

Table 35.1 The physiological changes experienced during the different stages of pregnancy.

Time period	Insulin sensitivity	Typical insulin requirement in pregestational diabetes
0−9 weeks	Decreased	Increased doses of basal and bolus insulin are needed[a]
9−16 weeks	Increased	Reduce basal and bolus doses to avoid hypoglycaemia
16−37 weeks	Decreased	Requirements gradually increase by 50%−150% (from preconceptual levels)
37−40 weeks	Increased	Stable/decreased insulin doses are generally needed from week 37−40
Postpartum	Increased	Reduce insulin doses to < preconceptual levels Take into account breastfeeding (25% further reduction)

[a]Titrate insulin to target a fasting glucose of ≤ 5.3 mmol/L and 1-hour postprandial glucose of ≤ 7.8 mmol/LT.

Handbook of Obesity in Obstetrics and Gynecology. DOI: https://doi.org/10.1016/B978-0-323-89904-8.00012-3

In addition, some of the medications used in pregnancy can have a marked impact on glycaemic control (Tables 35.2 and 35.3).

Important points

1. It is worth noting that an individualised approach to titration of insulin is necessary — pregnancy may be complicated by reduced appetite, nausea, and vomiting which can predispose to hypoglycaemia.
2. Many patients may have suboptimal glycaemic control entering pregnancy and may require a substantial increase in their prepregnancy insulin doses to achieve the tighter glycaemic control necessary in pregnancy
3. Insulin requirements vary between women with type 1 diabetes mellitus (T1DM) and type 2 (T2DM) — women with T2DM may need large increases in insulin doses per trimester compared to those with T1DM

---

1. Antihyperglycaemic agents
   a. Insulin
      i. Prandial insulin
         1. Human insulins (e.g. actrapid, regular Humulin — given 20–30 minutes before a meal) are the least immunogenic preparations and minimise the risk of insulin crossing the placenta. Human insulins were traditionally used in the management of hyperglycaemia in pregnancy and there is significant clinical evidence to support their efficacy and safety. However, there is now considerable evidence to support the use of certain analogue insulins, for example, lispro, aspart.
         2. Rapid acting insulin analogues (injected 10–15 minutes before a meal), for example, lispro and aspart have been studied in pregnancy. They show low

Table 35.2 Diabetes medications in pregnancy.

Diabetic medications approved for pregnancy	Diabetic medications to avoid in pregnancy	Insufficient data
• Insulin (regular, NPH, lispro, aspart, detemir) • Metformin • Glyburide	• Glucagon Like Peptide-1 (GLP-1) agonists • Dipeptidyl Peptidase (DPP)-4 inhibitor • Sodium Glucose Co-Transporter-2 (SGLT2) inhibitors • Thiazolinediones • Rapid acting insulin analogue glulisine — Food and Drug Administration (FDA) Category C	• Long-acting insulin analogues — lantus/degludec (frequently continued in pregnancy but generally not started de novo)

Table 35.3 Treatment of hypertension in pregnancy.

Medications	Changes to dosage in pregnancy versus nonpregnant population
Labetolol	Increased dose in pregnancy
Nifedepine	Increase dose and frequency in pregnancy, for example, (10 mg twice daily modified release preparation); 40 mg in 2—3 divided doses often used in pregnancy
Methyldopa	Reduced doses compared to nonpregnant adults — 500 mg per day in 2 divided doses
Drugs to avoid/stop in pregnancy	• ACEi/ARBs/DRI • MRA • Diuretics

ACEi, Angiotension converting enzyme inhibitors; *ARBs*, angiotensin receptor blockers; *DRI*, direct renin inhibitors; *MRA*, mineralocorticoid receptor antagonists.

rates of immunogenicity, do not (or minimally) cross the placenta and are not teratogenic. Pregnancy decreases the metabolic clearance of these insulins. These changes are most obvious in aspart in late pregnancy.

3. From the second trimester onwards insulin absorption is delayed. To prevent postprandial glucose spikes premeal insulin should be given 10—15 minutes before food in the first trimester, 20 minutes before food in the second trimester, and 30 minutes before food in the third trimester.

ii. Intermediate acting insulin

1. Neutral protamine Hagedorn (NPH insulin/Isophane) — this human insulin is well studied and has a good safety profile in pregnancy. It is frequently the first choice in pregnancy for basal requirements however, it often requires more frequent dosing (2—3 times per day) when compared to longer acting insulin analogues.

iii. Long-acting insulin analogues (e.g. detemir, glargine, degludec) — insulin detemir has been studied and is noninferior to NPH in pregnancy; insulin glargine has not been studied in a randomised controlled trial but observational data suggests it is safe; insulin degludec is not currently approved, however a number of case series suggest it is safe for use in pregnancy. The EXPECT trial which will compare degludec to detemir and aspart is completed and publication is awaited (NCT03377699).

b. Oral hypoglycaemic agents

i. Metformin — commonly used outside of pregnancy and remains the first-line oral therapy for many patients with gestational diabetes mellitus (GDM). Readily crosses the placenta and low quantities are found in breast milk. The bioavailability of metformin is significantly higher in pregnancy, however no dose adjustment is needed in pregnancy as renal clearance is also increased. The starting dose of metformin is 500 mg once daily and can be increased to 2500 mg total daily dose.

ii. Sulphonylureas

1. Glyburide (Glibenclamide) — crosses the placenta, however cord blood levels are highly variable among exposed. Long-term effects on the offspring are

unknown; however, there is no evidence to suggest increased congenital anomalies.

Pregnancy increases the clearance of glyburide and twice daily dosing is frequently needed.

In 2015 NICE removed glyburide from its guidelines on the treatment of diabetes in pregnancy; however it is reasonable to consider glyburide for patients who are or intolerant to or unwilling to consider metformin or insulin.

The dose of glibenclamide is 2.5 mg increasing up to 20 mg daily.

Disadvantages of glibenclamide include:

- higher rates of neonatal macrosomia and large for gestational age (LGA) births when compared to insulin; and
- higher rates of hypoglycaemia and maternal weight gain when compared to metformin

   iii. Agents to avoid in pregnancy

      1. Glucagon-like peptide receptor analogues — injectable weekly or daily therapies for T2DM. Commonly used outside of pregnancy due to cardiovascular and renoprotective effects. They may have a beneficial effect on pregnancy rates in obese patients with polycystic ovarian syndrome (PCOS). Teratogenicity has been reported in animal models and they are currently contraindicated in pregnancy and are not recommended for use during breastfeeding.

      2. Dipeptidyl peptidase-4 inhibitor — oral agents commonly used outside of pregnancy. These medications are also contraindicated in pregnancy and breastfeeding

      3. Sodium-glucose cotransporter-2 inhibitors — oral agents which are common used in T2DM outside of pregnancy. Have replaced metformin as first-line in some guidelines due to their cardio- and renoprotective effects. Associated with modest weight loss. Should be avoided in pregnancy - renal abnormalities observed in animal models.

      4. Thiazolinediones — second/third line in the treatment of patients with T2DM outside of pregnancy however fewer than 4% of patients now receive these drugs due to concerns over weight gain, fluid retention, and urological malignancies. These drugs are occasionally used in PCOS and can lead to a return of ovulation. They have been associated with growth restriction in animal models and should be discontinued in pregnancy. They are not recommended in breastfeeding.

2. Medications used to treat the complications of diabetes

   a. Statins — outside of pregnancy statins are commonly used to reduce the risk of cardiovascular events in high-risk patients with diabetes.

They are contraindicated in pregnancy and breastfeeding. A paper published by the National Institute of Health in 2004 reported 20 cases of central nervous system and limb anomalies in 52 patients exposed to statins in the first trimester.

Later studies have suggested that statins themselves do not confer a greater risk of congenital anomaly and rather that patients taking statins are naturally higher risk, for example, due to diabetes and this may account for higher rates of congenital anomalies.

   b. Fibrates — effective at lowering triglycerides outside of pregnancy however they have been associated with teratogenicity in animal models. A 2014 expert analysis by the American College of Cardiology recommends they should be avoided in pregnancy.

c. Antihypertensives

The NICE Guidelines advise that clinicians

- use labetalol as a first-line agent for hypertension in pregnancy;
- use nifedipine if labetolol is not tolerated or contraindicated; and
- use methyldopa if both of the above medications are unsuitable

Choice of antihypertensive

i. Beta-blockers

1. Labetolol: has both alpha and beta-blockade action; oral clearance is increased by nearly 30% in pregnancy.

2. Atenolol: no alpha activity; suitable if labetolol is unavailable and there are contraindications to the other medications.

3. Metoprolol — enhanced clearance in pregnancy; rarely used in pregnancy.

ii. Calcium channel blockers

1. Nifedipine: pregnancy-induced changes in its oral clearance and half-life mean dosing is higher and more frequent than in the nonpregnant population.

2. Amlodipine — widely used outside of pregnancy as a second-line agent. Very little data on its use in pregnancy, should only be used it the benefits outweigh the risks (FDA).

iii. Alpha-blockers — fourth-line antihypertensive therapy outside of pregnancy

1. Methyldopa: weak antihypertensive effect when compared to labetolol; safety is well established in pregnancy.

iv. Angiotensin Converting Enzyme Inhibitors (ACEi): commonly used outside of pregnancy in hypertensive patients and in those with persistently abnormal albumin:creatinine ratios.

1. Exposure in the first trimester has been associated with foetal ventriculo-septal and atrio-septal defects and central nervous system anomalies.

2. In the second and third trimesters exposure to ACEi is associated with renal tubular dysgenesis, anuria and oligohydramnios (and its resultant complications). Neonatal renal failure has been observed and though there is limited long-term followup data chronic renal impairment has been observed.

v. Angiotensin Receptor Blockers and Direct Renin Inhibitors (ARBS/DRIs) have similar mechanisms of action and are also contraindicated in pregnancy.

ACEi/ARBs/DRI should also be used with extreme caution in breastfeeding as trace amounts are secreted in breast milk and infants are extremely sensitive to the hypotensive effects of these agents.

vi. Mineralocorticoid receptor antagonists (MRAs): spironolactone (reports of ambiguous genitalia in humans and feminisation of males in animal models) — measures should be taken to avoid pregnancy when used in PCOS; epleneron (not conclusively safe in pregnancy); amilorde (insufficient data). As a class should be avoided in pregnancy.

vii. Diuretics

1. Loop diuretics: risk of neonatal jaundice however they have been used to treat pulmonary oedema in pregnancy (FDA group C).

2. Thiazide diuretics: associated with cases of foetal and neonatal jaundice and thrombocytopenia, FDA contraindicated in pregnancy.

Advantages and disadvantages of metformin.

Advantages	Disadvantages
• Reduces maternal weight gain • Reduces macrosomia when compared to insulin • Reduces neonatal hypoglycaemia when compared to insulin • Reduces dose of supplemental insulin (if required) and is highly acceptable to women	• Higher rates of small for gestational age infants when compared to placebo in insulin-treated patients with type 2 DM • May affect body composition and BMI of offspring both in patient with GDM and those exposed to metformin for PCOS • Further studies are ongoing to evaluate both the short-term efficacy (EMERGE study- NCT02980276) and long-term effects of metformin in GDM

Hypertension in breastfeeding mothers
Safe: propranolol, metoprolol, labetolol, nifedipine, verapamil, nicardipine, captopril, and enalopril (with extreme caution); hydrochlorothiazide at <50 mg /day, methyldopa.
Avoid: atenolol (can result in bradycardia, hypothermia, and cyanosis in the infant); arvediolol and bisoprolol (insufficient data); all other ACEi and ARBs.

ACEi, Angiotension converting enzyme inhibitors; *ARBs*, angiotensin receptor blockers.

d. Antihypertensives
   i. Aspirin — used at a dose of 75—150 mg in women thought to be at high risk of pre-eclampsia (including diabetes); teratogenicity reported in animal models and in humans at analgesic doses (which are significantly higher than those used in pregnancy).
   ii. Clopidogrel — used as a part of dual antiplatelet therapy postcardiac stenting, cardio-vascular accident, etc. The prevalence of advanced maternal age and subsequent complications including ischaemic events means these medications are likely to become more common. Clopidogrel works by inhibiting platelet aggregation via inhibition of the glycoprotein IIb/IIIa complex formation. Clopidogrel is a FDA class B drug — no known teratogenicity in animal models however there is insufficient data in humans to definitely rule out harmful effects. It is advised to stop clopidogrel at least 7 days before any procedures due to bleeding risk. Women using clopidogrel should be cared for in a large centre with obstetricians and cardiologists experienced in the management of these complex cases.
   iii. Prausgrel/Ticagrelor — used post percutaneous coronary intervention. Inhibits the P2Y12 receptor on platelets. Should be stopped 7 days before procedures. These are FDA class C medications. Women using these medications should be cared for in a large centre with obstetricians and cardiologists experience in managing complex cases.

In the event of an unintended pregnancy
• 40%—50% of all pregnancies are unintended with higher rates observed in women aged 18—24 with lower incomes and lower levels of education (Center for Disease Control).

- Unintended pregnancies are further complicated if a women has or goes on to develop diabetes − rates of preeclampsia, caesarean delivery, macrosomia, congenital anomaly, and stillbirth are all increased in women with diabetes.
- Many of these women also take teratogenic medications to reduce the cardiovascular risks associated with diabetes.

If a patient attends your service with an unintended pregnancy the following steps should apply:

  o Stop all teratogenic medications: ACEi, ARBs, DRI, thiazides, spironolactone, eplenerone, statins. Also be aware of common coexisting or related conditions, for example, thyrotoxicosis in T1DM treated with carbimazole (should be avoided in the first trimester and switched to propylthioracil).
  o If blood pressure control is needed switch to (or initiate) labetolol, nifedipine, or methyldopa.
  o Check Haemoglobin A1c (HbA1c) level (preconceptual HbA1c of <48 mmol/mol associated with reduced risk of congenital anomaly).
  o Check a urinary albumin:creatinine ratio.
  o Refer for urgent retinal screening due to the risk of deterioration in retinopathy during pregnancy.
  o Check thyroid function tests, Rubella status, viral hepatitis, and Human Immunodeficiency Virus status as standard.
  o Advise to check capillary blood glucose 7 times per day, before and 1−2 hours after each meal and before bed, or use continuous glucose monitoring system.
  o If satisfactory glycaemic control on an approved agent consider continuing same.
  o If uncontrolled consider initiating treatment with either an oral hypoglycaemic agent or insulin therapy in line with local guidelines.
  o Initiate folic acid (5 mg) if still within the first trimester.
  o Refer to a centre with a dedicated diabetes team.

3. Medications used to treat the complications of diabetes
   a. Steroids: administered to help foetal lung maturity. Steroids promote insulin resistance and hepatic gluconeogenesis, resulting in hyperglycaemia a number of hours later and profound hyperglycaemia 12 hours after the first steroid dose. In the nonpregnant state fasting glucose levels are usually unaffected and postprandial excursions are seen. However due to lower targets of 5.3 mmol/L in pregnancy, fasting hyperglycaemia is often seen with steroid use.

   Roughly 40% of women with diet-controlled GDM will go on to require insulin during the course of steroid therapy and women already on insulin will frequently need increases of 50%−100%.

   There are a number of different approaches and different protocols for controlling hyperglycaemia during steroid administration. A general approach to monitoring hyperglycaemia includes:

- Check a renal profile to ensure the patient has normal electrolytes.
- Check a capillary blood glucose (CBG) every hour aiming for a glucose level 4−7.8 mmol/L.
- CBGs may be required more regularly if the patient experiences symptoms of hypoglycaemia.
- If a patient is using a continuous glucose monitor this is sufficient, however any readings suggestive of hypoglycaemia may need to be rechecked.

- Continue the patient's long-acting insulin (where relevant).
- Start a variable rate insulin infusion (VRII) − 50 units of actrapid/Humulin S and 49.5 mL of 0.9% NaCL (made up to 50 mL) in a syringe driver and adjust the dose of insulin using a sliding scale to keep glucose levels in the above target.
- If a patient is also taking meal time subcutaneous insulin this can be continued or held (and glucose controlled with VRII) depending on the hospital's protocol.
- Stop the VRII if the patient becomes hypoglycaemic and treat as per the hospital's protocol.
- To avoid hypoglycaemia, hyponatraemia, and hypokalaemia consider administering fluids with dextrose and potassium, for example, 0.9% NaCL and 5% glucose with 20 mmol/L KCl at 50 mL/hour (evaluate the patients fluid status first).
- For women on insulin pump therapy a 40% increase in insulin doses may be needed and the local diabetes service should be consulted immediately.

Hyperglycaemia from steroids use can result in Diabetic Ketoacidosis or Hyperosmolar Hyperglycaemic State, so regular monitoring is essential and ketones should be checked in all women with diabetes who are hyperglycaemic.

  b. Beta-blockers cause hyperglycaemia in obese individuals outside of pregnancy; however, in pregnancy clinical significance is less.

35.2   Obesity and pregnancy

Globally the prevalence of obesity is rising − the Royal College of Obstetrics and Gynaecology has estimated that over 21% of all patients attending antenatal services have a BMI in the obese range (> 30 kg/m^2). Providing care for these patients is increasingly complex and requires a multidisciplinary approach.

In addition to the risks posed by obesity during pregnancy and around the time of delivery (preeclampsia, caesarean birth) there are a number of other components of the patient's health to consider.

Obesity and menstruation/contraception

1. PCOS − up to 80% of women with a diagnosis of PCOS will have obesity, however many of the drugs used to treat PCOS are either not used (e.g. oral contraceptives, letrozole, and clomiphene for ovulation induction) or contraindicated (spironolactone, cyproterone acetate, finasteride (antiandrogens which risk feminising a male foetus) and statins (for cardiovascular protection)) in pregnancy.

   Metformin which is commonly used in PCOS has been discussed above.

2. Contraception

   a. For women wishing to avoid pregnancy or taking an oral contraceptive for menstrual health, the choice may be limited by obesity.

   b. For women with a BMI >30 kg/m^2 but who are otherwise healthy contraceptive choice does not need to be limited; indeed the risk of an unintended pregnancy in a patient with obesity may pose a greater risk to their overall health.

   c. Venous thromboembolism (VTE) − for women with a BMI 30−34.9 kg/m^2 with no other risk factor for VTE a combined oestrogen-progestin pill (COCP) can be used. For those with a BMI ≥ 35 kg/m^2 (or 30 kg/m^2 with additional risk factors) who desire an oral form of contraception, progestin-only is an acceptable choice. Patients should be advised that progesterone-only pills (POPs) have a higher failure rate than other forms of contraception and window for taking POP is smaller than COCP.

   d. Contraception and bariatric surgery:

      i. Preoperative: oestrogen-containing oral contraceptives should be stopped 1 month prior to surgery. POPs can be continued until surgery and may be a suitable alternative for someone previously on oestrogen-containing preparations.

      ii. Postoperative: women undergoing restrictive bariatric interventions can continue oral contraceptive postoperatively. There is insufficient data on the efficacy of oral hormonal contraceptives in malabsorptive bariatric interventions, however the CDC recommends that 'bariatric surgical procedures involving a malabsorptive component have the potential to decrease oral contraceptive effectiveness, perhaps further decreased by postoperative complications such as long-term diarrhoea, vomiting, or both'; as such other forms of contraception should be considered in such patients.

VTE in Pregnancy

1. VTE risk − all women should have a VTE risk assessment done by 12 weeks; obesity is considered a risk factor (along with smoking, parity ≥ 3, and a previous provoked VTE) and should be score as follows:

   a. ≥ 4 risk factors = prophylactic low-molecular-weight heparin (LMWH) throughout pregnancy and for 6 weeks postpartum.

   b. ≥ 3 risk factors − LMWH from 28 weeks of gestation until 6 weeks postpartum

   c. ≥ 2 risk factors − LMWH for at least 10 days postpartum.

   d. A weight-based approach has been advised by the RCOG given the risk of underdosing patients with a high BMI. In high-risk obese patients where accurate dosing may be difficult, anti-Xa monitoring (under the advice of a haematologist) may be appropriate.

Testing for GDM after bariatric surgery

- Approximately half of all bariatric surgeries in the United States are performed in women aged 18−45 years.
- Bariatric surgery often improves fertility in these patients resulting in pregnancy.
- Pregnancies in women with previous bariatric surgery can be complicated by malabsorption of multiple micronutrients and it is generally recommended to avoid pregnancy for at least 12−24 months. This allows the woman to achieve her weight loss goals and also protects the foetus from the effects of rapid maternal weight loss.
- For women who successfully achieve pregnancy specialist dietician input is essential.
- These women also have a higher risk of SGA and preterm deliveries (greater for malabsorptive than restrictive procedures); though LGA rates are reduced.
- Traditional oral glucose tolerance test is unreliable in these patients.
- Half of all patients will have hypoglycaemia at 120 minutes during Oral Glucose Tolerance Test (OGTT).
- High rates of reactive hypoglycaemia is seen during the OGTT (most especially in women who have undergone roux-en-y-gastric bypass); however, despite this 50% of women with a history of bariatric surgery meet the International Association of Diabetes in Pregnancy Study Groups criteria for GDM versus 0% of patients using the Carpenter and Coustan criteria.

- While there are no specific and approved guidelines for the investigation of GDM in these patients, one suggested approach is:
  ○ HbA1c and fasting glucose in early pregnancy to rule out overt diabetes. A HbA1c of ≥ 48 mmol/mol or a fasting glucose level of ≥ 7 mmol/L indicate overt diabetes.
  ○ Self-monitoring of blood glucose 7 times/day in weeks 24−28.
    Fasting levels ≥ 5.3 mmol/L: (indicates need for treatment)
    − 1 hour postprandial readings ≥ 7.8 mmol/L;
    − 2 hour postprandial readings ≥ 6.7 mmol/L.

Further reading

American College of Obstetricians and Gynecologists. ACOG practice bulletin no. 105: bariatric surgery and pregnancy. *Obstet Gynecol* 2009;113:1405−1413.

Brenner B, Arya R, Beyer-Westendorf J, et al. Evaluation of unmet clinical needs in prophylaxis and treatment of venous thromboembolism in at-risk patient groups: pregnancy, elderly and obese patients. *Thrombosis J*. 2019;17:24.

Freitas C, Araujo C, Caldas R, et al. Effect of new criteria on the diagnosis of gestational diabetes in women submitted to gastric bypass. *Surg. Obes. Relat. Dis.* 2014;10 (6):1041−1046.

García-Patterson A, Gich I, Amini SB, et al. Insulin requirements throughout pregnancy in women with type 1 diabetes mellitus: three changes of direction. *Diabetologia*. 2010;53 (3):446−451. Available from: https://doi.org/10.1007/s00125-009-1633-z. Epub 2009 Dec 15. PMID: 20013109.

NICE. Diabetes in pregnancy: management from pre-conception to the postnatal period. *National Institute for Health and Care Excellence Guidelines NG3*; 2015.

Royal College of Obstetricians and Gynaecologists. Reducing the risk of venous thromboembolism during pregnancy and the puerperium. *Green-top Guideline No 27a*; 2015. Available from: https://www.rcog.org.uk/media/qejfhcaj/gtg-37a.pdf.

Rowan JA, Hague WM, Gao W, et al. MiG Trial Investigators. Metformin vs insulin for the treatment of gestational diabetes. *N Engl J Med*. 2008;358(19):2003−2015.

Rowan JA, Rush EC, Plank LD, et al. Metformin in gestational diabetes: the offspring follow-up (MiG TOFU): body composition and metabolic outcomes at 7−9 years of age. *BMJ Open Diabetes Res Care*. 2018;6(1):e000456. Available from: https://doi.org/10.1136/bmjdrc-2017-000456. Published 2018 Apr 13.

Short- and long-term effects of gestational diabetes and foetal outcomes

Laura Stirrat[1] and Tahir Mahmood[2]
[1]Royal Infirmary of Edinburgh, Edinburgh, Scotland, [2]Department of Obstetrics & Gynaecology, Victoria Hospital, Kirkcaldy, United Kingdom

- Gestational diabetes (GDM) affects around 3.5% of pregnancies in the United Kingdom and has a higher prevalence in obese pregnant women.
- Maternal hyperglycaemia increases risk of adverse pregnancy outcomes and GDM is associated with adverse outcomes for women and their offspring both in the short and long term.

36.1 Maternal diabetes and insulin resistance

- Pregnancy is a state of increasing insulin resistance (up to 40%−50%), with most of this increase occurring in the third trimester.
- In obese pregnant women, the prepregnancy insulin resistance associated with obesity is exacerbated such that obese pregnant women tend to have reduced peripheral insulin sensitivity even in early pregnancy.
- Increased adipose tissue in obese women also leads to a greater release of inflammatory markers and free fatty acids, which further contribute to insulin resistance, and inhibits the action of insulin on suppressing lipolysis.
- In the 1950s Pederson first postulated that in utero exposure to hyperglycaemia due to maternal diabetes may cause long-term changes to the foetus including congenital malformations, increased birthweight, and increased risk of diabetes and obesity later in life.
- An increasing body of evidence has characterised associations of maternal hyperglycaemia with offspring size and cardiovascular health.
- There are longer term risk of maternal cardiovascular disease, chronic kidney disease, and cancer risk.

36.2 Short-term risks

36.2.1 Foetal risks

36.2.1.1 Increased foetal growth

- Insulin resistance in later gestations helps facilitate nutrient transfer to the foetus and in turn promotes foetal growth.

Handbook of Obesity in Obstetrics and Gynecology. DOI: https://doi.org/10.1016/B978-0-323-89904-8.00036-6

- In states of maternal overnutrition, such as maternal obesity and diabetes, further increased insulin resistance may lead to a greater supply of nutrients to the foetus and result in increased foetal growth, particularly of adipose tissue.
- Studies of placentas from women with diabetes have identified increased expression of placental glucose transporters even with near-normal glycaemia at term.
- This may have a significant impact on nutrient transfer to the foetuses of women with diabetes, even in the presence of good glycaemic control.
- It has also been suggested that the resultant foetal hyperinsulinaemia may exacerbate the glucose concentration gradient across the placenta, further increasing glucose transfer to the foetus.
- It is well established that there is a linear relationship between the level of maternal glycaemia during pregnancy with indices of growth such as offspring birthweight, adiposity, and BMI.
- The Hyperglycaemia and Adverse Pregnancy Outcome study demonstrated raised maternal blood glucose at fasting, 1 and 2 hours after oral glucose tolerance test (OGTT) at 24—32 weeks' gestation was associated with approximately 1—1.5-fold increased risk for neonatal macrosomia (birthweight >90th centile) and neonatal hyperinsulinaemia (via cord blood C-peptide).

36.3   Short-term clinical consequences of increased foetal size include

- Polyhydramnios leading to increased risk of preterm labour, malpresentation, unstable lie, and cord prolapse.
- Obstructed labour and increased likelihood of caesarean section delivery.
- Foetal injury (including shoulder dystocia) or damage to the birth canal (including perineal or obstetric anal sphincter injury).

36.3.1   Congenital anomalies

Maternal hyperglycaemia can cause defects in the developing organs leading to congenital anomalies. Cardiac and neural tube defects are the most common.

36.3.2   Preterm birth

- Women with GDM are more likely to deliver at preterm gestations. This is in part due to earlier induction of labour for iatrogenic reasons and premature rupture of membranes due to polyhydramnios.
- Preterm neonates are at an increased risk of complications including difficulties with feeding and respiration, jaundice, infection, neonatal unit admission, and perinatal death.

36.3.3   Neonatal hypoglycaemia

- Neonates of women with GDM are at risk of developing hypoglycaemia. This occurs due to hyperinsulinaemia of the foetus which develops in the presence of maternal hyperglycaemia in utero.

- Neonatal hypoglycaemia can lead to more serious complications including severe CNS and cardiopulmonary disturbances.
- In the longer term, sequelae can include mental retardation, seizures, developmental delay, and personality disorders.

36.3.4 Neonatal jaundice

- Factors such as prematurity, immature hepatic conjugation of bilirubin, and increased enterohepatic circulation of bilirubin due to poor feeding may contribute to neonatal jaundice in offspring of women with GDM.
- Macrosomic infants have a higher oxygen demand which causes increased erythropoiesis and polycythaemia.
- The subsequent break down of these cells causes more bilirubin to increase and results in neonatal jaundice.

36.3.5 Stillbirth

The relationship between GDM and risk of stillbirth is not well understood, but women with raised fasting blood glucose appear to be at higher risk of late stillbirth. The increased risk is dependent on BMI and parity of the women. It can be as high as 4 times when BMI is >40.

36.4 Long-term (foetal)

- Offspring of women with GDM have been identified as being at risk of longer term complications.
- The abnormal in utero environment is thought to induce changes in gene expression by epigenetic mechanisms such that these predispose the offspring to metabolic complications later in life.
- Such changes have also been identified in second generation offspring.
- Passing a disease from one generation to the next via epigenetic changes is described as transgenerational transmission.

36.4.1 Childhood size and obesity

- The impact of size on offspring of women with GDM appears to extend beyond birth.
- Longitudinal studies of offspring of mothers with diabetes have demonstrated a significantly different growth pattern, with higher gestational-adjusted birthweight, and a greater rate of weight gain over the first few years of life.
- Such differences persist despite adjustment for factors including birthweight, gestational age at birth, maternal smoking during pregnancy, socioeconomic factors, current diet, and physical activity, but are attenuated after adjustment for maternal BMI.
- This suggests offspring of obese women with GDM are at higher risk than offspring of women with GDM and normal BMI.

- This highlights the important contribution of maternal obesity in the intergenerational effects of obesity and maternal diabetes.
- Compared with large for gestational age (LGA) offspring of women who had normal glucose tolerance, LGA offspring of women with GDM have been found to have increased fat mass and decreased lean body mass.

36.4.2 Impaired glucose tolerance in the offspring

- Offspring of women with GDM have been shown to have significantly higher fasting insulin and 2-hour glucose during OGTT compared with offspring of nondiabetic mothers, up to a mean age of 12 years (5.4% at 5−9 years and 19.3% at 10−16 years).
- Both childhood obesity and hyperinsulinaemia in utero (as reflected by amniotic fluid insulin concentration) have been identified as independent predictors of impaired glucose tolerance in childhood.
- These results suggest that at least part of the long-term consequences of maternal diabetes and obesity are related to hyperinsulinaemia in utero and its effects on adipogenesis.

36.4.3 Cardiometabolic risk in the offspring

- Offspring of women with GDM are thought to have a sustained increase in cardiovascular risk.
- Contributing factors are likely to include the combination of significantly higher systolic and diastolic blood pressure as well as lower HDL cholesterol which remains elevated after adjustment for age and gender, increased adiposity, and increased impaired glucose tolerance.
- A retrospective study of children aged 6−13 years found that offspring of women with diabetes during pregnancy had significantly increased markers of endothelial dysfunction; this was substantially attenuated by adjusting for maternal BMI.

36.5 Long-term maternal risks

Pregnancy has been described as a cardiometabolic "stress test" such that the physiological changes of pregnancy unmask subclinical conditions including diabetes and hypertension (HTN).

36.5.1 Type 2 diabetes

- During the 1950s it was first recognised that women with hyperglycaemia during pregnancy were more likely to develop type 2 diabetes (T2DM) following pregnancy.
- It is now recognised that up to 50% of women with GDM go on to develop T2DM with the cumulative incidence of T2DM at its highest in the first 3−6 years after the affected pregnancy.
- In a meta-analysis of individual risk factors, higher rates of converting to T2DM were associated with increased BMI, non-White ethnicity, family history of T2DM, insulin use during pregnancy, higher OGTT values, and earlier gestational age at diagnosis.

Gestational weight gain and birthweight were not associated with higher conversion to T2DM.

- As the supporting body of evidence has become more established this has informed clinical guidelines.
- NICE recommend that women with GDM should be offered lifestyle advice (including weight control, diet, and exercise) and offered a fasting plasma glucose measurement at 6 weeks postnatal and annually thereafter.

36.5.2   Cardiovascular risk

- Women with GDM are at risk of developing HTN later in life, with a twofold increased risk of developing HTN requiring medication within 10 years after the affected pregnancy, and are at increased risk for coronary artery disease and stroke.
- Cerebrovascular events appear to be more common more than 10 years after the affected pregnancy suggesting that efforts to reduce risk through lifestyle changes need to be sustained.
- It is not yet known whether the increased cardiovascular risk in women with GDM is related to the development of T2DM, which itself is a risk factor for cardiovascular disease.

36.5.3   Chronic kidney disease

- Chronic kidney disease is estimated to affect 13.4% of the global population.
- Early evidence suggests that women with GDM are at increased risk of developing micro-albuminuria and having subsequent renal morbidity.

36.5.4   Cancer

- Systematic reviews have reported conflicting evidence about the relationship of GDM and future development of cancer.
- One cohort study has reported that women with GDM were at increased risk of cancers of the kidney, lung, breast, thyroid, and nasopharynx.

36.6   Short-term maternal risks

- Women with GDM are more likely to have a caesarean delivery (35% vs 31% in recent studies). The most common reason for this is a suspected large for gestational infant, or delay in labour.
- The risk of gestational hypertensive disorders is approximately three times higher in women with GDM. Evidence suggests that insulin resistance may contribute to sodium retention and vasoconstriction, both of which are thought to contribute to the pathogenesis of hypertensive disorders during pregnancy.
- Both antepartum haemorrhage (APH) and postpartum haemorrhage (PPH) are more likely in women with GDM. It has been suggested that the increased risk of PPH might be due to complications of GDM which are independently associated with PPH. These include fatal macrosomia or large for gestational age, birth trauma, shoulder dystocia and operative deliveries.

36.7   Evidence for preventing type 2 diabetes after gestational diabetes

- Longer breastfeeding is associated with a reduction in the incidence of progression to T2DM following pregnancies affected by GDM.
- There is limited data to support the use of intensive diet and exercise or pharmacological intervention for reducing the risk of developing T2DM.

36.8   Implications for the obstetrician and primary care provider

- Obstetricians have an opportunity for future health promotion by providing accurate information about the longer term implications of GDM to patients, colleagues, and primary care providers.
- Specifically, obstetricians can promote breastfeeding, communicate to women and primary care providers the importance of postnatal fasting blood glucose (6−13 weeks postnatally) and annually thereafter.
- In 2016, only 18.5% of women with GDM were screened for GDM at 6 months postnatal. Further research is needed to target interventions to prevent the longer term complications of GDM for women and their offspring.

Further reading

Beloushi M.A., et al. Obesity, insulin resistance, and placental dysfunction − fetal growth. In: *Elsevier Obesity and Obsterics. 2nd ed.*; 2020:191−197.
Ching Wan Ma R., et al. Maternal obesity and developmental priming of risk of later disease. In: *Elsevier Obesity and Obstetrics. 2nd ed.*; 2020:149−163.
NICE. Diabetes in pregnancy: management from preconception to the postnatal period. *NICE guideline*, 2020. https://www.nice.org.uk/guidance/ng3.
Shou C, et al. Updates in long-term maternal and foetal adverse effects of gestational diabetes mellitus. *MFM*. 2019;1(2):91−94.
The HAPO Study Cooperative Research Group. Hyperglycaemia and adverse pregnancy outcomes (HAPO). *NEJM*. 2008;353:1991−2002.

Obesity and preeclampsia

37

Caroline Brewster[1], Chu Chin Lim[2] and Tahir Mahmood[2]
[1]Department of Obstetrics and Gynaecology, Royal infirmary of Edinburgh, Edinburgh, Scotland,
[2]Department of Obstetrics & Gynaecology, Victoria Hospital, Kirkcaldy, United Kingdom

37.1 Introduction

- Pregnant women who are obese are at risk of a multiplicity of pregnancy-related complications, in particular preeclampsia and gestational diabetes.
- It has been suggested that similar to gestational diabetes and development of type 2 diabetes, gestational hypertension and preeclampsia are revealing those with a tendency to develop cardiovascular disease.
- Obesity is known to predispose to cardiovascular disease (CVD) out with pregnancy, therefore it is important to consider that those who are obese and develop preeclampsia in pregnancy may be at a larger risk in the future.

37.2 Definitions

- Preeclampsia is defined as new onset hypertension ($>140/>90$ mmHg) after 20 weeks of pregnancy and the coexistence of one or more of the following new onset conditions: proteinuria (>300 mg in 24 hours or 2^+ on urine dipstick), uteroplacental dysfunction and organ dysfunction such as renal insufficiency, liver involvement, neurological complications or haematological complications.
- Whereas gestational hypertension is defined as new onset hypertension after 20 weeks without proteinuria.
- Preeclampsia is described as superimposed if it occurs in a women known to have preexisting hypertension or hypertension present in the first half of pregnancy (essential or secondary hypertension with a known underlying cause).
- It is well-documented that obesity is associated with chronic hypertension. Therefore it is possible that some diagnosis of 'preeclampsia' in the obese is in fact unmasking of the underlying hypertension by pregnancy.

Classification of BMI as per WHO includes overweight as BMI ≥ 25 kg/m^2, preobese 25.00−29.99 kg/m^2, obese class I 30.00−34.99 kg/m^2, class II 35.00−39.99 kg/m^2 and class III ≥ 40.00 kg/m^2.

Handbook of Obesity in Obstetrics and Gynecology. DOI: https://doi.org/10.1016/B978-0-323-89904-8.00041-X

37.3   Predisposition to preeclampsia

- The higher the BMI, the higher the patient's risk of developing preeclampsia!
- One metaanalysis showed that with increasing BMI, there was an increase in risk ratio for preeclampsia with overweight being 1.70, obese 2.93 and severely obese 4.14, respectively.
- Interestingly, a study of more than half a million women showed that short stature itself was associated with an increased risk of all types of preeclampsia occurring before 32 weeks gestation.
- The UK Obstetric Surveillance System (UKOSS) study showed an incidence of preeclampsia of 9% in severely obese patients compared to 2% of nonobese controls.
  - Obesity-related metabolic abnormalities, in particular an increase in circulating insulin, raised triglycerides (TG) concentrations, and inflammation TG concentrations associated with gestational diabetes also further increase the risk of developing preeclampsia.

37.4   Pathophysiological basis of pregnancy induced hypertension (PIH) during pregnancy

- Obesity is a type of chronic inflammatory condition. During pregnancy, fat deposition is predominantly central in obese women, therefore they are at increased risk of gestational hypertension and late onset preeclampsia.
- White adipose tissue is increased in obesity and is responsible for triglyceride storage. Visceral white adipose tissue is strongly associated correlated with insulin resistance, plasma lipids, and CVD both during pregnancy as well as in the nonpregnant state.
- Physiological hyperlipidaemia is exaggerated in obese pregnant women with a greater increase in total and VLDL TG and cholesterol and small dense and low-density lipoproteins and lower high-density lipids.
- The small dense LDL particles are increased in both obesity and preeclampsia and are both atherogenic, promoting endothelial dysfunction. It is possible that dyslipidaemia in obese women triggers the development of placental bed atherosis and preeclampsia.
- It is alleged that metabolic factors including increased leptin, tumour necrosis factor-alpha (TNF-α), interleukin 1 and 8, reduced levels of adiponectin, raised insulin, glucose and lipids exacerbate the proinflammatory and antiandrogenic mechanisms of 'placenta ischaemia-induced vascular dysfunction', thereby increasing the risk of preeclampsia.
  - Several mechanisms have been proposed as to why obesity increases your risk of preeclampsia. Preeclampsia is suspected to be as a result of ischaemic insults in the placenta increasing antiandrogenic and proinflammatory factors; soluble fms-like tyrosine kinase-1 and TNF-α, into maternal circulation which causes maternal endothelial dysfunction.
  - It is hypothesised that as obesity is itself a chronic inflammatory condition, obese women are predisposed to an amplified inflammatory response.

37.5 The placenta

- The placenta expresses virtually all known cytokines, including TNF, resistin, and leptin, which are also produced by adipose cells and their levels are more than twofold higher in obese women.
- Adipokines also modulate placental function. Leptin regulates placental angiogenesis, protein synthesis, and growth and causes immunomodulation. Thus increased local levels of leptin may modulate placental inflammation and function,
- In obese women, a higher circulating concentration of leptin and insulin leads to impaired glucose intolerance or gestational diabetes, and possibly development of syncytiotrophoblast.
- Placentae from obese women showed a two- to threefold increase in placental macrophages as compared with nonobese women. The macrophage population was characterised by the increased expression of IL-1, TNF and IL-6. These cytokines may also increase transport of amino acids to the foetus.
- Insulin stimulates foetal aerobic glucose metabolism, reduced oxygen delivery to intervillous space, leading to thickening of the placental basement membrane and reduced uteroplacental or foetoplacental blood flow, thus causing foetal hypoxaemia. Such a change will have a compounding effect in association with cytokines-led proinflammatory changes within placental structure.

37.6 Risk of miscarriage

- Miscarriage is also associated with an inflammatory response. Obesity is associated with an increased risk of miscarriage and recurrent miscarriage.
- It has also been proven that women who miscarry in their first pregnancy are three times more likely to develop preeclampsia in subsequent pregnancies compared to those who reached term in their first pregnancy.
- A heightened inflammatory response in early pregnancy is linked to both miscarriage and preeclampsia and therefore as previously discussed, may be heightened in obesity, a chronic inflammatory condition.

37.7 Chronic hypertension

- Obesity has long been known to be associated with chronic hypertension. It therefore raises the question if the development of preeclampsia in the obese pregnant population is revealing those who already have a degree of underlying hypertension.
- Interestingly the GOPEC study showed a significant increase in booking blood pressure as well as level of proteinuria in patients with a higher BMI. However, it is important to note that not all obese patients develop preeclampsia and thus these hypotheses can still be questioned.

37.8   Effect of obesity-associated PIH on pregnancy outcomes

- The rate of obesity in pregnancy is increasing with as estimated prevalence of 8.7 cases of extreme obesity (BMI >50 kg/m^2) per 10,000 deliveries in the United Kingdom. Obesity in pregnancy is associated with higher maternal morbidity and mortality.
- In the 2016–18 MBRACE review into maternal deaths, 29% of women who died were obese and 26% were overweight.
  - Compared with patients with a normal BMI, obese women are more at risk of preeclampsia as well as miscarriage, gestational diabetes, venous thromboembolism and complications in labour.
  - Foetal complications as a result of obesity include macrosomia, prematurity, congenital anomalies and stillbirth. Infants born to obese women are also at risk of obesity and metabolic disorders later in childhood.
  - It is imperative that weight is addressed prepregnancy and optimised prior to conception where possible. All women should have their height, weight and BMI calculated at their booking appointment.

37.9   Care during pregnancy

- Women with a BMI >30 kg/m^2 should be advised of the risks of pregnancy and childbirth. They should be commenced on high dose folic acid at least 1 month prior to conception and for the first trimester.
- Every obese women should be screened for risk factors (previous preeclampsia, antiphospholipid antibody syndrome, preexisting medical conditions such as renal disease or diabetes mellitus, previous pregnancy complicated with growth restriction, history of recurrent miscarriages and multiple pregnancy) for eclampsia at first antenatal visit to plan an individualised care plan,
- For women with preexisting hypertension, serum creatinine, fasting blood glucose, serum potassium and urinalysis should be performed in early pregnancy,
- Women should be advised about warning signs and symptoms of preeclampsia and the importance of timely reporting them to their healthcare providers,
- Calcium supplementation of at least 500 mg/day orally is recommended for women with a low dietary intake of calcium.
- They should be screened for gestational diabetes according the local protocols.
- Women deemed to be at high risk of preeclampsia: those with more than one additional risk factor for preeclampsia (BMI ≥ 35 kg/m^2, primiparity, family history, age over 40 and multiple pregnancy), should be commenced on aspirin 150 mg from 12 weeks and continued until delivery.

Additional interventions which may be of help are:

- Metformin in women with polycystic ovaries.
- Increased rest at home in the third trimester.
- Reduction of workload or stress.
- Low-molecular-weight heparin in the management of women with antiphospholipid syndrome.

- Regular blood pressure assessment at hospital day care units for those with nonsevere preeclampsia.
- Antihypertensive therapy of severe and nonsevere hypertension with oral and parenteral route.
- Magnesium sulfate for foetal neuro protection for preterm births (<34 weeks).
- Women with preexisting hypertension and those suspected at developing severe pre-eclampsia should undergo a schedule of blood tests regularly for maternal well-being. This would involve assessing haematological, renal and hepatic parameters as follows:
  - o Haematological: elevated white cell count, elevated INR or APTT, low platelet count.
  - o Renal: elevated serum creatinine, elevated serum uric acid.
  - o Hepatic: nausea or vomiting, RUQ or epigastric pain, elevated serum AST, ALT, LDH, or bilirubin, low plasma albumin.
- Women with a BMI of over 35 kg/m^2 should also undergo serial growth ultrasounds and Doppler velocimetry-based assessment of the foetal circulation as symphysial fundal height measurements are less accurate for the diagnosis of intrauterine growth assessment. This should be complemented by maternal foetal movement awareness and foetal heart rate assessment
- An anaesthetic assessment is useful in preparation for pain relief choices during labour and anaesthetics for interventions for delivery are important for forward planning.
- Planned induction of labour at term has been shown to reduce the incidence of caesarean section.
- Monitoring of blood pressure can be less accurate in obese patients and readings can vary depending on the cuff size used. It is important that larger cuff sizes are available in ante-natal clinics. A clean urine sample can also be more difficult to collect in obese patients who are heavily pregnant.

37.10   Place of delivery

- All obese women with or without a high blood pressure should be delivered in a hospital with facilities for emergency obstetric and neonatal care.
- Close monitoring of blood pressure should be continued and if on antihypertensive treat-ment, it should be continued during labour.
- Provision of effective analgesia also helps with the control of blood pressure. Epidural analgesia benefits the foetus by decreasing maternal respiratory alkalosis, compensatory metabolic acidosis, and release of catecholamines.
- Epidural or spinal should be preferred methods for operative delivery as there are signifi-cant challenges associated with general anaesthesia.
- The third stage should be actively managed as obese women are at higher risk of postpar-tum haemorrhage, and intravenous oxytocin should be used.

37.11   Postpartum

- During immediate postpartum period, there should be close monitoring of blood pressure, and antihypertensive therapy should be continued postpartum,

- Nonsteroidal antiinflammatory should be cautiously used if hypertension is difficult to control as there is evidence of kidney injury in there is elevated creatinine ($>90\,\mu mol/L$), or the platelet count is $<50\times10^9\,L^{-1}$.
- Postpartum thromboprophylaxis should be instituted based on individual risk factors.

37.12 Long-term implications

- Approximately one in five women who suffer hypertensive disease in pregnancy will develop it again in future pregnancies.
- Beyond pregnancy, those with a history of preeclampsia are up to three times more likely to have a major adverse cardiac event, twice the risk of cardiovascular mortality, three times an increase in risk of stroke, and up to five times the risk of developing hypertension.
  - Support should be given to obese women to achieve a healthier BMI following pregnancy. One study has shown that in patients unaffected by hypertensive disorders in their first pregnancy an increase in BMI of $2-4\,kg/m^2$ between deliveries is associated with the development of hypertension in the subsequent pregnancy.
  - Achieving a healthy BMI will therefore not only lessen risks in subsequent pregnancies but also reduce their risks of hypertension, type 2 diabetes, stroke and CVD in later life.
  - It raises the issue that obese women who develop preeclampsia in pregnancy are at a significant increase of long-term morbidity, in particular hypertension and CVD. It is therefore particularly important to encourage these women to lose weight, have a healthy diet, and maintain a healthier lifestyle.

37.13 Practical consideration for obesity and pregnant women

37.13.1 Challenge of accurately measuring blood pressure

- There is a direct relationship between BMI and upper arm circumference in the pregnant population, and therefore the increasing incidence will directly affect the accuracy of blood pressure measurements in pregnancy and of preeclampsia, if correct cuff size is not used.
- Therefore large-sized cuff should be available in the antenatal clinics which can encircle at least 80% of the arm. In a series of patients undergoing bariatric surgery with BMI of $66.7+/- 13.8\,kg/m^2$, the mean upper arm circumference was $48.6+/-7.5\,cm$. Such findings should be considered when using an appropriate sized cuff.
- The mainstay of management targets is prevention of hypertension in pregnant women.
- Preconception counselling encourage women to enter pregnancy with a lower BMI, limiting weight gain, and taking low-dose aspirin to prevent preeclampsia from before 16 weeks' gestation.
- Low-dose aspirin has been recommended for use in women with a BMI over $35\,kg/m^2$ who have at least one other moderate risk factor for preeclampsia.

- A systematic review and metaanalysis has been published including 15 studies; no statistically significant difference was demonstrated in treating hypertensive diseases in pregnancy with the use of metformin. The metaanalysis was able to extrapolate that there was a high probability of over 90% that metformin has a beneficial effect in preventing preeclampsia, gestational hypertension, and any hypertensive disease in pregnancy, when compared to placebo or other treatments.
- Home blood pressure monitoring can not only enable closer and more accurate monitoring, but can also help to enable appropriate diagnosis and initiation of antihypertensive medication. According to the NICE guidance, the first-line medication in the treatment of hypertension in pregnancy is labetalol, followed by nifedipine and methyldopa. There is no particular mention regarding any difference in the choice of antihypertensive therapy in the obese pregnant woman.
- With the adjunct of a home blood pressure monitor that has been validated for use in pregnancy and a specially designed Smartphone app, women's blood pressure can be monitored remotely and antihypertensive treatment can be titrated appropriately without compromising safety

37.14   Conclusion

- It is clear that obesity is a risk factor for hypertension and subsequent preeclampsia in pregnancy. The basis of physiology is around the chronic inflammatory condition of obesity and this augmenting the inflammation within the placenta associated with preeclampsia.
- Both preeclampsia and obesity themselves are associated with a greater risk of CVD in later life. It is unclear how much this is exaggerated if a patient suffers from both conditions.
- Ultimately, it is important to address BMI preconception as well antenatally and postpartum.
- A reduction in BMI between pregnancies can help to reduce risks in subsequent pregnancies. For obese patients who develop gestational hypertension or preeclampsia, it is important they are followed up to support them in reducing their risks of diabetes and CVD in later life.

Further reading

Al Beloushi M, Doshani A, Konje JC. Obesity, insulin resistance, and placental dysfunction-fetal growthAvailable from: https://doi.org/10.1016/B978-0-12-817919-5.00020-5.

Broughton-Pipkin F, Loughna P. Obesity, hypertension, and preeclampsiaAvailable from: https://doi.org/10.1016/B978-0-12-817919-5.00013-8.

Denison F.C., Aedla N.R., Keag O., Hor K., Reynolds R.M., Milne A., Diamond A., On behalf of the Royal College of Obstetricians and Gynaecologists. Care of women with obesity in pregnancy. *Green-top Guideline No. 72 BJOG*; 2018

Dude AM, Shahawy S, Grobman WA. Delivery-to-delivery weight gain and risk of hypertensive disorders in a subsequent pregnancy. *Obstet Gynecol.* 2018;132(4):86874.

GOPEC Consortium. Disentangling fetal and maternal susceptibility for pre-eclampsia: a British multicenter candidate-gene study. *Am J Hum Genet*. 2005;77(1):12731.

Knight M, Kurinczuk JJ, Spark P, Brocklehurst PUK Obstetric Surveillance System. Extreme obesity in pregnancy in the United Kingdom. *Obstet Gynecol*. 2010;115(5):98997.

NICE. National Collaborating Centre for Women's and Children's Health. Hypertension in pregnancy: the management of hypertensive disorders during pregnancy. *NICE Clinical Guideline, NG133*: London: RCOG;2019.

Knight M, Bunch K, Tuffnell D, Shakespeare J, Kotnis R, Kenyon S, Kurinczuk JJ, (eds.), Saving lives, improving mothers' care lessons learned to inform maternity care from the UK and Ireland Confidential Enquiries into Maternal Deaths and Morbidity 2016−18; December 2020.

Oxford: National Perinatal Epidemiology Unit, University of Oxford.

Spradley FT. Metabolic abnormalities and obesity's impact on the risk for developing pre-eclampsia. *Am J Physiol Regul Integr Comp Physiol*. 2017;312(1):R512.

Wang Z, Wang P, Liu H, He X, Zhang J, Yan H, et al. Maternal adiposity as an independent risk factor for pre-eclampsia: a meta-analysis of prospective cohort studies. *Obes Rev*. 2013;14:508. 21.

Venous thromboembolism in obese mother

Nithiya Palaniappan[1] and Nirmala Mary[2]
[1]Department of Obstetrics and Gynaecology, Victoria Hospital, Kirkcaldy, Scotland, [2]NHS Lothian, Edinburgh, Scotland

38.1 Introduction

Venous thromboembolism (VTE) is the leading cause of direct maternal deaths in the United Kingdom. Obesity is a significant contributory factor for VTE in pregnancy and the risk increases by fourfold when compared with women of normal BMI.

- Recent evidence has shown obesity as an important contributory factor for inappropriate dosage of low-molecular-weight heparin for thromboprophylaxis or VTE treatment as the dosage is mainly weight-based.
- With increasing weight, exact dosing may be a challenge to assess.
- The classification of overweight and obesity according to the National Institute of Clinical Excellence is:
  o Obesity I—BMI 30−34.9 kg/m^2
  o Obesity II—BMI 35−39.9 kg/m^2
  o Obesity III—BMI 40 kg/m^2 or more.

38.2 Pathophysiology

- Physiological increase in coagulation factors VIII, IX, X and fibrinogen with simultaneous decreased antifibrinolytic activity.
- Anatomical changes caused by the gravid uterus favours venous stasis and marked reduction in the usual blood flow in the lower limbs.
- Obesity causes exacerbation of the inflammatory changes causing further increase in coagulation factors, prothrombin and fibrinogen levels, increased levels of plasminogen activator inhibitor Type I.
- Exaggerated proinflammatory environment which increases endothelial dysfunction and oxidative stress, which in turn exacerbates the risk factors for VTE.
- Increased maternal abdominal adiposity associated with obesity further restricts venous return and also increases intraabdominal pressure especially during the intrapartum period.

Handbook of Obesity in Obstetrics and Gynecology. DOI: https://doi.org/10.1016/B978-0-323-89904-8.00025-1

38.3   Other contributory factors

- Increased risk of preeclampsia which further increases risk of VTE.
- Increased risk of interventions during pregnancy and delivery including failed inductions, operative delivery, and prolonged labour further multiplies the risk of VTE.
- Increased risk of sepsis associated with obesity further increases risk of VTE.
- Risk of reduced mobility associated with obesity during pregnancy can increase risk of VTE.

38.4   Symptoms of venous thromboembolism

38.5   Prevention of venous thromboembolism in obese pregnant women

- Robust guidelines for VTE prophylaxis in obese women.
- Risk scoring strategies and auditing continuously to improve the risk scoring is beneficial in the overall reduction of VTE in pregnancy.
- Correct assessment of BMI so that appropriate weight-based dosage is prescribed
- Universal and individualised risk assessment at booking, start of third trimester and repeated during any admission to the hospital or intercurrent illness.
- Repeat risk assessment at delivery.
- Education of women for prompt self-referral if any signs or symptoms of VTE.
- Early mobilisation after delivery.
- Postnatal thromboprophylaxis regardless of mode of delivery for women over BMI 40.
- Avoid unnecessary hospital admission.
- Multidisciplinary input from metabolic and haematology team in extremes of obesity.

38.6   Agents for thromboprophylaxis

Low-molecular-weight heparin (LMWH) is the agent of choice.

Table 38.1 Symptoms of deep vein thrombosis and pulmonary embolism.

DVT	PE
Unilateral leg pain and swelling	Dyspnoea
Unequal calf size	Raised JVP
Pain in the groin/flank/buttock area	Haemoptysis
Pain in the lower abdomen	Syncope
Leucocytosis	Chest pain/pleuritic pain

- Predictable pharmacokinetics with LMWH.
- Once daily dosage for prophylaxis.
- Weight-specific dosage to be followed although lack of data for exact dosage.
- Reduced risk of bleeding due to enhanced ratio of anti-Xa to anti-IIa.
- Reduced risk of heparin-induced thrombocytopenia, skin reactions and osteoporosis.
- Monitoring anti-Xa levels not required unless there is extremes of weight or abnormal renal function.
- Monitoring platelet levels are not required unless there is a previous exposure to unfractionated heparin.
- Tinzaparin, Dalteparin, and Enoxaparin are generally used in the UK.
- Doses to be reduced for women with renal impairment.
- Evidence suggests increased bruising episodes when women on LMWH undergo caesarean section.

Unfractionated heparin:

- Not used as first line.
- Usually considered during peripartum period where women are at high risk of haemorrhage or to cover regional anaesthesia.
- Platelets levels to be checked every 2−3 days from day 4−14.
- Hospital or local guidelines to be referred for administration.

Fondaparinux:

- Not used as first line.
- Considered reasonable alternative for women who are allergic to Heparin despite limited evidence.
- In pregnancy only to be used with expertise from haematologists.
- Considered safe for foetus.
- Half-life is 18 hours.
- 36−42 hours to be allowed prior to considering regional anaesthesia in the intrapartum period.

Danaparoid:

- Not used as first line.
- Used in conjunction with haematologists who have special interest in pregnancy.
- Limited evidence in pregnancy.
- Half-life is 24 hours and can therefore pose difficulties for women requiring regional anaesthesia.
- Not contraindicated in breastfeeding although evidence is limited for this.

Warfarin:

- Ideally avoided in pregnancy
- Causes warfarin embryopathy in 5% of foetuses exposed between 6 and 12 weeks.
- Reported complications include increased risk of miscarriage, bleeding, stillbirth, and foetal and maternal haemorrhage.
- Suitable during postnatal period from day 5−7 onwards as this does not increase risk of haemorrhage.
- Suitable during breastfeeding.

Antiembolism stockings/thromboembolic deterrent stockings:

- Appropriate sizing is required.
- Graduated compression with a calf pressure of 14−15 mmHg is recommended in pregnancy and the puerperium for women who are hospitalised and have a contraindication to LMWH.
- Women with extreme obesity may not be able to get appropriate sized stockings. Prior planning and procurement is essential.

Aspirin:

- Not considered for prophylaxis for VTE.

38.7   Investigation for venous thromboembolism

- Doppler ultrasound of both the lower limbs is the investigation of choice, if a deep vein thrombosis is suspected.
- If Doppler is negative and clinical suspicion is high, continue treatment dose of anticoagulant and repeat Dopplers on day 4 and day 7.
- Ileofemoral DVTs are common in pregnancy.
- The ileofemoral veins may be particularly difficult to image in obese pregnant women due to tissue density; caution to be exercised when interpreting Doppler in obese pregnant women.
- D-dimers are high in pregnancy due to physiological changes and therefore are not an investigation of choice for suspected VTE in pregnancy.
- Chest X-ray and ECG are the initial investigations when a pulmonary embolism is suspected.
- Chest X-Ray is performed to rule out concurrent infection.
- Ventilation quotient scan or computerised tomography pulmonary angiogram is the investigation of choice if a pulmonary embolism is suspected.
- Manual handling during imaging has to be considered cautiously for obese pregnant women.
- Imaging may be affected due to obesity and limitations of equipment will have to be addressed.
- Baseline bloods for a full blood count, renal functions can be done to assess basic parameters.
- CT scan of the brain with contrast is the investigation of choice if a cerebral venous sinus thrombosis is suspected.

38.8   Management of venous thromboembolism in obese pregnant women

- Treatment with low-molecular-weight heparin should be immediately commenced even prior to any investigation until the diagnosis confirms or refutes VTE.
- Investigate as above with appropriate imaging.
- Multidisciplinary team involving the haematology team and obstetric team are crucial.

- The treatment dose of low-molecular-weight heparin should be titrated against weight and given in single or twice daily doses.
- RCOG recommends anti-Xa levels monitoring if the weight of the patient is >90kg.
- If massive pulmonary embolism is confirmed, thrombolysis is the treatment of choice.
- Treatment is recommended until 6 weeks following delivery.

38.8.1   RCOG charts for dosage of low-molecular-weight heparin

Thromboprophylaxis (Table 38.2):
   Treatment doses of low-molecular-weight heparin:
   (Table 38.3)
   (Table 38.4)
   (Table 38.5)

38.9   Intrapartum care in obese pregnant women with regard to venous thromboembolism

- Clear instruction to be documented for intrapartum care when women are on prophylactic or treatment dose of LMWH antenatally.
- Multidisciplinary input involving obstetrics, haematology, and anaesthetic specialities are mandatory.
- Prior discussion regarding the course of LMWH should be discussed and documented.
- Thromboprophylaxis dosage should be stopped if spontaneous labour commences.
- Thromboprophylaxis to be stopped 12 hours prior to induction of labour or elective procedure.
- Treatment dose to be stopped 24 hours prior to elective procedure.
- If patients are high risk then unfractionated heparin may be considered during intrapartum period or for regional anaesthesia.
- LMWH can be restarted in the puerperal period, the timing of which is determined by the mode of delivery, additional procedures like epidural and other complications.

38.10   Postnatal care in obese pregnant women

- Prothrombotic changes continue for several weeks following delivery.

Table 38.2 RCOG chart for dosage of low molecular weight heparin for thromboprophylaxis.

Booking weight	Enoxaparin	Dalteparin	Tinzaparin
91–130 kg	60 mg daily	7500 units daily	7000 units daily
131–170 kg	80 mg daily	10,000 units daily	9000 units daily
>170 kg	0.6 mg/kg/day[a]	75 u/kg/day[b]	75 u/kg/day[c]

[abc]Can be given in divided doses.

Table 38.3 RCOG chart for treatment dose of enoxaparin in venous thromboembolism.

Booking or early pregnancy weight	Enoxaparin
90–109 kg	100 mg twice daily or 150 mg once daily
110–125 kg	120 mg twice daily or 180 mg once daily
>125 kg	Discuss with haematologist

Table 38.4 RCOG chart for treatment dose of dalteparin in venous thromboembolism.

Booking or early pregnancy weight	Dalteparin
90–109 kg	10,000 units twice daily or 20,000units once daily
110–125 kg	12,000 units twice daily or 24,000units once daily
>125 kg	Discuss with haematologist

Table 38.5 RCOG chart for treatment dose of tinzaparin in venous thromboembolism.

Tinzaparin
175 mg/kg/once daily

- Risks increase with operative delivery and complications like postpartum haemorrhage.
- Postnatal thromboprophylaxis is recommended for 10 days postdelivery regardless of mode of delivery.
- With additional risk factors and or extreme obesity thromboprophylaxis can be extended to 6 weeks.
- Antiembolism stockings which are measured for appropriate calf size are recommended for use in women in the puerperium.
- Early mobilisation and hydration reduces the risk of VTE.
- In women who are being treated for VTE in pregnancy, treatment should continue for 6 weeks and at least 3 months of total treatment.
- Warfarin may be given after the 5th day of delivery to avoid postpartum haemorrhage.
- Both heparin and warfarin are safe for breastfeeding.
- Women will need a postnatal review to assess the medications and plans discussed and documented for discontinuing medications.

38.11   Contraception in obese women with a history of venous thromboembolism

- Barrier methods can be considered, however they carry a high failure rate.

- Nonhormonal intrauterine device is the preferred contraception of choice with a history of VTE.
- Levonorgestrel intrauterine system, *Mirena-IUS*, progesterone-only pill, or the progesterone implant may be considered, however only if advantages of using the method generally outweigh the theoretical or proven risks.
- Combined oral contraceptive pills are contraindicated when there is a history of VTE.

38.12 Future advice in obese pregnant women with history of venous thromboembolism in index pregnancy

- Thrombophilia testing should be considered after discontinuing treatment but only if it will alter future management for the woman.
- Prepregnancy planning in future pregnancies is advised to assess any new risk factors that have developed in the interval period.
- Antenatal and postnatal thromboprophylaxis in any future pregnancy will need to be discussed and documented.
- Early referral to the combined obstetric haematology clinic is recommended to commence antenatal thromboprophylaxis.
- Advice for thromboprophylaxis has to be given in case of new risk factors like long-haul air travel or major surgery.

Further reading

Reducing the Risk of Venous Thromboembolism during Pregnancy and the Puerperium. *Green-top Guideline No.37a. RCOG April* 2015.

Thromboembolic Disease in Pregnancy and the Puerperium: Acute Management. Green-top Guideline No.37b. RCOG. April; 2015.

RCOG. Care of. Women with Obesity in Pregnancy. *Green-top Guideline No.72.* RCOG November 2018.

NIH. Obesity in Pregnancy, ACOG Practice Bulletin, Number 230. *Obstet Gynecol*; 137(6): e128−e144. June 2021.

Ellison J, Thomson A, Chapter 14 Obesity and Obstetrics. *2020 Obesity and Venous Thromboembolism and management plan.* Elsevier; June. Second Edition Elsevier; June.

NPEU. MBRRACE-UK: Saving Lives, *Improving Mothers' Care 2020: Lessons to inform maternity care from the UK and Ireland Confidential Enquiries in Maternal Death and Morbidity 2016−18.*

Management of Prepregnancy. Pregnancy, and Postpartum Obesity from the FIGO Pregnancy and Non-Communicable Diseases Committee: A FIGO. International Federation of Gynecology and Obstetrics Guideline.

Induction of labour in obese pregnancies

39

Kahyee Hor
Department of Obstetrics and Gynaecology, Royal Infirmary of Edinburgh and University of Edinburgh, Scotland

- Induction of labour (IOL) involves artificially stimulating the onset of labour through chemical and/or mechanical methods, with the aim of achieving a vaginal delivery, prior to the onset of spontaneous labour.
- Approximately 33.0% of deliveries in the United Kingdom (UK) in 2019−20 were induced, whereas data from the Centres for Disease Control and Prevention demonstrated that 29.4% of deliveries in the United States (US) in 2019 were induced. These figures have increased over the last decade where in the UK, only one in five women had IOL in 2009−10.
- Induction of labour is offered when continuing with the pregnancy confers more risks to the mother and/or the foetus than delivery. There are various clinical indications which support IOL including maternal diabetes, preeclampsia and hypertension, as well as foetal growth restriction, to name a few.
- One of the key indications for IOL is to prevent prolonged pregnancy which has been shown to be associated with increased perinatal morbidity and mortality.
- Data from the Cochrane Database suggest IOL at or beyond 37 weeks' gestation in 'low-risk' pregnancies associates with a reduction in perinatal deaths, stillbirths, and neonatal intensive care admissions.
- Additionally, there is a modest reduction in caesarean section rates in the induction of labour group compared to expectant management, with no evidence of differences in the rate of instrumental deliveries.
- The National Institute for Health and Care Excellence (NICE) in the UK as well as the Society of Obstetricians and Gynaecologists of Canada suggest that women with 'uncomplicated pregnancies should usually be offered induction of labour between 41^{+0} and 42^{+0} weeks'.
- Regular antenatal assessments with twice-a-week amniotic fluid volume measurements and 'nonstress test' with cardiotocography should be offered to women who choose to delay induction.
- The American College of Obstetrics and Gynaecology (ACOG), on the other hand, have suggested IOL may be considered at 41^{+0} to 42^{+0} weeks' gestation and should be recommended beyond 42^{+0} weeks' gestation.
- Given the global increase in obese pregnant women, it is crucial to have clear guidance on IOL in this group of women. However, there remains a lack of consensus in optimum gestation for IOL and choice of induction agents. Nevertheless, this chapter aims to collate the available information to aid in decision-making with respect to induction of labour in obese women.

Handbook of Obesity in Obstetrics and Gynecology. DOI: https://doi.org/10.1016/B978-0-323-89904-8.00002-0

39.1 Indications of induction of labour

1. Postdates pregnancy
   - Denison et al. have demonstrated that maternal obesity is associated with a reduced likelihood of spontaneous onset of labour and greater risk of postdates pregnancies.
   - Maternal obesity during pregnancy is also associated with an increased risk of complications including maternal diabetes and preeclampsia, which may in themselves warrant earlier delivery.
   - This is reflected in an increased risk of induction of labour among obese pregnancies (odds ratio 1.70, 99% confidence interval 1.64−1.76).
   - In general, IOL among obese pregnant women with prolonged pregnancy is considered safe with no difference in Apgar scores and cord blood pH between obese and lean pregnancies.
   - However, it has also been reported that the likelihood of vaginal delivery following IOL in obese pregnant women is reduced when compared to those with normal body weight (rate of unassisted vaginal delivery in obese vs lean pregnancies, 55.0% vs 57.9%).
   - Obese pregnancies are associated with an increased risk of delivery by caesarean section (rate of caesarean section delivery in obese vs lean pregnancies, 28.5% vs 18.9%, $P < .001$) and prolonged second stage of labour.
   - Interestingly, when compared to IOL, elective caesarean sections in obese pregnant women have not been shown to improve maternal and neonatal outcomes, further supporting IOL as a safe and reasonable intervention for obese pregnancies.
2. Diabetes, including gestational diabetes and preexisting diabetes
   - Both gestational diabetes mellitus (GDM) and preexisting diabetes is more common among obese pregnant women, and associated with an increased risk of maternal and foetal morbidity and mortality.
   - Delivery between before 39^{+0} weeks' gestation is recommended in women with preexisting type 1 or 2 diabetes, whereas those with uncomplicated GDM should be offered delivery before 40^{+0} weeks' gestation.
   - Both NICE and ACOG have suggested that the mode of delivery should be determined by maternal diabetic control and foetal well-being.
3. Hypertensive disorders of pregnancy, including preeclampsia
   - Maternal obesity is associated with a higher incidence of preexisting hypertensive disorders, pregnancy-induced hypertension and preeclampsia.
   - Early delivery is commonly indicated in cases where there is evidence of maternal and/or foetal deterioration.
   - Further, results from the HYPITAT trial indicated that induction of labour in women with pregnancy-induced hypertension and mild preeclampsia was associated with a lower composite risk for poor maternal outcome secondary to progression of the disease.
   - While this study did not perform a subgroup analysis for obese pregnant women, there is evidence suggesting that obese pregnant women with preeclampsia are less likely to progress to a vaginal delivery following IOL.
   - The decision with regards to mode of delivery in this cohort of patients therefore requires a personalised and multidisciplinary approach.

4. Foetal macrosomia
- While maternal obesity, with or without concurrent diabetes, associates with an increased risk of foetal macrosomia and shoulder dystocia.
- However, prediction of large-for-gestational age or foetal macrosomia during the antenatal period remains challenging due to inaccuracies with symphyseal-fundal height measurements in obese pregnant women.
- Ultrasound assessment of estimated foetal weight during the third trimester is also subject to a degree of variability and increased maternal adiposity can impact on the quality of the images – thus resulting in inaccurate measurements.
- A Cochrane review on induction of labour in pregnancies with suspected foetal macrosomia did not demonstrate a clear role of IOL at or near term in reducing the risk of brachial plexus injuries.
- Therefore, in the absence of other risk factors, IOL for suspected foetal macrosomia in obese pregnancies should be considered with caution.

5. Elective induction of labour
- The risk of stillbirth is greater in pregnancies affected by maternal obesity and there appears to be a linear relationship between risk of stillbirth and advancing gestation in obese pregnancies.
- Crucially, there is evidence that elective IOL in obese pregnancies may improve neonatal outcomes without increasing maternal morbidity.
- Indeed, the Royal College of Obstetricians and Gynaecologists (RCOG) has recommended that while maternal obesity in itself is not an indication for elective IOL, it should still be considered and discussed on an individual basis.

39.2 Counselling for induction of labour

1. Gestation at the time of IOL
- In the absence of maternal and/or foetal complications, consider offering elective IOL at term.
- Pregnancies complicated by diabetes:

	NICE	ACOG
Gestational diabetes	If uncomplicated, offer delivery between 39^{+0} to 40^{+6} weeks' gestation	Diet/exercise control: expectant management until 40^{+6} weeks' gestationWell-controlled on medications: offer delivery between 39^{+0} to 39^{+6} weeks' gestationPoorly controlled in spite of medications: offer delivery between 37^{+0} to 38^{+6} weeks' gestationOnly consider delivery 34^{+0} to 36^{+6} weeks' gestation if poor glycaemic control in spite of in-hospital treatment and/or evidence of foetal compromise

(Continued)

(Continued)

	NICE	ACOG
Preexisting diabetes	Offer delivery between 37^{+0} to 38^{+6} weeks' gestation	Well-controlled diabetes: offer delivery between 39^{+0} to 39^{+6} weeks' gestationPoor glucose control, evidence of vasculopathy/ nephropathy or history of previous stillbirth: offer delivery between 36^{+0} to 38^{+6} weeks' gestation.

2. Assessment prior to IOL
- Confirm gestation and presence of any risk factors (e.g. diabetes or hypertensive disorders of pregnancy) which may influence timing of IOL.
- Consider referral to Anaesthetic team − the RCOG recommends that women with BMI ≥ 40 kg/m^2 should be referred to the anaesthetic team for early discussion of options for analgesia and to assess and plan for potential challenges with regional and general anaesthesia.
- Foetal lie and presentation − this may be challenging with increased maternal adiposity and therefore confirmation of presentation with ultrasound may be indicated.
- Cervical assessment − the Bishop scoring system is commonly used to assess the cervix and may influence the choice of induction agent.

Score	Dilatation (cm)	Position of cervix	Effacement (%)	Cervical consistency	Station (−3 to +3)
0	Closed	Posterior	0−30	Firm	−3
1	1−2	Mid-position	40−50	Medium	−2
2	3−4	Anterior	60−70	Soft	−1 to 0
3	5−6		>80		+1 to +2

3. Previous caesarean section
- Individualised care, taking into consideration severity of obesity, parity and previous history of vaginal deliveries, cervical assessment and maternal preference.
- Increased risk of unsuccessful trial of labour in obese pregnant women with previous caesarean section
  o Risk of unsuccessful trial of labour in morbidly obese (i.e. BMI ≥ 40 kg/m^2) versus those with normal BMI − 39.3% versus 15.2%.
  o Successful trial of labour in obese (i.e. BMI ≥ 30 kg/m^2) versus those with normal BMI − 54.6% versus 70.5%.
- Risk of uterine scar dehiscence or rupture is increased with maternal obesity (2.1% in morbidly obese pregnancies).
- Risk of neonatal injuries, including brachial plexus injuries, fractures, and lacerations (fivefold increase in risk, 1.1% in morbidly obese pregnancies versus 0.2% in normal weight pregnancies).
4. Foetal monitoring
- Continuous electronic foetal monitoring should be performed following administration of IOL measures. Once there is confirmation of foetal well-being with a normal

cardiotocogram, intermittent auscultation is reasonable unless there are indications for continuous electronic foetal monitoring.
- Once active labour is established, continuous electronic foetal monitoring should be offered.
- Palpation of uterine activity and accurate recording of foetal heart using transabdominal probes may be challenging with increased maternal adiposity. Women should be counselled about the possible need for a foetal scalp electrode to monitor foetal heart rate.

5. Home versus inpatient IOL
- Evidence is inconclusive although there is evidence suggesting that there is no difference in maternal and foetal morbidity and mortality in home versus inpatient IOL.
- Outpatient IOL may associate with shorter duration of hospital stay and reduced caesarean section rates. However further studies are required to confirm the validity of these findings, especially in pregnancies complicated by maternal obesity.

39.3 Methods of induction of labour

1. All women should be offered a vaginal examination with membrane sweep where possible.
2. Pharmacological methods for cervical ripening:
- Prostaglandin E_2
  o Available in vaginal preparation as a controlled-release pessary, gel or tablet.
  o Controlled-release pessary: 1 dose over 24 hours (maximum).
  o Gel or tablet: usually divided into multiple doses given at 6-hour intervals. NICE guidance recommends maximum of 2 doses.
  o Risk of tachysystole or hyperstimulation.
- Prostaglandin E_1
  o Not recommended for IOL under NICE guidance, unless it was for the use of induction of labour following stillbirth or in the context of a clinical trial.
  o Misoprostol can be used in IOL in viable pregnancies under the SOGC and ACOG guidance and is recognised to have a rapid onset of action. It carries a risk of tachysystole and therefore is not recommended for use in an outpatient setting or in women with a previous Caesarean section due to risk of uterine rupture.
  o In Europe, the use of vaginal misoprostol for IOL was approved in 2013 and Varlas et al. demonstrated that MVI-Misodel is an effective pharmacological IOL agent in high risk (including postdates pregnancy, hypertension and diabetes) obese pregnant women with no evidence of difference in Caesarean section rates, as well as maternal and neonatal outcomes.
- Consider adjusting prostaglandin (and oxytocin) dose as maternal obesity has been shown to require high doses of induction agents.
3. Mechanical methods:
- Transcervical double-balloon or Foley catheters
  o NICE does not recommend routine use of catheters for IOL. However, there appears to be a renewed interest in this IOL method, particularly in women who have had a previous caesarean section, as it does not directly stimulate uterine contractions, and thus should in theory have a reduced risk of uterine rupture. It may

also be a favoured option for outpatient IOL due to lower risk of uterine hyperstimulation.

o The PROBAAT and PROBAAT-II randomised controlled trial has shown that IOL in an unfavourable cervix with pharmacological methods (prostaglandin, PROBAAT and misoprostol, PROBAAT-II) was not superior to the use of Foley catheter. Additionally, the use of Foley catheters may associate with a lower risk of maternal morbidity.

o There is evidence supporting the use of intrauterine devices such as transcervical catheters in conjunction with oral misoprostol in obese pregnant women. Kehl et al. have shown that sequential use of the double-balloon catheter followed by oral misoprostol associated with higher rates of vaginal deliveries when compared to oral Misoprostol alone. However, this is a small study with only 400 participants and further studies are required to explore neonatal outcomes and patient experience from this method.

4. Amniotomy:
   - NICE guidance does not recommend the use of amniotomy as the primary method for IOL, unless there are contraindications for pharmacological agents.
   - Reserved for women with favourable cervix.
   - Consider use of oxytocin following amniotomy. There is a lack of consensus about the interval between amniotomy and commencement of the oxytocin infusion, although the SOGC suggests 'early' use of oxytocin following amniotomy.

Further reading

Clinical Practice Obstetrics Committee. Induction of Labour. SOGC clinical practice guideline. *J Obstetr Gynaecol Can.* 2013;35(9):840−857.

Denison FC, Aedla NR, Keag O, Hor K, Reynolds RM, Milne A, Diamond A. Care of women with obesity in pregnancy (Green-top Guideline No. 72) *BJOG.* 2018;126(3). e62−106.

Dong S, Khan M, Hashimi F, Chamy C, D'Souza R. Inpatient vs outpatient induction of labour: a systematic review and meta-analysis. *BMC Pregnancy Childbirth.* 2020;20:382.

Ellis JA, Brown CM, Barger B, Carlson NS. Influence of maternal obesity on labor induction: a systematic review and meta-analysis. *J Midwifery Women's Health.* 2019;64 (1):55−67.

NIH. Practice bulletin no. 146: Management of late-term and postterm pregnancies. *Obstet. Gynecol.* 2014;124(2 Pt 1):390−396.

Middleton P., Shepherd E., Morris J., Crowther C.A., Gomersall J.C. Induction of labour at or beyond 37 weeks' gestation. *Cochrane Database Syst.* Rev. 2020;(7) CD004945. Available from: https://doi.org/10.1002/1461858.CD004945.pub5. Accessed 01.05.21.

National Institute for Health and Care Excellence (NICE). Inducing Labour. Clinical Guideline [CG70]. Published 23 July 2008, last updated June 2012. <https://www.nice. org.uk/guidance/cg70>; 2008 Accessed 01.05.21.

Sebire NJ, Jolly M, Harris JP, et al. Maternal obesity and pregnancy outcome: a study of 287 213 pregnancies in London. *Int. J. Obes.* 2001;25:1175−1182.

Subramaniam A, Jauk VC, Goss AR, Alvarez MD, Reese C, Edwards RK. Mode of delivery in women with class III obesity: planned caesarean compared with induction of labour. *Am. J. Obstetr. Gynaecol.* 2014;211(6):700. e1−9.

Robinson CJ, Hill EG, Alanis MC, Chang EY, Johnson DD, Almeida JS. Examining the effect of maternal obesity on outcome of labour induction in patients with preeclampsia. *Hypertens Pregnancy.* 2010;29(4):446−456.

Varlas VN, Bostan G, Nasui BA, Bacalbasa N. AL Pop. Is misoprostol vaginal insert safe for the induction of labor in high-risk pregnancy obese women? *Healthcare.* 2021;9(4):464. Basel.

Yao R, Schuh BL, Caughey AB. The risk of perinatal mortality with each week of expectant management in obese pregnancies. *J Maternal-Fetal Neonatal Med.* 2019;32 (3):434−441.

Intrapartum care for obese women

Diogo Ayres-de-Campos and Andreia Fonseca
Obstetrics Department, North Lisbon University Hospital Center, Lisbon Academic Medical Center, Lisbon, Portugal

40.1 Introduction

- The prevalence of obesity in high-resource countries is rising in all age groups.
- It is estimated that over a third of women in the reproductive age are obese.
- Obesity is both a metabolic and an inflammatory disorder.
- Obesity is defined as a body mass index (BMI) of 30 or above, and the following classification is used in this chapter:
  o class I: BMI 30.0−34.9.
  o class II: BMI 35.0−39.9.
  o class III or morbid obesity: BMI ≥ 40.0.
- Obesity carries with it a higher risk of adverse pregnancy outcomes, and this risk increases with the severity of obesity.
- Obese women often present with comorbidities, which ultimately amplify the risk of maternal and perinatal complications.
- Labour progress is slower in obese women and intrapartum care is often more challenging. This does not mean that obesity is an indication for caesarean delivery.
- Planning and a multidisciplinary approach are required for an adequate surveillance of pregnancy and delivery in obese women.

40.2 The impact of obesity in labour

- Maternal obesity is an established risk factor for iatrogenic preterm birth. Some studies also report an increased risk of preterm prelabour rupture of membranes. Although data is conflicting, obesity does not seem to increase the risk of spontaneous preterm birth.
- Spontaneous onset of labour is less likely in obese women, either due to altered labour onset mechanisms or to inaccurately dated pregnancies. The likelihood of labour induction is therefore increased in this population.
- Obesity is associated with a slower progress of labour. The underlying mechanism is yet unknown, but several theories have been proposed:
  o The deposition of adipose tissue in the maternal pelvis may interfere with foetal progression.
  o The effectiveness of uterotonic agents may be adversely affected by a greater volume of distribution.

Handbook of Obesity in Obstetrics and Gynecology. DOI: https://doi.org/10.1016/B978-0-323-89904-8.00024-X

o Myometrial contractility may be inhibited by increased levels of leptin, cholesterol and apelin.
- Obese women have a higher risk of complications during labour delivery and the immediate postpartum period:
  o These include failed labour induction, prolonged labour, failure to progress, operative vaginal delivery, failed operative vaginal delivery, shoulder dystocia, foetal trauma, perineal lacerations, emergency caesarean delivery, use of general anaesthesia, and anaesthetic complications.
  o Postpartum complications include postpartum haemorrhage, wound infection, wound dehiscence, and thromboembolic events.

40.3   Labour management

- Prenatal assessment and planning are of the utmost importance in preparation of labour.
  o The patient's BMI and comorbidities must be clearly documented in the medical records.
  o If the resources and staff available are insufficient to ensure a safe labour and delivery, the patient should be transferred to another health facility.
  o Appropriate monitoring equipment should be available, including appropriately sized blood pressure cuffs.
  o Material capable of supporting the patient's weight and size must be accessible. This includes high-weight capacity chairs, wheelchairs, motorised lifts, birthing beds, and operating tables.
  o Procedures such as epidural placement, intubation, and surgery, are often challenging. Longer spinal needles and difficult intubation kits, including a fibre optic bronchoscope, should be available. Longer surgical instruments and self-retaining retractors should also be accessible.
  o The team must be prepared to deal with emergent situations, such as emergency caesarean delivery, shoulder dystocia, and postpartum haemorrhage. Additional staff members may need to be available, if necessary.
  o Obese patients are prone to neural injury and to pressure sores, not only due to their weight and difficult positioning, but also because procedures often take longer. Belts and gel pads help to properly secure the patient and to protect pressure areas.
- Antenatal counselling and assessment of expectations are important. Possible complications associated with maternal obesity and comorbidities should be discussed.
- The success rate of a trial of labour after caesarean delivery is inversely correlated with maternal BMI. It is less than 50% in women with no previous vaginal deliveries. Conversely, maternal and neonatal morbidities increase with BMI, two and fivefold, respectively. Women with class III obesity who undergo a trial of labour after caesarean delivery have a higher rate of morbidity compared with those submitted to elective caesarean section. However, the absolute rate of complications is low, so individual counselling and shared decision-making are essential.
- Obesity and obesity-related comorbidities increase the likelihood of caesarean delivery. This risk is proportional to maternal BMI, with a threefold increase in class III obese women. The decision to perform a caesarean delivery should consider intra- and postoperative risks, but also the decreased success rate of a subsequent trial of labour.
  o Infectious morbidity is increased, and preoperative antibiotic prophylaxis may be less effective. Some authors recommend increasing antibiotic doses, particularly if maternal

weight exceeds 120 kg, if the procedure outlasts two times the antibiotic's half-life, or if blood loss is greater than 1500 mL. However, current evidence is insufficient to evaluate the efficacy of such adjustments.

o If adequate retractors are unavailable, taping of the panniculus to the patient's thorax or anaesthesia screen should be considered, ensuring that the weight of the pannus does not compromise maternal ventilation.

o There is no evidence to suggest that the Pfannenstiel or Misgav-Ladach skin incisions should not be preferred in these women.

o The risk of surgical wound complications, such as dehiscence, can be minimised by closure of the subcutaneous layer when its thickness exceeds 2 cm. Intraabdominal or subcutaneous are not recommended.

o Prophylactic use of negative pressure wound therapy has not been shown to be beneficial.

o Healthcare providers should be aware that longer times may be necessary for transfer of the patient to the operating theatre, and from abdominal incision to delivery of the newborn. Both factors may affect newborn outcome in emergency caesarean deliveries.

• Maternal and foetal assessment and monitoring are challenging.

o Maternal habitus may make vaginal examination more difficult.

o The panniculus may limit clinical assessment of the foetal presentation. Ultrasound confirmation on admission may be needed.

o Continuous intrapartum cardiotocography is recommended, as obese women are at higher risk of foetal hypoxia. However, reliable external monitoring with external Doppler and tocodynamometry may be difficult.

o After rupture of membranes, foetal electrode placement should be considered if external Doppler does not provide an adequate quality signal.

o Cardiotocographic monitoring with continuous transabdominal ECG and electrohysterography is an alternative in centres where this recent technology is available.

• Because of the higher morbidity associated with obese parturients, particularly those with class III obesity, an intravenous (IV) access placement upon admission should be recommended. An additional peripheral IV catheter may be considered in class III women or those with important comorbidities.

• If ambulation is limited, use of compression stockings is recommended.

• Cross-matching of blood should be considered, particularly if other risk factors for postpartum haemorrhage or anaemia are present.

• Obese women are more likely to experience anaesthetic complications, so additional care should be considered:

o An antenatal consultation with an anaesthesiologist may be considered in the late third trimester.

o Obstructive sleep apnoea is particularly common in this population and increases the risk of hypoxaemia, hypercapnia, and sudden death.

o If antepartum anaesthesiology assessment did not take place, early evaluation after admission should be provided.

o Anaesthetic complications, such as nonfunctioning epidural catheter, failed intubation, and aspiration, should be anticipated.

o Early placement of an epidural/spinal catheter should be recommended, as this is often technically challenging, and a working catheter may decrease the need for general anaesthesia should an emergency caesarean section be required.

o There is an increased risk of intensive care unit (ICU) admission. Consider contacting ICU team on admission to secure a bed if the woman has several comorbidities.

1. Cervical Ripening and Induction of Labour
   a. Obese women are more likely to require labour induction, either due to postterm pregnancy or to medical or obstetrical complications.
      i. Obesity alone is not an indication for labour induction, but it is associated with an increased risk of macrosomia and stillbirth at term. Potential risks and benefits must be individually discussed.
      ii. Induction of labour at term for suspected macrosomia is warranted to decrease the risk of shoulder dystocia, both in women with normal and high IMC. Likewise, labour induction due to comorbidities or pregnancy-related complications remains unchanged.
      iii. Some studies suggest labour induction at 40 weeks to all women with class II and class III obesity should be considered, to reduce the risk of stillbirth and the likelihood of emergency caesarean delivery.
   b. Stripping of the foetal membranes at term may be offered to increase the possibility of spontaneous onset of labour.
   c. Evidence is scarce regarding the best labour induction method in this population. Cervical ripening is often required. This process takes longer and is also more likely to fail in this population. If cervical ripening is successful, there is still an increased induction-to-delivery time. Failed labour induction occurs two to three times more frequently in the obese population and is a common indication for caesarean delivery.
2. First Stage of Labour
   a. The first stage of labour is on average longer in the obese population, both in spontaneous and induced labours. Prolonged first stage of labour and failure to progress are up to four times more likely to occur, particularly after induction of labour and in the more severe cases of obesity. In women successfully delivered vaginally, active labour lasts up to 4 hours longer in obese paturients.
   b. Inadequate uterine contractility occurs more often in obese women, so labour augmentation is more frequently necessary. When compared to women with a normal BMI, obese women tend to have higher oxytocin requirements, including dosage and duration of infusion, which are directly proportional to maternal BMI. This may be due to the increased volume of distribution, receptor dysfunction associated with metabolic, hormonal and inflammatory changes, and decreased myometrial response.
   c. Studies report a particularly slow progress of labour until 6 cm of cervical dilation. Therefore obese women should be granted increased time to allow progression of dilatation during the first stage of labour, as long as maternal and foetal states remain reassuring.
3. Second Stage of Labour
   a. The duration of the second stage of labour does not appear to be affected by obesity. Studies evaluating the quality of maternal expulsive efforts report that intrauterine pressure generated by the Valsalva manoeuvre is comparable between obese and normal-weight women.
   b. An experienced obstetrician should be available at the time of delivery, as:
      i. Maternal obesity is a risk factor for operative vaginal delivery, which is up to two times more frequent in class III obesity. The likelihood of failed instrumental vaginal delivery is also increased, consequently increasing associated maternal and foetal morbidity.
      ii. The decision to attempt an instrumental vaginal delivery increases the risk of shoulder dystocia, birth trauma, perineal lacerations, and postpartum haemorrhage.
      iii. Shoulder dystocia is nearly three times more likely in the obese population.

4. Immediate Postpartum Period
   a. Metabolic, hormonal, and inflammatory mechanisms associated with obesity also hinder uterine contractility after delivery. Postpartum haemorrhage is more frequent in this population, mostly due to uterine atony. The need for blood transfusions is also increased.
   b. The panniculus limits the evaluation of uterine contraction and reduces the efficiency of uterine massage.
   c. Preventive administration of uterotonics, such as oxytocin, is strongly recommended, as they reduce the risk of postpartum haemorrhage.
   d. Obese women should be kept under close surveillance of vital signs and vaginal bleeding during the early postpartum period.

40.4 Summary

- Obesity increases the likelihood of intrapartum complications and the need for intervention during labour. Both maternal and foetal monitoring may be challenging. Specialised equipment and additional staff are often needed to ensure patient safety.
- Labour induction is more frequent, mostly due to postterm pregnancy, medical and obstetrical complications. There is also an increased rate of failed labour induction.
- The first stage of labour usually takes longer, so longer time intervals are needed to diagnose protracted or arrested labour.
- The duration of the second stage of labour remains unchanged. Healthcare providers must be prepared to deal with a higher rate of instrumental vaginal deliveries and shoulder dystocia.
- Obese women have an increased risk of postpartum haemorrhage. Preventive administration of uterotonics is strongly recommended.

Further reading

American College of Obstetrics and Gynecologists (ACOG). Practice Bulletin No 495: obesity in pregnancy. *Obstet Gynecol*. 2015;126(6). e112-e126.

Bogaerts A, Witters I, van den Bergh BRH, Jans G, Devlieger R. Obesity in pregnancy: altered onset and progression of labour. *Midwifery*. 2013;29:1303–1313.

Carpenter JR. Intrapartum management of the obese gravida. *Clin Obstet Gynecol*. 2016;59:172–179.

Denison FC, Aedla NR, Keag O, et al.on behalf of the Royal College of Obstetricians and Gynaecologists (RCOG) Care of women with obesity in pregnancy. *Green-top Guideline No. 72. BJOG*. 2018;126(3):e62–e106.

Ghaffari N, Srinivas SK, Durnwald CP. The multidisciplinary approach to the care of the obese parturient. *Am J Obstet Gynecol*. 2015;213:318–325.

Gunatilake RP, Perlow JH. Obesity and pregnancy: clinical management of the obese gravida. *Am J Obstet Gynecol*. 2011;204:106–119.

Harper A. Reducing morbidity and mortality among pregnant obese. *Best Pract Res Clin Obstet Gynaecol*. 2015;29:427–437.

Kither H, Whitworth MK. The implications of obesity on pregnancy. *Obstet Gynaecol Reprod Med*. 2012;22:362–367.

Liat S, Cabero L, Hod M, Yogev Y. Obesity in obstetrics. *Best Pract Res Clin Obstet Gynaecol*. 2015;29:79–90.

Lim CC, Mahmood T. Obesity in pregnancy. *Best Pract Res Clin Obstet Gynaecol*. 2015;29:309–319.

Maxwell C, Gaudet L, Cassir G, et al. Guideline No. 391-pregnancy and maternal obesity part 1: pre-conception and prenatal care. *J Obstet Gynaecol Can*. 2019;41:1623–1640.

Maxwell C, Gaudet L, Cassir G, et al. Guideline No. 392-pregnancy and maternal obesity part 2: team planning for delivery and postpartum care. *J Obstet Gynaecol Can*. 2019; 41:1660–1675.

McAuliffe FM, Killeen SL, Jacob CM, et al. Management of prepregnancy, pregnancy, and postpartum obesity from the FIGO Pregnancy and Non-Communicable Diseases Committee: a FIGO (International Federation of Gynecology and Obstetrics) guideline. *Int J Gynaecol Obstet*. 2020;151:16–36.

Mission JF, Marshall NE, Caughey AB. Pregnancy risks associated with obesity. *Obstet Gynecol Clin N Am*. 2015;42:335–353.

RANZCOG. The Royal Australian and New Zealand College of Obstetricians and Gynaecologists. *Management of Obesity in Pregnancy*; 2017.

Assisted vaginal delivery in obese women

Inês Martins[1] and Diogo Ayres-de-Campos[2]
[1]Department of Obstetrics and Gynecology, Medical School, University of Lisbon, Portugal, [2]Department of Obstetrics and Gynecology, Santa Maria University Hospital, Lisbon, Portugal

41.1   Introduction

- Although the duration of the second stage of labour is not affected by obesity, and the intrauterine pressure generated by the Valsalva manoeuvre is similar to normal-weight women, maternal obesity is a known risk factor for operative vaginal delivery, which is up to two times more frequent in class III obesity.
- The overall success rate of assisted vaginal delivery has been reported to be 86%−91%, but lower rates are found when maternal body mass index exceeds 30 kg/m^2.
- Similarly to nonobese patients, the choice of instrument is mainly dependent on operator experience, with vacuum delivery being more likely to fail (OR 1.7; 95% CI 1.3−2.2), to be associated with cephalohaematoma (OR 2.4; 95% CI 1.7−3.4), and retinal haemorrhage (OR 2.0; 95% CI 1.3−3.0), but less likely to be associated with important maternal perineal and vaginal trauma (OR 0.4; 95% CI 0.3−0.5).
- A prophylactic dose of intravenous amoxicillin plus clavulanic acid at the time of assisted vaginal delivery was associated, in a single large randomised controlled trial, with a 42% reduction in postpartum infection, and this measure should be considered in obese women, particularly if episiotomy or lacerations have occurred.

41.2   Indications

- The indications for assisted vaginal delivery are the same for obese and nonobese women:
  - o Maternal conditions:
    - ■ The inability to push effectively, because of exhaustion or uncoordinated pushing efforts.
    - ■ A medical condition which may be aggravated by the Valsalva manoeuvre (e.g. cardiac or neurologic diseases).
  - o Presumed foetal hypoxia:
    - ■ Suspicious or pathologic cardiotocographic tracings.
  - o Prolonged or arrested second stage of labour
    - ■ There is no worldwide consensus in these definitions, and the total duration of the second stage of labour (from full dilatation to birth) is frequently confused with the duration of active pushing.

Handbook of Obesity in Obstetrics and Gynecology. DOI: https://doi.org/10.1016/B978-0-323-89904-8.00020-2

- The World Health Organisation states that in nulliparous women, birth is usually completed within 3 hours, whereas in subsequent labours it is usually completed within 2 hours. The American College of Obstetricians and Gynaecologists (ACOG) and the Royal College of Obstetricians and Gynaecologists (RCOG) define prolonged second stage when it lasts more than 3 hours in nulliparous women with regional anaesthesia; more than 2 hours in nulliparous women without regional anaesthesia. For multiparous women, these limits are 2 hours with regional anaesthesia, and 1 hour without regional anaesthesia.
- The International Federation of Obstetrics and Gynaecologists (FIGO) defines a delay in the active second stage when it has lasted 2 hours in nulliparous women and birth is not imminent, 1 hour in multiparous women and birth is not imminent.
- The recommendation that the second stage of labour should not last more than 3−4 hours is based on strong evidence of a significant increase in maternal and neonatal infection after that time has elapsed.
- The maximum duration of maternal pushing should take into consideration the frequency and duration of contractions, but most women find it difficult to maintain adequate expulsive efforts for periods greater than 60 minutes.
- The definition of arrested/protracted labour in the active second stage of labour needs to be adapted to the frequency and duration of uterine contractions, to the mother's expectations, her physical capabilities, and to the progress of foetal descent.
- However, action is usually taken if little or no descent is documented by 30 minutes, or if pushing exceeds 60 minutes.
- Oxytocin augmentation or manual rotation is usually considered at this stage, but an operative delivery should be recommended if little or no progress is observed in the following 30−45 minutes.

41.3 Prerequisites

- The prerequisites for an assisted vaginal delivery are the same for obese and nonobese women:
  - o Fully dilated cervix, ruptured membranes, vertex presentation, known foetal position (intrapartum ultrasound may be considered in this context), engaged foetal head, adequate-sized maternal pelvis.
  - o Empty maternal bladder.
  - o Informed maternal consent.
  - o All additional prerequisites for a vaginal delivery (adequate maternal analgesia, conditions to perform an emergency caesarean delivery, conditions to perform neonatal resuscitation, etc.).

41.4 Contraindications

- The contraindications for performing an assisted vaginal delivery are the same for obese and nonobese women:
  - o Suspected foetal−pelvic disproportion.

o Vacuum extractors should be avoided before 34 weeks of gestation, because of increased risks of intraventricular haemorrhage.

o Some rare foetal demineralising diseases and bleeding disorders.

41.5   Specific considerations in obese women

- Because of impaired uterine contractility in obese patients, ongoing labour augmentation with oxytocin is more frequent at the start of the second stage of labour.
- Cardiotocographic monitoring is more difficult in these patients, conditioning poor quality foetal heart rate and contraction signals.
- Use of the foetal electrode should be strongly considered at the start of the second stage of labour, in the absence of contraindications, if foetal heart rate signal is not good.
- An effort should be made to avoid inserting the electrode in the area of subsequent vacuum cup placement.
- There are no specific recommendations regarding the use of assisted vaginal delivery in obese pregnant women, but it is frequently useful to ask women to hyperflex the thighs in order to have better accessibility to the presenting part. Stirrups may be considered, if women find them more comfortable.
- Obese women have an increased risk of gestational diabetes and of having large-for-gestational-age foetuses. The latter are associated with prolonged second stages of labour, and consequently an increased need of assisted vaginal delivery.
- It is necessary to select judiciously which large-for-gestational-age foetuses should be submitted to a trial of assisted vaginal delivery, based on ultrasound-estimated foetal weight, abdominal examination, and vaginal examination of the maternal pelvis, foetal head size and position.
- All clear-cut cases at high-risk of procedure failure should be offered a caesarean delivery, it is reasonable in all others to recommend an attempted assisted vaginal delivery, which can be abandoned if there is no foetal descent.
- It is important to remember and accept that the likelihood of failed instrumental vaginal delivery is increased in obese women.
- Decisions are made more difficult because of the increased risk of shoulder dystocia associated with maternal obesity, large-for-gestational-age foetuses, and assisted vaginal delivery, but which remains mainly an unpredictable condition.
- This situation is best handled by recommending caesarean delivery to clear-cut cases at increased risk of shoulder dystocia, and in the remaining cases to have on standby an experienced team in management of the situation.

41.6   Summary

- Obese women have an increased risk of assisted vaginal delivery.
- It is useful to ask women to hyperflex the thighs for instrument placement in order to have better accessibility to the presenting part.
- As in nonobese women, it is essential to guarantee a good quality foetal heart rate signal throughout the whole procedure.

- If a large-for-gestational-age foetus is suspected, ultrasound-estimated foetal weight, abdominal examination, vaginal examination of the maternal pelvis, foetal head size and position are important to select cases where a trial of assisted vaginal delivery should be recommended. Clear-cut cases at high-risk of procedure failure or shoulder dystocia should be offered caesarean delivery.
- The likelihood of failed instrumental vaginal delivery is increased in the obese population.
- It is important to have on standby an experienced team in management of shoulder dystocia.
- A prophylactic dose of intravenous amoxicillin plus clavulanic acid at the time of assisted vaginal delivery was associated, in a single large randomised controlled trial, with a 42% reduction in postpartum infection, and this measure should be considered in obese women, particularly if episiotomy or lacerations have occurred.

Further reading

Denison FC, Aedla NR, Keag O, et al. on behalf of the Royal College of Obstetricians and GynaecologistsCare of Women with Obesity in Pregnancy *BJOG.* 2019;126(3): e62−e106.

NICE. Intrapartum care for women with existing medical conditions or obstetric complications and their babies. *NICE Guideline, No. 121.* National Institute for Health and Care Excellence (UK); 2019. ISBN-13: 978-1-4731-3296-2.

Johanson RB, Menon V. Vacuum extraction vs forceps for assisted vaginal delivery. *Cochrane Database Syst Rev.* 1999;(2):CD000224.

Macrosomia. ACOG practice bulletinNumber 216 *Obstet Gynecol.* 2020;135(1):e18−e35.

Murphy DJ, Strachan BK, Bahl R. on behalf of the Royal College of Obstetricians Gynaecologists Assisted Vaginal Birth *BJOG.* 2020;127:e70−e112.

Nassar A, Visser G, Ayres-de-Campos D, et al. FIGO Statement: restrictive use rather than routine use of episiotomy. *Int J Gynecol Obstet.* 2019;146:17−19.

Obesity in Pregnancy. ACOG practice bulletin, number 156. *Obstet Gynecol.* 2020;135(1): e18−e35.

Operative Vaginal Birth. ACOG practice bulletin, number 219. *Obstet Gynecol.* 2020;135(4): e149−e159.

Royal Australian and New Zealand College of Obstetricians and Gynaecologists. *Instrumental vaginal delivery. College Statement C-Obs 16.* East Melbourne, Australia: RANZCOG; 2012.

WHO Recommendations. *Intrapartum Care for a Positive Childbirth Experience.* Geneva: World Health Organization; 2018. Licence: CC BY-NC-SA 3.0 IGO.

Wright A, Nassar A, Visser G, et al. FIGO good clinical practice paper: management of the second stage of labor. *Int J Gynaecol Obstet.* 2021;152(2):172−181.

Sepsis in obese pregnant women (concise version)

42

Hannah Waite, Katrine Orr and Ailie Grzybek
Ninewells Hospital, Dundee, Scotland

42.1 Introduction

The terms "overweight" and "obesity" are defined by the World Health Organisation (WHO) as an abnormal or excessive fat accumulation that presents a risk to health.[1] Obesity is further classified into three categories: Class 1 (BMI 30.0–34.9 kg/m^2), Class 2 (BMI 35–39.9 kg/m^2) and Class 3 or morbid obesity (BMI >40.0 kg/m^2).[1,2] Animal, epidemiological and limited human studies have reported that obesity increases susceptibility to both bacterial and viral infections.[3,4] There is increasing evidence of the increased risk of sepsis in association with obesity in pregnancy.[5]

Obesity is now one of the most frequently occurring obstetric comorbidities with 21.3% of the UK antenatal population obese at booking.[2] Maternal sepsis is defined by the WHO as a "life threatening condition defined as organ dysfunction resulting from infection during pregnancy, childbirth, postabortion, or postpartum period."[6] The most recent Mothers and Babies: Reducing Risk through Audit and Confidential Enquiries Across the UK (MBRRACE-UK) Report (November 2018) highlights that more than a third of maternal deaths occurred in women classified as obese.[7] Obesity has been recognised as an independent risk factor for both infection and sepsis in pregnancy and this chapter will discuss the epidemiology, immunology, infection sites, management and specific intra and postpartum care of obese pregnant woman with infection.

42.2 Epidemiology of obesity and sepsis

Obesity is a global epidemic affecting both developed and developing countries. The rate of obesity worldwide has tripled since 1975 and the United Kingdom has one of the highest obesity rates in the world at 27%.[1,2,8]

Risk factors for obesity include social deprivation, age >35 years, ethnic minorities, regional location and diabetes mellitus.

Obesity in pregnancy[2,9–11]: 21.3% are obese at booking with one in two overweight and one in 20 having BMI >35.

The higher a woman's booking BMI, the greater the risk of adverse perinatal outcomes, including the risk of ICU admission and death.[12]

Handbook of Obesity in Obstetrics and Gynecology. DOI: https://doi.org/10.1016/B978-0-323-89904-8.00033-0

Table 42.1 Compilation of data regarding obesity and sepsis in pregnant women taken from CEMACH, MBRRACE-UK and UKOSS reports.

Report	Findings
MBBRACE-UK (2017)[13]	Sepsis remains a key theme. At examination of all direct and indirect causes of death 34% of women were obese. Decreasing numbers of maternal deaths from genital tract sepsis and influenza. Same pattern not observed in cases of indirect sepsis (such as pneumonias). In 47% of cases improvements in care that would have altered the outcomes could have been made.
CEMACH (2003–05)[14]	Obesity a risk factor for maternal death. 33% of those with sepsis were obese. Regarding deaths of pregnant women in this report, 28% were obese (compared to a background of 16%–19%). Substandard care of sepsis identified in 71% of deaths.
UKOSS (2014)[15]	365 cases of severe maternal sepsis to 757 controls. Interestingly did not associate obesity with development of severe sepsis.
MBRARACE-UK (2018)[7]	Decrease in deaths related to indirect sepsis and improved care for these women. Worrying increase in the number dying from direct causes.
CEMACH (2006–08)[16]	Substandard care reduced only slightly to being a factor in 69% of sepsis deaths compared with the previous 71%.

Table 42.1 summarises MBRRACE-UK, UK Obstetric Surveillance System (UKOSS) and Confidential Enquiry into Maternal and Child Health (CEMACH) reported findings regarding the epidemiology of sepsis in obese patients.

Potential for improvements in care leading to better outcomes and avoidable harm is emphasised by these reports: clinical guidelines, care bundles and annually audited skills and drills sessions have been employed in the hope to improve care. Multidisciplinary care is crucial.

42.3 Immunology

Adipose tissue is known to have both endocrine, steroidogenesis and immunological function. Mechanism of dysregulation of the immune system in obesity is yet to be established but there is clear evidence that adipose tissue causes immunosuppression.[17] Mechanisms include suppressed function of both CD4 and CD8 T cells, T cell diversity, impaired NK cells, decreased cytokine production, disrupted lymphoid tissue architecture[18], alterations in leucocyte coordination and function[18], failure to coordinate

Table 42.2 Effects and results of obesity on the immune system.

Immune system effect	Result
Lower levels of NK and CD8 cells (responsible for apoptosis of infected cells—particularly viruses)	Increased morbidity/mortality to viruses—seen in obese pregnant women in H1N1 influenza pandemic[19]
Higher baseline levels of cytokines/monocytes	Chronic inflammatory state—desensitising the immune system to acute infection
Increased systemic inflammation and oxidative stress	Increases B cell proliferation at the expense of T cells
Change to memory T cell response	Increased risk of repeat infections, even from previously exposed pathogens

innate and adaptive immune responses[18], dysfunction of thymus and spleen[18] and inhibited effectiveness of vaccines.[18]

Table 42.2 summarises these effects,

42.4   Specific infections

There are numerous associated risk factors that put an obese woman at risk of an infection[16], including anaemia, diabetes mellitus, operative deliveries, cervical cerclage, amniocentesis, Group B streptococcus (GBS) carriage, manual removal of placenta, black or ethnic minorities, young age, multiparity, preterm birth, induction of labour, immunosuppression and ruptured membranes.

42.4.1   Genital tract

Signs and Symptoms: offensive, green vaginal discharge, preterm labour, foetal distress (may be a sign of chorioamnionitis and foetal infection), meconium stained liquor, premature rupture of membranes, stillbirth.

Investigations: Microbiology swab of vagina, placenta and baby for culture and sensitivities (if CS include uterine cavity swab), continuous CTG if maternal temperature $>38°C$ once or $>37.5°C$ twice, foetal scalp electrode good practice in obese patients, placenta to histopathology.

Management: 10 days oral erythromycin if preterm premature rupture of membranes (higher incidence in obese patients due to subclinical infection[20,21]), prophylactic antibiotics against GBS to all women in preterm labour, inform neonatal team at delivery if chorioamnionitis suspected.

Postnatal:

1. Consider endometritis if postnatal vaginal discharge, heavy lochia and/or delayed uterus inversion are present. Commonly associated with streptococcal and Gram-negative anaerobes. Endometritis more common in obese women.
2. Perineum should be inspected as a site of infection postnatally

42.4.2   Breast

Breast pain can be a symptom of: breast abscess, necrotising fasciitis and toxic shock syndrome.

Causative organisms: *Staphylococcus*, *Streptococcus*, methicillin-resistant *Staphylococcus aureus* (MRSA).

Hospital review if no response to first line antibiotics within 48 hours, recurrence of symptoms or signs of sepsis.[22]

42.4.3   Surgical site infection

Obese women are twice as likely to develop a surgical site infection following caesarean section.[23]

Causative organisms: polymicrobial (24%)[24]. *Staphylococcus aureus*, anaerobes (23.2%), MRSA (17%), Enterobacteriaceae (13.3%), Streptococci (7.4%).

Signs and symptoms: wound exudate, swelling, high dose of opiate pain relief required.

Necrotising fasciitis signs and symptoms: no visible skin changes in deep infection, severe pain, eventual skin blistering and necrosis.

Investigations: MRI to diagnose deep soft tissue infection.

Management: swabs of exudate/infected tissue, antibiotics to cover likely causative organisms and then directed at any pathogens from swabs, surgical drainage or debridement in cases where worsening or no improvement, involvement of plastic surgeons early if necrotising fasciitis suspected.

Infection at regional anaesthesia sites is rare but should be considered. Resiting of epidural cannulas is more common in the obese patient; multiple attempts at resiting can predispose to an infection. Urgent neurology review if suspected.

42.4.4   Hospital acquired infections

Obesity increases susceptibility to nosocomial infections: MRSA, carbapenemase producing enteroabacteiaceae (CPE) and *Clostridium difficile* (*C. diff*).[25,26]. See Table 42.3.

42.4.5   Urinary tract infection

Source of sepsis in just over a third of women who died of pregnancy-related sepsis in the 2013−15 triennium.[13]

Causative organisms: Gram-negative bacteria, these may produce β-lactamases making them resistant to many antibiotics (approx. 12% of coliform infections[22]).

42.4.6   Pneumonia

Obese pregnant woman are at an increased risk of viral pneumonia.[30] Liaise with microbiology and infectious diseases to target investigations and treatment.

Table 42.3 Summary of nosocomial infections.

	Description/symptoms	Associated with	Risk factors	Investigations	Management
MRSA	*Staphylococcus aureus* strains with resistance to β-lactam antibiotics	Mastitis Cellulitis Breast abscess Pelvic thrombophlebitis Pneumonia Wound infection UTI Sepsis	Previous MRSA colonisation/infection Multiparity[27]	Whole genome sequencing Surveillance swabs in outbreaks on wards[28]	Discuss with microbiology Isolation until negative proven, barrier nursing, good hand hygiene Vancomycin, teicoplanin[28]
CPE	Part of normal bowel flora but are highly resistant to most antibiotics. Increasingly common in Europe[29]	UTI Intra-abdominal infections Sepsis	Hospital stay abroad in last 12 months	Rectal swab or Faecal culture	Isolation until negative result returned
C. diff	Diarrhoea, Abdominal distension, abdominal pain, fever, vomiting	Foetal loss, ICU admission, colectomy, septic shock, death	Inflammatory bowel disease, recent/current antibiotic use, long term antibiotic therapy	Stool sample	Oral metronidazole or vancomycin Frequent hand hygiene (alcohol hand sanitisers do not eradicate *C. diff* spores) Avoidance of unnecessary antibiotics

Typical causative organisms: *Streptococcal pneumoniae, Staphylococcus aureus*, mycoplasma, *Haemophilus influenza*.

Atypical causative organisms: *Mycoplasma pneumoniae, Chlamydia pneumoniae, Legionella pneumoniae*.

42.4.7 Pharyngitis

A sore throat can be a symptom of a streptococcus infection. 10% caused by group A streptococcus (GAS). Consider GAS if three out of four of the following Centor Criteria are present: fever, tonsillar exudate, absence of cough and tender cervical lymphadenopathy.

Risk factors: Infection between December and April and contact with young children.

GAS associations: sepsis, streptococcal toxic shock syndrome and necrotising fasciitis.

Management: staff to use full PPE including visors, nurse in a single room with ensuite, inform infection control, neonatologist input as baby will require antibiotic regime and prophylactic antibiotics to all household contact.

NB: Asymptomatic cases require aggressive treatment in pregnancy.

42.4.8 Rare infections

Summarised in Table 42.4.

42.5 Viral infections

Pregnancy and obesity are major risk factors for developing respiratory complications of H1N1 influenza A viral infection[17]; postulated to be partially due to a restricted lung

Table 42.4 Rarer infections which can lead to sepsis or severe symptoms in pregnant women.

Infection	Associations and symptoms
Salmonella, Listeria, Campylobacter	Gastrointestinal infection from intake of unpasteurised milk
Coxiella burnetti	Contact with birthing animals
	May cause pneumonia, hepatitis, endocarditis and placentitis
Chlamydia psittaci	Contact with infected birds or aborting sheep
Recreational IV drug use	Streptococcus, staphylococcus and MRSA infections
	Risk of septic seeding → endocarditis and septic pelvic thrombosis
Staphylococcal toxic shock syndrome	Wide spread macular rash, conjunctival hyperaemia or suffusion, fevers >39.9°C, hypotension, multisystem involvement, skin desquamation after 10–14 days[22]

volume in association with pregnancy exacerbated by obesity. Vaccination should be strongly encouraged. However, in the obese population, IgG response by 12 months is significantly lower and correlates positively with increasing BMI.[31]

42.5.1 Pandemics—influenza

In June 2009 the H1N1 strain of influenza A led to a pandemic; pregnant women were offered vaccination.[32] Obese pregnant women were more likely to require hospital treatment.[32]

Signs and symptoms: fever, cough, shortness of breath, muscle and joint aches, headache, sore throat, vomiting, diarrhoea, coryzal symptoms.

Investigations: viral PCR—may be falsely negative and should be repeated if viral influenza suspected, sputum culture, legionella urine antigen test.

Management: do not withhold treatment to await PCR results if influenza suspected. If other diagnoses not ruled out, empirical antibiotics and thromboprophylaxis until results are available. Barrier nursing and appropriate PPE. Antiretrovirals: Zanamir and Oseltamivir commenced within 48 hours; seek specialist advice if greater than 48 hours have elapsed since contact[33]. Delay elective delivery unless maternal condition necessitates delivery. Exclude pulmonary embolus, bacterial pneumonia and atypical pneumonia.

Prophylaxis: Annual vaccination, antiretrovirals if infected close contact.

42.5.2 COVID-19

In December 2019, a new coronavirus variant (SARS-CoV-2) was identified and led to a global pandemic being declared in March 2020; coronavirus disease (COVID-19) is the infectious disease caused by the new variant.[34] Obese pregnant patients should be considered as higher risk for severe/critical COVID-19[35]

Signs and symptoms: fever, cough, myalgia, anosmia, ageusia, headache, chills, vomiting, nausea and diarrhoea[36]

Risk factors: obesity, age >35 years, preeclampsia, black and ethnic minorities, preexisting diabetes.[37–39]

Management: Oxygen if required, dexamethasone therapy may be used for those with severe symptoms (hypoxaemia) where benefits outweigh the risks[40,41], preterm delivery if maternal condition requires[35,42], vaginal delivery not contraindicated if maternal condition is stable.[43]

Prophylaxis: strict adherence to hand hygiene and social distancing, particularly in second and third trimester, vaccination.[44]

There does not appear to be any obvious neonatal or developmental adverse outcomes in relation to maternal COVID-19 infection during pregnancy, especially in those with mild/moderate infection.[39,42,45,46] However, adverse neonatal events are more likely to occur in those with worse symptoms[46], therefore there is the potential for them to be more common in obese patients, but this has not been studied to any depth.

42.6 Sepsis

Sepsis can be insidious and difficult to recognise and women in particular will often appear well prior to collapse. Sepsis is an emergency and all patients who are suspected should be assessed and treatment started within an hour as every hour without treatment results in a significant increase in mortality rate of 7.9%.[16]

Sepsis signs and symptoms: swinging or persistent fever, hypothermic, Leucocytosis or leucopenia (neutropenia an ominous sign), respiratory rate >20 breaths per minute, hyperglycaemia in absence of diabetes.

Septic shock: vasopressors needed to maintain mean arterial blood pressure above 65 mmHg. Serum lactate >2 mmol/L in absence of hypovolaemia.[47]

Sepsis identification tools

It has now been proposed that organ dysfunction can be assessed using the Sequential [Sepsis-related] Organ Failure Assessment (SOFA) score described by Singer et al.[47] It should be noted that this is yet to be validated within the obstetric population but may be used in the future. It is a scoring system taking into account clinical and biochemical variables and assigning a score based on deviation from baseline. A 2-point increase from baseline indicates significant organ dysfunction and is associated with a mortality of 10%.[47] The clinical variables include PaO_2/FiO_2, MAP, Glasgow Coma Scale and urine output.[47] The biochemical variables include platelet count, bilirubin level and serum creatinine.[47]

An alternative to this is the Quick SOFA score which can be carried out at the bedside. The presence of two out of three of respiratory rate >22/minute, altered mentation and systolic blood pressure <100 mmHg is associated with a greater likelihood of poor outcome.[47]

The Surviving Sepsis Campaign highlights that the initial identification of patients with infection should continue to be based on signs and symptoms; organ dysfunction should be identified in the same way.[48]

The use of any scoring system based on normal adult physiology is inherently challenging within the obstetric population due to the effect of the change in normal pregnancy physiology such as: white cell increases up to 25×10^9/L, maternal heart rate elevated to 90–100 bpm in active labour, occasional febrile episodes in labour.

The RCOG criteria for diagnosis of suspected sepsis should be used.[22,49] This is summarised in Table 42.5.

Investigations: FBC, U + E, LFT, CRP, glucose, coagulation screen and lactate, blood cultures, culture for other infection sources including urine, stool, sputum, breast milk, cerebrospinal fluid, wound swab, high vaginal swab, throat swab, baby and placenta swab, uterine swab, and nasal swab for MRSA, consider unusual pathogens if no source found, MRI for deep tissue infection

No investigation should delay sepsis management.

Management of sepsis: "Hour-1 bundle" (see Table 42.6). Blood cultures prior to antibiotic therapy but if unattainable by 45 minutes begin antibiotics.[50,51] Broad spectrum empirical antibiotics immediately—guided by local policy. Liaise with microbiology as needed. If GAS infection suspected clindamycin should be prescribed as it inhibits

Table 42.5 RCOG Criteria for diagnosis of sepsis in pregnant women.[18]

Infection, documented or suspected, and some of the following
General variables:
• Fever ($>38°C$) • Hypothermia (core temperature $<36°C$) • Tachycardia (>100 beats per minute) • Tachypnoea (>20 breaths per minute) • Impaired mental state • Significant oedema or positive fluid balance (>20 mL/kg over 24 h) • Hyperglycaemia in the absence of diabetes (plasma glucose >7.7 mmol/L)
Inflammatory variables:
• White blood cell count (WBC) $>12 \times 10^9$/L (note that transient leucocytosis in labour is normal) • Leucopenia (WBC count $<4 \times 10^9$/L) • Normal WBC count with $>10\%$ immature forms • Plasma C-reactive protein >7 mg/L
Haemodynamic variables:
• Arterial hypotension (systolic blood pressure <90 mm Hg; mean arterial pressure <70 mmHg or systolic blood pressure decrease >40 mmHg)
Tissue perfusion variables:
• Raised serum lactate > 4 mmol/L • Decreased capillary refill or mottling
Organ dysfunction variables:
• Arterial hypoxaemia (PaO_2 (arterial oxygen partial pressure)/FiO_2 (fraction of inspired oxygen) <40 kPa). Sepsis is severe if <33.3 kPa in the absence of pneumonia or <26.7 kPa in the presence of pneumonia. • Oliguria (urine output <0.5 mL/kg/h for at least 2 h, despite adequate fluid resuscitation) • Creatinine rise of >44.2 μmol/L. Sepsis is severe if creatinine level >176 μmol/L • Coagulation abnormalities (International Normalised Ratio (INR) >1.5 or activated partial thromboplastin time (APTT) >60 s) • Thrombocytopaenia (platelet count $<100 \times 10^9$/L) • Hyperbilirubinaemia (plasma total bilirubin >70 μmol/L) • Ileus (absent bowel sounds)

Source: Reproduced from Royal College of Obstetricians and Gynaecologists. "Bacterial sepsis in pregnancy." Green-top Guideline No. 64a. London: RCOG, April 2012, with the permission of the Royal College of Obstetricians and Gynaecologists; Anderson CJ, Murphy KE, Fernandez ML. Impact of obesity and metabolic syndrome on immunity. Adv Nutr 2016;7(1):66−75.

endotoxins and avoid NSAIDS.[22] Consider viral/fungal coverage.[50] Repeat a lactate level after 2−4 hours if initially >2 mmol/L to assess resuscitation response. VTE prophylaxis if not contraindicated—especially in obese patients.[7,50]

Table 42.6 Surviving Sepsis Campaign Hour-1 Bundle.[51]

• Measure lactate level. Remeasure if initial lactate is >2 mmol/L
• Obtain blood cultures prior to administration of antibiotics
• Administer broad-spectrum antibiotics
• Begin rapid administration of 30 mL/kg crystalloid for hypotension or lactate >4 mmol/L
• Apply vasopressors if patient is hypotensive during or after fluid resuscitation to maintain MAP >65 mm Hg

Table 42.7 Specific considerations for septic patients.

Special consideration	Actions
Diabetic patient with infection	Tight glucose control can lead to hypoglycaemia Commence insulin infusion when there are 2 consecutive readings of >180 mg/dL Aim to avoid hypoglycaemia and an upper target of 180 mg/dL[50]
Closed space infection	Swinging temperature/thrombocytosis → abscess or infected haematoma Will require surgery 6−12 h after patient stabilised[50] Expect significant blood loss in surgery
Delivery of baby	An individualised, multidisciplinary decision when regarding a septic patient

Fluid resuscitation in sepsis: 30 mL/kg of crystalloid if shocked or signs of hypoperfusion (lactate >4 mmol/L).[51] Beware potential overload. Consider need for vasopressor therapy and ICU/HDU admission if MAP fails to exceed 65 mmHg after resuscitation.[51] Aim urine output >0.5 mL/kg/hour. Assess initial response via heartrate, blood pressure, oxygen saturations, respiratory rate and temperature.[50]

Source: Reprinted by permission from Springer Nature Customer Service Centre GmbH: Springer Nature, Intensive Care Medicine; Levy MM, Evans LE, Rhodes A. The Surviving Sepsis Campaign Bundle: 2018 Update. Intensive Care Med 2018;44(6): 925−928.

Special considerations: Specific considerations for septic patients have been described in Table 42.7.

There is a risk of disseminated intravascular coagulation in sepsis therefore haemoglobin and platelet levels need to be carefully considered, summarised in Table 42.8.

Consider ICU admission if:[50] refractory hypotension or lactic acidosis, need for respiratory support, need for dialysis, decrease in conscious level, hypothermia, evidence multiorgan failure.

Table 42.8 Haemoglobin and platelet targets in septic patients.

Haemoglobin	Platelets
Aim for Hb level >7.0 g/dL[50,52]	Maintain levels above 10 × 10⁹/L if no bleeding Maintain levels above 20 × 10⁹/L if bleeding risk identified Maintain levels >50 × 10⁹/L if requiring surgical management[50]

42.7 Intrapartum care

Steps to reduce risk of sepsis: limiting vaginal examinations.[53] Oxytocin infusions in a timely fashion to also reduce genital tract infection. Avoidance of long labour and caesarean sections.[53]

Caesarean section

Obese women are twice as likely to require delivery by caesarean section.[54]

The chance of successful vaginal birth declines with increasing BMI. Therefore, obese women may be considered for elective caesarean section to avoid the significant morbidity associated with an emergency caesarean section based on individualised approach.

Vaginal birth after caesarean section (VBAC)

Obesity increases risk of[55] failed VBAC, uterine rupture and emergency caesarean section. Chance of successful VBAC is inversely proportional to pre-pregnancy BMI.[54]

Pressure sores: Risk is greatest in BMI >40 kg/m². Worsened by periods of immobility.[2] Potential infection entry source.

42.8 Operative issues

Factors complicating operation and postoperative healing: technically more difficult requiring good operative technique and experienced assistant.[2] Poor vascular perfusion and decreased immunological response contributing to increased risk of wound infection. Apron of adipose tissue delays healing and impedes cleaning and hygiene.

42.9 Caesarean sections

As BMI increases there is an increasing risk of caesareans section, however previous successful vaginal delivery in the absence of comorbidities is protective.[56,57] Obese women who undergo a caesarean section are at an increased risk of: intra-abdominal infections, wound infections (doubled in obesity and continue to double with every increment of 5 units of BMI[58]) and endometritis.

Further risks in obese patients: anaemia, smoking, older age, use of subcutaneous drains.[58]

42.10   Preoperative considerations

Obesity in pregnancy, particularly those with a BMI >40 kg/m^2, increases the risk of anaesthetic complications including[59,60] failed epidural placement, difficult/failed intubation, respiratory sensitivity to opiates, gastric aspiration, difficult IV access, failure of previously functioning epidural, patchy epidural anaesthesia, increased risk of hypotension in spinal/epidural anaesthesia. Potential complications are summarised in Table 42.9.

Preoperative considerations for obese patients: Antenatal anaesthetic review to plan care.[2] Early intrapartum epidural. Table specifications and/or adjustments. Hair should only be clipped (if needed for incision) to reduce risk of infection from microabrasions. Patient should be in a left lateral tilt position >15 degrees as aortocaval compression more profound. Consideration of good temperature and glycaemic control: increased surgical site infections in hypothermia and hyperglycaemia.[62,63]

42.11   Antibiotic prophylaxis

All patients undergoing a caesarean section should have a single dose of antibiotics administered prior to the skin incision.[64]

There has been a suggestion that standard antibiotic dosing in the obese patient may be inadequate because of altered pharmacokinetics and inadequate tissue penetration.[58] Therefore, morbidly obese patients are often given a prolonged course of prophylactic antibiotics however, there appears to be a limited evidence base.[65]

42.12   Incision

The abdominal wall anatomy can be distorted by the panniculus in obese patients. Careful attention should be paid to the symphysis pubis (which can be hard to palpate) to avoid injury to the bladder. Some studies suggest a vertical incision gives better access in obese patients.[66] However, these incisions were found to produce[67,68] difficulty accessing the lower segment potentially leading to use of the

Table 42.9 Complications associated with anaesthetic challenges in obese patients.[61]

Anaesthetic challenge	Complication
Failed intubation/aspiration Difficult regional anaesthetic placement	Aspiration pneumonia Epidural haematoma Abscess (1 in 1000) Meningitis (1 in 50,000)

classical caesarean section technique, increased postoperative pain, atelectasis, wound infections and fascial dehiscence.

Alternative incisions to consider in obese patients: suprapubic above pannus, transverse above pannus, low transverse incision with traction on the pannus (may compromise respiratory function; fat necrosis has been reported).[69]

42.13   Closure

Closure of subcutaneous space if the wound depth is >2 cm is recommended to reduce the risk of subsequent wound complication.[2,59,70]

There is no clear evidence of best practice for skin closure in obese patients following a caesarean section.[59] Current NICE Guidance advises considering closure using suture rather than staples at the time of caesarean section as this is associated with a decreased incidence of wound dehiscence, however the limited evidence for this practice in the obese population is highlighted.[71]

42.14   Postoperative care

1. Regular assessment of tissues for pressure sores, wound infection and wound breakdown[2]
2. Continuous care by the same medical team
3. Removal of urinary catheters and other indwelling lines as soon as clinically possible
4. Encouraging mobilisation as soon as clinically possible
5. Consider use of PICO dressings in patients with a BMI >35 kg/m[2][72,73]

42.15   Care in wound dehiscence

Most wounds will only partially dehisce and in nearly all cases, the rectus sheath will remain intact. Occasionally the wound will completely dehisce superficially.

Options for management of full/partial dehiscence: **primary reclosure of the wound**: Reduces healing time, requires antibiotic administration beforehand.[74] **Delayed wound closure:** increased rate of wound infections[75], higher rates of repeat dehiscence.[75] **Secondary intention**: partial dehiscence only; longer healing times.

42.15.1   Rectus sheath dehiscence

Signs and symptoms: marked abdominal pain, copious amount of serous discharge from wound

Investigations: Abdominal ultrasound and/or CT

Management: Surgical repair

42.15.2 Evisceration due to dehiscence

This is a surgical emergency.

Management: Contact general surgeons. Keep bowel warm presurgery with warmed sterile packs. Bowel inspection, washout and debridement in surgery. Wound closure. Prolonged course of broad spectrum antibiotics.

Risks: Bowel necrosis, sepsis.

42.16 Patient education

Obese women should be educated to the benefits of a normal weight prior to conceiving to both herself and her baby. They should be supported to avoid excessive weight gain in pregnancy. The target for the next pregnancy should be achieving a normal BMI prior to conception.

References

1. WHO. Obesity and overweight. <https://www.who.int/en/news-room/fact-sheets/detail/obesity-and-overweight>; 2019 Accessed 30.05.19.
2. Denison F.C., Aedia N.R., Keag O., et al. on behalf of the Royal College of Obstetricians and Gynaecologists. Care of women with obesity in pregnancy. Green-top Guideline No. 72. BJOG 2018.
3. Smith AG, Sheridan PA, Harp JB, Beck MA. Diet-induced obese mice have increased mortality and altered immune responses when infected with influenza virus. *J Nutr.* 2007;137(5):1236–1243.
4. Milner JJ, Beck MA. The impact of obesity on the immune response to infection. *Proc Nutr Soc.* 2012;71(2):298–306.
5. Axelsson D, Blomberg M. Maternal obesity, obstetric interventions and post-partum anaemia increase the risk of post-partum sepsis: a population-based cohort study based on Swedish medical health registers. *Infect Dis (Lond).* 2017;49(10):765–771.
6. World Health Organisation (WHO). WHO statement on maternal sepsis. WHO reference number: WHO/RHR/17.02. <https://www.who.int/reproductivehealth/publications/maternal_perinatal_health/maternalsepsis-statement/en/>; 2017 Accessed 30.05.19.
7. on behalf of MBRRACE-UK Knight M, Bunch K, Tuffnell D, et al., eds. *Saving lives, Improving Mothers' Care—Lessons learned to inform maternity care from the UK and Ireland Confidential Enquiries into Maternal Deaths and Morbidity 2014-16.* Oxford: National Perinatal Epidemiology Unit, University of Oxford; 2018.
8. NHS Digital. Statistics on Obesity, Physical activity and diet. Available from: <https://files.digital.nhs.uk/publication/0/0/obes-phys-acti-diet-eng-2018-rep.pdf>; 2018, Accessed 30.05.19.
9. Centre for Maternal and Child Enquiries (CMACE). *Maternal Obesity in the UK: Findings from a National Project.* London: CMACE,; 2010.
10. NMPA Project Team. National Maternal and Perinatal Audit: Clinical Report 2017. RCOG London; 2017.

11. Lisonkova S, Muraca GM, Potts J, et al. Association between prepregnancy body mass index and severe maternal morbidity. *JAMA*. 2017;318(18):1777−1786.
12. McCall SJ, Li Z, Kurinczuk JJ, Sullivan E, Knight M. Maternal and perinatal outcomes in pregnant women with BMI >50: an international collaborative study. *PLoS ONE*. 2019;14(2):e0211278.
13. on behalf of MBRRACE-UK Knight M, Nair M, Tuffnell D, Shakespeare J, Kenyon S, Kurinczuk JJ, eds. *Saving Lives, Improving Mothers' Care—Lessons learned to inform maternity care from the UK and Ireland Confidential Enquiries into Maternal Deaths and Morbidity 2013−15*. Oxford: National Perinatal Epidemiology Unit, University of Oxford; 2017.
14. The Confidential Enquiry into Maternal and Child Health (CEMACH). Saving Mothers' Lives: reviewing maternal deaths to make motherhood safer 2003−2005. In: Lewis G, ed. *The Seventh Report on Confidential Enquiries into Maternal Deaths in the United Kingdom*. London: CEMACH; 2007.
15. Acosta CD, Kurinczuk JJ, Nuala Lucas D, Tuffnell DJ, Sellers S, et al. Severe maternal sepsis in the UK, 2011−2012: a national case-control study. *PLoS Med*. 2014;11(7):e1001672.
16. Cantwell R, Clutton-Brock T, Cooper G, et al. Saving mothers' lives: reviewing maternal deaths to make motherhood safer: 2006−08. the eighth report on confidential enquiries into maternal deaths in the United Kingdom. *BJOG*. 2011;118(Suppl. 1):1−203.
17. Karlsson E, Marcelin G, Webby RJ, Schultz-Cherry S. Review of the impact of pregnancy and obesity on influenza virus infection. *Influenza Other Respir Viruses*. 2012;6(6):449−460.
18. Anderson CJ, Murphy KE, Fernandez ML. Impact of obesity and metabolic syndrome on immunity. *Adv Nutr*. 2016;7(1):66−75.
19. Sen S, Iyer C, Klebenov D, Histed A, Aviles JA, Meydani SN. Obesity impairs cell-mediated immunity during the second trimester of pregnancy. *Am J Obstet Gynecol*. 2013;208(2):139. e1−8.
20. Nohr EA, Bech BH, Vaeth M, Rasmussen KM, Henriksen TB, Olsen J. Obesity, gestational weight gain and preterm birth: a study within the Danish National Birth Cohort. *Paediatr Perinat Epidemiol*. 2007;21(1):5−14.
21. Faucett AM, Metz TD, DeWitt PE, Gibbs RS. Effect of obesity on neonatal outcomes in pregnancies with preterm premature rupture of membranes. *Am J Obstet Gynecol*. 2016;214(2):287. e1−5.
22. Morgan M, Hughes RG, Kinsella SM. Bacterial sepsis following pregnancy. Green-top guideline No. 64b on behalf of the Royal College of Obstetricians and Gynaecologists *RCOG, Lond*. 2012.
23. Krieger Y, Walfisch A, Sheiner E. Surgical site infection following cesarean deliveries: trends and risk factors. *J Matern Fetal Neonatal Med*. 2017;30(1):8−12.
24. Wloch C, Wilson J, Lamagni T, Harrington P, Charlett A, Sheriden E. Risk factors for surgical site infection following caesarean section in England: results from a multicentre cohort study. *BJOG*. 2012;119(11):1324−1333.
25. Cantürk Z, Cantürk NZ, Çetinarslan B, Utkan NZ, Tarkun I. Nosocomial infections and obesity in surgical patients. *Obes Res*. 2003;11(6):769−775.
26. Reush M, Ghosh P, Ham C, Klotcho A, Singapuri S, Everett G. Prevalence of MRSA colonization in peripartum mothers and their newborn infants. *Scand J Infect Dis*. 2008;40(8):667−671.
27. Wang B, Suh KN, Muldoon KA, et al. Risk factors for Methicillin-resistant Staphylococcus aureus (MRSA) colonization among patients admitted to obstetrical units: a nested case-control study. *J Obstet Gynaecol Can*. 2018;40(6):669−676.

28. Longmore M., Wilkinson I.B., Baldwin A., Wallin E. Oxford Handbook of Clinical Medicine. 9th ed.; 2014. Chapter 6 + 9.

29. Carmeli Y, Akova M, Cornaglia G, et al. Controlling the spread of carbapenemase-producing Gram-negatives: therapeutic approach and infection control. *Clin Microbiol Infect.* 2010;16(2):102−111.

30. Modder J. Review of maternal deaths in the United Kingdom related to A/H1N1 2009 Influenza. CMACE, 2010.

31. Sheridan PA, Paich HA, Handy J, et al. Obesity is associated with impaired immune response to influenza vaccination in humans. *Int J Obes (Lond).* 2012;36(8):1072−1077.

32. Yates L, Pierce M, Stephens S, et al. Influenza A/H1N1v in pregnancy: an investigation of the characteristics and management of affected women and the relationship to pregnancy outcomes for mother and infant. *Health Technol Assess.* 2010;14(34):109−182.

33. Public Health England guidance on the use of antiviral agents for the treatment and prophylaxis of seasonal influenza. Public Health England, London. Version 9.1. <https://assets.publishing.service.gov.uk/government/uploads/system/uploads/attachment_data/file/773369/PHE_guidance_antivirals_influenza.pdf>; January 2019 Accessed 30.05.19.

34. Satpathy HK, Lindsay M, Kawwass JF. Novel H1N1 virus infection and pregnancy. *Postgrad Med.* 2009 Nov;121(6):106−112. Available from: https://doi.org/10.3810/pgm.2009.11.2080.

35. Khoury R, Bernstein PS, Debolt C, et al. Characteristics and outcomes of 241 births to women with severe acute respiratory syndrome Coronavirus 2 (SARS-CoV-2) Infection at Five New York City Medical Centers. *Obstet Gynecol.* 2020;136(2):273−282. Available from: https://doi.org/10.1097/AOG.0000000000004025.

36. Umakanthan S, Sahu P, Ranade AV, et al. Origin, transmission, diagnosis and management of coronavirus disease 2019 (COVID-19). *Postgrad Med J.* 2020;96(1142):753−758. Available from: https://doi.org/10.1136/postgradmedj-2020-138234. Dec.

37. Knight M, Bunch K, Vousden N, et al. UK Obstetric Surveillance System SARS-CoV-2 Infection in Pregnancy Collaborative Group. Characteristics and outcomes of pregnant women admitted to hospital with confirmed SARS-CoV-2 infection in UK: national population based cohort study. *BMJ.* 2020;8(369):m2107. Available from: https://doi.org/10.1136/bmj.m2107.

38. Kayem G, Lecarpentier E, Deruelle P, et al. A snapshot of the Covid-19 pandemic among pregnant women in France. *J Gynecol Obstet Hum Reprod.* 2020;49(7):101826. Available from: https://doi.org/10.1016/j.jogoh.2020.101826.

39. Czeresnia RM, Trad ATA, Britto ISW, et al. SARS-CoV-2 and pregnancy: a review of the facts. *Rev Bras Ginecol Obstet.* 2020;42(9):562−568. Available from: https://doi.org/10.1055/s-0040-1715137. English.

40. Johnson RM, Vinetz JM. Dexamethasone in the management of Covid-19. *BMJ.* 2020;370:m2648. Available from: https://doi.org/10.1136/bmj.m2648. Jul 3.

41. Berton AM, Prencipe N, Giordano R, Ghigo E, Grottoli S. Systemic steroids in patients with COVID-19: pros and contras, an endocrinological point of view. *J Endocrinol Invest.* 2021;44(4):873−875. Available from: https://doi.org/10.1007/s40618-020-01325-2. Epub 2020 Jun 8.

42. Royal College of Obstetricians and Gynaecologists. 2021. Coronavirus infection and pregnancy: information for pregnant women and their families. <https://www.rcog.org.uk/en/guidelines-research-services/guidelines/coronavirus-pregnancy/covid-19-virus-infection-and-pregnancy/>.

43. Liao J, He X, Gong Q, Yang L, Zhou C, Li J. Analysis of vaginal delivery outcomes among pregnant women in Wuhan, China during the COVID-19 pandemic. *Int J Gynaecol Obstet.* 2020;150(1):53−57. Available from: https://doi.org/10.1002/ijgo.13188.
44. Public Health England. COVID-19: the green book. <https://www.gov.uk/government/publications/covid-19-the-green-book-chapter-14a>; 2020. Chapter 14a
45. Antoun L, Taweel NE, Ahmed I, Patni S, Honest H. Maternal COVID-19 infection, clinical characteristics, pregnancy, and neonatal outcome:a prospective cohort study. *Eur J Obstet Gynecol Reprod Biol.* 2020;252:559−562. Available from: https://doi.org/10.1016/j.ejogrb.2020.07.008. Sep.
46. Brandt JS, Hill J, Reddy A, et al. Epidemiology of coronavirus disease 2019 in pregnancy: risk factors and associations with adverse maternal and neonatal outcomes. *Am J Obstet Gynecol.* 2021;224(4):389. Available from: https://doi.org/10.1016/j.ajog.2020.09.043. e1−389.e9.
47. Singer M, Deutschman CS, Seymour CW, et al. The third international consensus definitions for sepsis and septic shock (Sepsis-3). *JAMA.* 2016;315(8):801−810.
48. Surviving Sepsis Campaign. Surviving sepsis campaign responds to sepsis-3. <http://www.survivingsepsis.org/SiteCollectionDocuments/SSC-Statements-Sepsis-Definitions-3-2016.pdf>; 1 March 2016. Accessed 30.05.19.
49. Pasupathy D., Morgan M., Plaat F.S., Langford K.S. on behalf of the Royal College of Obstetricians and Gynaecologists. Bacterial sepsis in pregnancy. Green-top Guideline No 64a. RCOG, Lond, 2012.
50. Rhodes A, Evans LE, Alhazzani W, et al. Surviving Sepsis Campaign: international guidelines for management of sepsis and septic shock: 2016. *Intensive Care Med.* 2017;43(3):304−377.
51. Levy MM, Evans LE, Rhodes A. The surviving sepsis campaign bundle: 2018 Update. *Intensive Care Med.* 2018;44(6):925−928.
52. Dellinger RP, Levy MM, Rhodes A, et al. Surviving sepsis campaign: international guidelines for management of severe sepsis and septic shock: 2012. *Crit Care Med.* 2013;41(2):580−637.
53. Tran TS, Jamulitrat S, Chongsuvivatwong V, Geater A. Risk factors for post caesarean surgical site infections. *Obstet Gynecol.* 2000;95(3):367−371.
54. Catalano PM, Shankar K. Obesity and pregnancy: mechanisms of short term and long term adverse consequences for mother and child. *BMJ.* 2017;360:j1.
55. Gupta J.K., Smith G.C.S. and Chodankar R.R. on behalf of the Royal College of Obstetricians and Gynaecologists. Birth after previous caesarean birth. Green-top Guideline No. 45. RCOG, Lond 2015.
56. Carpenter J. Intrapartum management of the obese gravida. *Clin Obstet Gynecol.* 2016;59(1):172−179.
57. Hollowell J, Pillas D, Rowe R, Linsell L, Knight M, Brocklehurst P. The impact of maternal obesity on intrapartum outcomes in otherwise low risk women: secondary analysis of the Birthplace national prospective cohort study. *BJOG.* 2014;121(3):343−355.
58. Alanis MC, Villers MS, Law TL, Steadman EM, Robinson CJ. Complications of caesarean delivery in the massively obese parturient. *Am J Obstet Gynecol.* 2010;203(3):271. e1−7.
59. Gaiser R. Anesthetic considerations in the obese parturient. *Clin Obstet Gynecol.* 2016;59(1):193−203.
60. Vricella LK, Louis JM, Mercer BM, Bolden N. Anaesthesia complications during scheduled caesarean delivery for morbidly obese women. *Am J Obstet Gynecol.* 2010;203(3):276. e1−5.

61. Moen V, Dahlgren N, Irestedt L. Severe neurological complications after central neuraxial blockades in Sweden 1990—1999. *Anesthesiology*. 2004;101(4):950—959.
62. Seamon MJ, Wobb J, Gaughan JP, Kulp H, Kamel I, Dempsey DT. The effects of intraoperative hypothermia on surgical site infection: an analysis of 524 trauma laparotomies. *Ann Surg*. 2012;255(4):789—795.
63. Ata A, Lee J, Bestle SL, Desemone J, Stain SC. Postoperative hyperglycemia and surgical site infection in general surgery patients. *Arch Surg*. 2010;145(9):858—864.
64. National Institute for Health and Care Excellence (NICE). Caesarean section. NICE Guideline 132; 23 November 2011.
65. Valent AM, DeArmond C, Houston JM, et al. Effect of post-cesarean delivery oral cephalexin and metronidazole on surgical site infection among obese women: a randomized clinical trail. *JAMA*. 2017;318(11):1026—1034.
66. Faucett AM, Metz TD. Delivery of the obese gravida. *Clin Obstet Gynecol*. 2016;59(1):180—192.
67. Ayres-de-Campos D. Obesity and challenges of caesarean delivery: prevention and management of wound complications. *Best Pract Res Clin Obstet Gynaecol*. 2015;29(3):406—414.
68. Bell J, Bell S, Vahratian A, Awonuga AO. Abdominal surgical incisions and perioperative morbidity among morbidly obese women undergoing caesarean delivery. *Eur J Obstet Gynecol Reprod Biol*. 2011;154(1):16—19.
69. Stirrat LI, Milne A, Paul A, Denison FC. Suprapannus transverse caesarean section incision in morbidly obese women. *Eur J Obstet Gynecol Reprod Biol*. 2015;186:110—111.
70. Chemmow D, Rodriguez EJ, Sabatini MM. Suture closure of subcutaneous fat and wound disruption after caesarean delivery: a *meta*-analysis. *Obstet Gynecol*. 2004;103(5 Pt 1):974—980.
71. National Institute for Health and Care Excellence (NICE). Surgical site infections: prevention and treatment. *NICE Guidel*. 2019;125. 11 April.
72. Searle RJ, Myers D. A survey of caesarean section surgical site infections with PICO™ single use negative pressure wound therapy system in high-risk patients in England and Ireland. *J Hosp Infect*. 2017;97(2):122—124.
73. Hyldig N, Joergensen JS, Wu C, et al. Cost-effectiveness of incisional negative pressure wound therapy compared with standard care after caesarean section in obese women: a trial-based economic evaluation. *BJOG*. 2019;126(5):619—627.
74. Dodson MK, Magann EF, Meeks GR. A randomized comparison of secondary closure and secondary intention in patients with superficial wound dehiscence. *Obstet Gynaecol*. 1992;80(3 Pt 1):321—324.
75. Duttaroy DD, Jitendra J, Duttaroy B, et al. Management strategy for dirty abdominal incisions: primary or delayed primary closure? A randomized trial. *Surg Infect*. 2009;10(2):129—136.

Immediate postnatal care in obese women

Rabia Sherjil[1] and Rashda Bano[2]
[1]Royal Infirmary of Edinburgh, Edinburgh, United Kingdom, [2]Zita West Assisted Fertility Clinic, London, United Kingdom

Postnatal care is preeminently about the provision of a supportive environment in which a woman, her baby, and the wider family can begin their new life together.

- Maternal overweight and obesity (body mass index $>25 \text{ kg/m}^2$) is an important modifiable risk factor in women of reproductive age.
- The rise in prevalence of obesity is seen in both high-income countries and low—middle-income countries.
- It is predicted that by 2025 more than 21% of women in the world will be overweight/obese.

43.1 Historical background

- Current models of postnatal care originated from the beginning of the 20th century, when there were concerns about the high maternal mortality rate in the postnatal period.
- Postnatal care provision is provided by both acute and primary healthcare sectors, with the majority of care taking place in the woman's home.
- Midwife-led postnatal care in the United Kingdom continues to be a statutory requirement and is usually provided until 6—8 weeks postnatal.
- There has been limited research into the provision and content of postnatal care, with even less information on effective framework of care in obese woman of child-bearing age.

43.2 Implications of obesity in planning of postnatal care

- The implications of being obese for a woman's postnatal health are likely to affect the content and planning of her individualised care, frequency of contact and what needs to be provided by the multiprofessional team.
- Overweight and obese women are at increased risk of starting pregnancy with existing medical comorbidity such as diabetes and hypertension and they may be already on medical treatments compared to women with normal weight.

Handbook of Obesity in Obstetrics and Gynecology. DOI: https://doi.org/10.1016/B978-0-323-89904-8.00034-2

- Furthermore new disorders such as gestational diabetes, new onset hypertension, and venous thrombosis (VTE) arise at a higher frequency when compared with normal weight women.
- Some of these existing prepregnancy morbidities are expected to worsen during pregnancy. There is an increased risk of operative intervention during labour, hence compounding risk of sepsis, VTE secondary to labour intervention and impaired mobility, impaired glucose metabolism, and difficulty in breastfeeding.
- The priorities for postnatal care should start during antenatal period of pregnancy and should continue to evolve throughout pregnancy, intrapartum period, and the postnatal phase.
- The key objective is to prevent and minimise potential postbirth complications and assist the woman and her baby to achieve the best possible outcome.
- Obstetric outcome data from a retrospective cohort of over 72,000 Australian women delivered in a tertiary centre are shown in Fig. 43.1, stratified by body mass index (BMI).
- These data provide BMI subgroups stratified risks, and could assist in planning of the whole journey of pregnancy especially during the postdelivery phase.

Variable	BMI (kg/m^2)					
	<18.5	18.5–<25	25–<30	30–<35	35–<40	≥40
Maternal outcome (%)						
Hypertension in pregnancy	1.1	1.7	3.3	5.1	7.0	9.6
GDM	1.0	1.2	2.1	3.4	5.5	6.9
Type 1 and 2 diabetes mellitus	0.2	0.5	3.3	1.7	2.8	4.1
Spontaneous vaginal birth	61	54.4	50.4	47.1	46.9	43.6
Assisted birth	13.3	12.9	10.0	8.4	5.9	4.9
Caesarean section	25.7	32.7	39.6	44.5	47.1	51.5
Neonatal outcomes (%)						
Perinatal death	0.5	0.7	1.0	1.1	1.5	1.8
Stillbirth	0.2	0.4	0.5	0.7	0.8	0.7
Neonatal death	0.3	0.3	0.5	0.5	0.7	1.1
Macrosomia	5.4	10.6	15.9	18.7	20.1	20.8
SGA	12.4	10.9	12.2	13.4	15.7	18.7
LGA	10.5	11.0	12.4	13.3	14.0	15.9
Pre-term birth < 37 weeks	8.5	6.7	7.5	8.5	9.5	11.3
Respiratory distress syndrome	4.2	4.3	5.3	5.7	6.4	7.3
Mechanical ventilation	5.9	4.7	5.8	6.5	8.6	10.4
Jaundice	6.4	4.7	5.4	6.4	7.5	9.3
Hypoglycaemia	1.1	0.9	1.3	1.8	3.0	2.5

SGA=small for gestational age, LGA=large for gestational age, ≥ greater than or equal to, < less than. Source: adapted from Mc Intyre HD, Giwons KS, Flenady VJ, Callaway LK. Overweight and obesity in Australian mothers: epidemic or endemic? Med J Aust. 2012;196(3);184-8..

Figure 43.1 Pregnancy outcomes in a cohort of 72,000 women who delivered in a Tertiary Maternity Hospital in Australia.

43.3   Postpartum care plan

- Anticipatory guidance for postpartum care plan should start following booking visit and should continue to evolve as the pregnancy progresses, taking account of new complicating factors which may affect new parents.
- Anticipatory guidance should be based upon comprehensive needs assessment, including discussion about infant feeding especially breastfeeding, 'baby blues', postpartum emotional health, and the challenges of parenting and recovery during postpartum period.
- Prenatal discussions also should address plans for long-term management of chronic health conditions, such as mental health, diabetes, hypertension, including identification of a primary healthcare provider who will care for the patient.
- Mode of delivery, past birth experience, and length of stay in maternity unit, particularly in obese parturient, further define postnatal care.
- The American College of Obstetrics and Gynecology has proposed to coin the term of fourth trimester to the postpartum period and a shift of paradigm to 12 weeks rather than 6 weeks of postpartum care (Table 43.1).

Overweight and obese women are at increased risk of experiencing the following complications significantly more than women with normal BMI.

43.4   Thromboembolism (venous thrombosis)

- Pulmonary embolism (PE) remains a leading direct cause of maternal death in the UK, particularly during the postpartum period.
- However, one aspect of note was the high risk of thrombosis as a consequence of obesity.
- Nearly all women dying from VTE following vaginal delivery in the 2003—05 Confidential Enquiry were overweight or obese.
- The relative risk of VTE during postpartum is fivefold higher when compared to antepartum period.
- A systematic review of risk of postpartum VTE found that the risk varied from 21—84-folds from the baseline nonpregnant, nonpostpartum state in studies that included an internal reference group. The absolute risk peaked in the first 3 weeks postpartum (421 per 100,000 person-years; a 22-fold increase in risk).
- There has been a marked decline in the number of deaths during pregnancy and following vaginal birth within the United Kingdom.
- This change most likely reflects routine use of thromboprophylaxis for VTE as recommended in national guidance, adopted from the RCOG and the ACOG, shown in Fig. 43.2.
- FIGO has emphasised the use of early mobilisation, compression stockings, and low-molecular-weight heparin according to local and national guidance for obese women post birth.
- RCOG has stratified postpartum risk factor scoring system for prevention of VTE, highlighting BMI ≥ 40 as an independent risk factor requiring prophylaxis.
- Women with obesity are at increased risk of interventions during labour and birth compared to women with lower BMIs and therefore mechanical thromboprophylaxis is considered essential for caesarean delivery.

Table 43.1 Suggested components of the postpartum care plan[a] ⇦.

Element	Components
Care team	Name, phone number, and office or clinic address for each member of care team
Postpartum visits	Time, date, and location for postpartum visit(s); phone number to call to schedule or reschedule appointments
Infant feeding plan	Intended method of infant feeding, resources for community support (e.g. WIC, Lactation Warm Lines, Mothers' groups), return-to-work resources
Reproductive life plan and commensurate contraception	Desired number of children and timing of next pregnancy Method of contraception, instructions for when to initiate, effectiveness, potential adverse effects, and care team member to contact with questions
Pregnancy complications	Pregnancy complications and recommended followup or test results (e.g. glucose screening for gestational diabetes, blood pressure check for gestational hypertension), as well as risk reduction recommendations for any future pregnancies
Adverse pregnancy outcomes associated with ASCVD	Adverse pregnancy outcomes associated with ASCVD will need baseline ASCVD risk assessment, as well as discussion of need for ongoing annual assessment and need for ASCVD prevention over lifetime.
Mental health	Anticipatory guidance regarding signs and symptoms of perinatal depression or anxiety; management recommendations for women with anxiety, depression, or other psychiatric issues identified during pregnancy or in the postpartum period
Postpartum problems	Recommendations for management of postpartum problems (i.e. pelvic floor exercises for stress urinary incontinence, water-based lubricant for dyspareunia)
Chronic health conditions	Treatment plan for ongoing physical and mental health conditions and the care team member responsible for follow-up

ASCVD, Atherosclerotic cardiovascular disease; *WIC*, Special Supplemental Nutrition Program for Women, Infants, and Children.
[a]A Postpartum Care Plan Template is available as part of the ACOG Pregnancy Record.

43.5 Sepsis risk

- According to the confidential enquiries from the maternal mortality report of 2006–08, genital tract sepsis was the largest cause of direct maternal death (CMACE, 2011).
- Numerous studies have shown that, compared with women of normal BMI, postpartum infection was considerably more common in obese pregnant women, independent of the

Protocols for Prophylaxis

Agent	LMWH Enoxaparin	Dalteparin	Tinzaparin	UFH Unfractionated heparin	
Weight based				**Gestational age-based**	
<50kg	20mg daily	2500 units daily	3500 units daily	First trimester	5000-7500 units Twice daily
50-90kg	40mg daily	5000 units daily	4500 units daily	Second trimester	7500-10000 units Twice daily
91-130kg	60mg daily*	7500 units daily*	7000 units daily*	Third trimester	10000 units Twice daily
131-170kg	80mg daily*	10000 units daily*	9000 units daily	Postpartum	5000 units twice daily
>170kg	0.6mg / kg / day*	75 units / kg / day	75 units / kg / day		

Hospitalized antepartum patients may receive 5000 units UFH twice daily for prophylaxis to facilitate regional anesthesia

*=may be given in two divided doses

Adapted from ACOG Practice Bulletin 123, ACCP Recommendations , RCOG Green Top Guideline 37a

Safe Motherhood Initiative

Figure 43.2 VTE prophylaxis adopted from the American College of Obstetrics and Gynecology (ACOG) Practice Bulletin and the Royal College of Obstetricians and Gynaecologists Green Top Guidelines (RCOG).

    form of delivery (vaginal, elective, or emergency caesarean delivery), and this is despite the use of prophylactic antibiotics in the majority of studied cases.
- Obesity is an independent risk factor for postcaesarean infection, and prophylactic use of antibiotics before caesarean section in women with class 1 obesity can help its reduction (RCOG).
- Of note, this increased infection rate has been partially attributed to the altered metabolic state of obesity, as well as poor vascularity of subcutaneous adipose tissue and incision dehiscence.
- RCOG has advised suturing of the subcutaneous tissue space more than 2 cm at caesarean section in order to reduce the risk of wound infection and wound separation.
- Women who are obese are at higher risk of experiencing other sites of infection. A systematic review (2008) of obstetric outcomes among obese women found almost 3.5-fold increased risk of infections (including wound, urinary tract, perineum, chest and breast).

43.6 Postpartum haemorrhage

- There were nine direct maternal deaths following major obstetric haemorrhage in the UK during 2006–08, and of these five deaths occurred postnatal (Lewis, 2007). Two of these women were overweight or obese.
- A population-based cohort study included 1,114,071 women with singleton pregnancies who gave birth in Sweden from January 1, 1997 through December 31, 2008, who were divided into six BMI classes,
- Obese women (class I–III) were compared with normal weight women concerning the risk for postpartum haemorrhage after suitable adjustments.

- The use of heparin-like drugs over the BMI strata was analysed in a subgroup. The risk of atonic uterine haemorrhage increased rapidly with increasing BMI. There was a twofold increased risk in obesity class III (1.8%).
- No association was found between postpartum haemorrhage with retained placenta and maternal obesity.
- There was an increased risk for postpartum haemorrhage for women with a BMI of 40 or higher (5.2%) after normal delivery compared with normal-weight women (4.4%) and even more pronounced (13.6%) after instrumental delivery compared with normal-weight women (8.8%).
- A retrospective population-based 10-year (2006−15) study from Australia concluded pre-obese and obese women are at significantly increased risk of PPH compared to women of normal weight (between 16%−54%) both in nulliparous and multiparous women.
- These findings emphasise the need for active management of third stage labour in preobese and obese women.

43.7 Preeclampsia

Obesity is a major risk for both gestational hypertension and preeclampsia.

The risk of preeclampsia typically doubles with each $5-7\,\text{kg/m}^2$ increase in prepregnancy.

It is however less clear whether or not an increasing BMI is associated with all types of PE, in particular EOP (early onset preeclampsia) or LOP (late onset preeclampsia).

The data of the 5600 healthy nulliparous pregnant women in the multicentre SCOPE study, just using a binary BMI cut-off of 30, appeared to indicate that obesity was associated with both EOP and LOP.

A recently published 18-year observational cohort study (2001−18) has reported that increasing BMI has a linear association with late onset preeclampsia.

Results from this study have further classified the risk factors for EOP and LOP separately.

The study found that following are the risk factors for EOP:

- chronic hypertension; and
- history of preeclampsia in multigravidas.

The study found that following are the risk factors for LOP:

- primiparity;
- age over 35 years; and
- BMI $\geq 35\,\text{kg/m}^2$
    - In a multivariate analysis with EOP or LOP as outcome variables compared with controls (normotensive), maternal age and prepregnancy BMI were independent risk factors for both EOP and LOP.
    - Evidence from a systematic review and metaanalysis proposed the benefit of prophylactic use of low-dose aspirin in women with BMI of 35 and above from 12 weeks of gestation until delivery.

- However, in postpartum period, use of aspirin when no longer in place, monitoring of signs and symptoms, along with blood pressure measurement using appropriate size cuff is currently in practice in U.K.
- Furthermore, FIGO (2020) has suggested appropriate postnatal followup in line with local resources' care pathways, in particular, those with preexisting hypertension or newly developed preeclampsia in antenatal period.

43.8 Breast feeding initiation and maintenance

- Maternal obesity is associated with a physiological delay in lactogenesis, lower rates of breastfeeding initiation, earlier cessation of breastfeeding, and earlier introduction of solids (RCOG, 2018).
- Breastfeeding is associated with nonexhaustive list of maternal health benefits in addition to improved infant health outcomes including lower risk of infection, obesity, diabetes, and sudden infant death syndrome.
- However, a number of studies have highlighted that women who are overweight or obese are less likely to commence breastfeeding than women of normal weight and if they do commence breastfeeding, they are more likely to cease earlier than women with a normal BMI.
- Another US study (2019) explored the impact of the ten steps to successful breastfeeding on supporting breastfeeding amongst mothers with obesity, which has disproportionately lower rates of exclusive breastfeeding than mothers who are not obese.
- Researchers found that mothers with obesity reported holding babies skin-to-skin significantly less often than other mothers.
- They concluded that interventions aimed at helping mothers with obesity to hold their babies skin-to-skin in the first hour and teaching them to breastfeed on demand have the potential to decrease the breastfeeding disparities in this population.
- Dedicated breastfeeding support during the postnatal period is needed as the onset of breastfeeding is likely to be more complicated than for other women.
- RCOG has suggested providing extra help to ensure frequent and effective milk removal to stimulate lactogenesis, and assistance with physical difficulties attaching the newborn infant to large breasts.
- In addition, as women who are obese are more likely to have had interventions during their labour and birth, it is imperative for clinicians to support women who wish to breastfeed and provide information on how to access local peer groups or lactation consultants following discharge.
- FIGO has further emphasised the need to advise all women who are obese that breastfeeding, including exclusive and mixed breast-feeding, is inversely related to postpartum weight retention.
- Psychological factors that are associated with breastfeeding behaviours in women include body image and social knowledge as well as beliefs on the nutritional adequacy and sufficiency of breast milk and the infant feeding preferences of others.
- Exploring these in women with obesity may support improved breastfeeding behaviours including improving body image and emphasising the adequacy of maternal milk.

43.9 Care plan for other health risks

- Immediate postpartum period is an ideal time for optimisation of weight, health conditions, and prevention of long-term metabolic conditions linked with obesity to decrease morbidity in future for both women and her baby.
- Appropriate referrals to a primary care physician who understands the intricacies of obesity should ideally be provided at the time of hospital discharge.
- That will include access to a comprehensive weight management program including dietary adjustments, physical activity, and behavioural modifications.
- Multidisciplinary care programs suitable for individual patients are available in most healthcare systems or community programs are recommended.
- Due to the risks associated with pregnancy for women with obesity, preventing unwanted pregnancy is an important consideration.
- Potential effects of various contraceptives have relevance to obesity and include increased risk of venous thromboembolism and weight gain.
- Based on this, the risk of contraceptive failure with combined oral contraceptives cannot be excluded, particularly in those with a BMI ≥ 35.
- Alternatives such as long-acting reversible contraceptives, including intrauterine devices and subdermal hormonal implants or progestin-only contraceptives may be preferable where available and acceptable to the woman.
- Postpartum health visits should be used as an opportunity to inform women about the benefits of weight loss between pregnancies, such as reduced risk of stillbirth, hypertensive complications, and foetal macrosomia, and increased chance of successful vaginal birth after caesarean delivery.
- Postpartum weight management, particularly in the first year after pregnancy, may therefore impact the long-term weight status and weight gain trajectory of women, influencing their BMI in subsequent pregnancies.
- Where postpartum weight loss is not possible or achieved, there should be an emphasis on the benefit of weight maintenance and the avoidance of additional weight gain after birth to reduce the risk of complications.
- Expanded counselling may involve the evaluation for bariatric surgery procedures and a proper referral provided.
- In cases where the pregnancy has been completed on a patient who had prepregnancy bariatric surgery, a return to the bariatric surgeon should be secured to allow for any necessary followup care, especially of those women with an adjustable device (lap band).
- Prepregnancy overweight and obesity are significant risk factors for the development of type 2 diabetes (T2DM).
- NICE recommends testing for diabetes at 6 weeks in primary care for all obese women diagnosed with gestational diabetes. CMACE/RCOG (2010) guidance recommends that women with a pregnancy booking BMI of ≥ 30 and gestational diabetes who have a normal glucose tolerance test after giving birth should be regularly followed up by their GP to screen for the development of T2DM.
- The importance of advising overweight and obese women who have Type 1 diabetes about weight loss as part of postnatal care has recently been highlighted to prevent future perinatal complications.
- Obese women have an increased risk of cardiac disease during pregnancy. This is not the only risk factor, as those who smoke, have hypertension, diabetes, are older or have a family history of cardiac disease are also at risk, and a woman may present with several risk factors.

- A recent American study examined the association between maternal obesity on left ventricular size and recovery in women with peripartum cardiomyopathy (PPCM) and found that obese women with PPCM had greater cardiac remodelling; higher leptin levels, and diminished cardiac recovery compared to nonobese controls.
- Healthcare professionals should be aware of the relationship between mental health and obesity and offer psychological support and referral where appropriate.
- Research has shown that prepregnancy obesity is associated with higher risk of postpartum mental health issues including depression.
- In addition, depressive symptoms in women with obesity have been shown to increase with increasing weight gain.
- It is therefore prudent to recommend that, where available, women with obesity should be screened for anxiety and depression in the postpartum period as part of routine postpartum care and using a standardised tool, appropriate to the setting, where available.

Further reading

Davis EM, Ewald G, Givertz MM, et al. Maternal obesity affects cardiac remodeling and recovery in women with peripartum cardiomyopathy. *Am J Perinatol*. 2019;36 (5):476−483. Available from: https://doi.org/10.1055/s-0038-1669439.

Dutton H, Borgengasser SJ, Gaudet LM, et al. Obesity in pregnancy. *Med Clin North Am*. 2018;102:87−106.

Kair L, Nickel N, Jones K, et al. Hospital breastfeeding support and exclusive breastfeeding by maternal pre-pregnancy BMI. *Matern Child Nutr*. 2019. Available from: https://doi. org/10.1111/mcn.12783.

Marshall N, Lau B, Purnell J, et al. Impact of maternal obesity and breastfeeding intention on lactation intensity and duration. *Matern Child Nutr*. 2018. Available from: https://doi. org/10.1111/mcn.12732.

Perlow JH, Morgan MA. Massive maternal obesity and perioperative cesarean morbidity. *Am J Obstet Gynecol*. 1994;170:560.

Robillard P-Y, Dekker G, Scioscia M, et al. Increased BMI has a linear association with late-onset preeclampsia: a population-based study. *PLoS ONE*. 2019;14(10):e0223888. Available from: https://doi.org/10.1371/journal.pone.0223888.

Postpartum weight management and future pregnancy planning

Mohamed ElMoursi
Consultant of Obstetrics and Gynaecology, Victoria Hospital, NHS Fife, Kirkcaldy, United Kingdom

Obesity and its associated metabolic transformations could constitute a significant burden in the postpartum period with its potential complications and outcomes. This chapter will highlight the effects of obesity beyond the immediate postpartum period and the best evidence to manage obesity as well as future fertility planning.[1]

44.1 Complications of obesity

- Obesity is associated with[2]
  - Increased risk of overall mortality
  - Lifetime impact of disability and morbidity[3]
  - Increased risk for[4,5]
    - cardiovascular disease, including coronary heart disease, stroke, heart failure, atrial fibrillation, and venous thromboembolism;
    - diabetes mellitus type 2;
    - dementia in middle-aged adults;
    - gastrointestinal diseases such as gastroesophageal reflux, gallstones, and nonalcoholic fatty liver disease.
- With obesity in pregnancy, associated complications include increased risks for pregnancy-induced hypertension, antepartum venous thromboembolism, labour induction, caesarean delivery, and wound infection, and foetal and neonatal mortality.
- Increasing body mass index (BMI) associated with increased risk of many cancers including colon, breast, corpus uteri, and kidney and leukaemia.
- Additional complications of obesity may include
  - increased risk of
    - chronic kidney disease;
    - kidney stones;
    - hot flashes;
    - obstructive sleep apnoea.
  - lower specificity of screening mammography;
  - lower efficacy of oral contraceptives;
  - orthopaedic issues such as increased risk for low back pain, lumbar disc degeneration, total hip replacement, and knee osteoarthritis;
  - dyspnoea and adult-onset asthma.

Handbook of Obesity in Obstetrics and Gynecology. DOI: https://doi.org/10.1016/B978-0-323-89904-8.00001-9

44.2   Complications of weight stigma

- 'Weight stigma'[6] defined as social devaluation/denigration of individuals due to excess body weight and may lead to negative attitudes, stereotyping, prejudice, and discrimination; weight bias (of patient themselves (internalised), other individuals patient interacts with, and healthcare provider) may be
  - o explicit − overt, consciously held negative attitudes that can be measured by self-report
  - o implicit − automatic negative attributions and stereotypes existing without conscious awareness
- Reported complications/associations of weight stigma include[6]
  - o increased obesity and weight gain over time
  - o increased risk of transitioning from overweight to obesity
  - o depressive symptoms
  - o higher anxiety levels
  - o low self-esteem
  - o social isolation
  - o perceived stress
  - o substance use
  - o unhealthy eating and weight-control behaviours, such as binge eating or emotional overeating
  - o paradoxically increased food intake (regardless of BMI)
  - o lower levels of physical activity and higher exercise avoidance
  - o increased sedentary behaviour
- Reported possible effects of obesity stigma in healthcare setting include
  - o avoidance of clinical care, such as age-appropriate cancer screening
  - o reduced adherence to prescribed treatment or self-care
  - o lower frequency of achieving weight loss goals.

44.3   Outcomes of obesity and its effects on future pregnancies

44.3.1   Maternal outcomes

- Pregestational BMI > 30 kg/m^2 may be associated with increased risk for cardiovascular hospitalisations after pregnancy.[7]
- Obesity during pregnancy may be associated with increased risk of premature death and cardiovascular disease.
- Excessive gestational weight gain during pregnancy in obese women is associated with increased risk of weight retention at 1-year postpartum.

44.3.2   Foetal/neonatal outcomes

- Increasing BMI is associated with increased foetal and neonatal mortality with higher risk of foetal death, stillbirth, and neonatal, perinatal, and infant death.[8,9]

- Prepregnancy maternal BMI >25 kg/m^2 associated with increased birth weight for gestational age and infant adiposity.[10]
- Maternal overweight and obesity is associated with small increased risk for congenital anomalies and birth defects.

44.3.3 Outcomes in childhood for children born to obese mothers

Children born to obese women have higher risk of[11]

- metabolic syndrome and childhood obesity
- development of autism spectrum disorders
- respiratory hospitalisation during childhood in offspring
- type 1 diabetes in offspring of parents without diabetes
- development of affective disorders in offspring
- adverse mental and motor development assessed at 2 years of age in preterm offspring
- developmental delay in offspring among children.

44.3.4 Outcomes in adulthood being born to obese mothers

- Maternal overweight and obesity is associated with increased risk of premature death in adult offspring.[12]
- Maternal gestational weight gain exceeding recommended guidelines may be associated with overweight or obesity in daughters during middle age.

44.4 General postpartum management

44.4.1 Management

- Routine postpartum care as discussed before in the immediate postpartum period.

44.4.2 Thromboprophylaxis

- Maternal obesity is associated with increased risk of postpartum venous thromboembolism.[13]
- Offer thromboprophylaxis to women having caesarean section in labour as they score 2 points in the RCOG scoring system due to increased risk of venous thromboembolism.
- BMI >30 kg/m^2 scores 1 point and BMI> 40 kg/m^2 scores 2 points in the RCOG scoring system for thromboprophylaxis.
- $\geq$ 2 current risk factors, administer for $\geq$ 10 days postpartum (RCOG Grade D).
- $\geq$ 3 current risk factors, starting from 28 weeks gestation and continuing for 6 weeks postnatally (RCOG Grade D).
- $\geq$ 4 current risk factors (excluding previous VTE and thrombophilia), starting in first trimester and continuing for 6 weeks postnatally (RCOG Grade D).

44.4.3   Postpartum contraception

- Discuss postpartum contraception (strong recommendation). Progestin-only methods including injection, 'mini-pill', intrauterine device, or implant may be initiated before discharge, and do not appear to affect breastfeeding or infant growth (strong recommendation).
- Overweight or obesity does not appear to reduce contraceptive efficacy of intrauterine device or etonogestrel subdermal contraceptive implants.[14]
- Obesity does appear to lower efficacy of oral contraceptives.

44.4.3.1   Intrauterine devices

- Most women are good candidates for hormonal and nonhormonal Intrauterine devices (IUDs).
- Both hormonal and nonhormonal IUDs may be inserted any time during cycle when pregnancy can be excluded [American College of Obstetricians and Gynecologists (ACOG) Level B], including
   o immediately postpartum (within 10 minutes after placental delivery in vaginal and caesarean births) (ACOG Level B).
- Immediate postpartum insertion (within 10 minutes of placental separation) is considered ideal for both breastfeeding and nonbreastfeeding women[15]
   o with copper IUD (USMEC Category 1; UKMEC Category 1);
   o with levonorgestrel-releasing intrauterine system (LNG-IUS) (USMEC Category 1 and 2; UKMEC Category 1; ACOG Level B).
- Benefits of immediate postpartum insertion include
   o convenience for patient;
   o patients may become pregnant prior to followup appointments;
   o patients may fail to attend followup appointments.
- Immediate postpartum insertion contraindicated in women with (USMEC Category 4)
   o peripartum chorioamnionitis;
   o endometriosis;
   o puerperal sepsis;
   o ongoing postpartum haemorrhage.
- IUD placement during caesarean delivery associated with higher rate of IUD use at 6 months compared with placement delayed $\geq$ 6 weeks after delivery
- Routine postpartum insertion (4−6 weeks postpartum) with copper IUD or LNG-IUS is acceptable for breastfeeding and nonbreastfeeding women (USMEC Category 1; UKMEC Category 1).

44.4.4   Nutrition and physical exercise

- Postnatal dietary guidelines are similar to those established during pregnancy.
- Calorie requirements for breastfeeding women:
   o on average, $\geq$ 1800 kcal/day required for adequate milk production;
   o usually, additional 500 kcal/day recommended while lactating.
- Routine vitamin/mineral supplementation not indicated.
- ACOG recommendations for physical activity and exercise in postpartum period:[16]
   o encourage women to engage in aerobic and strength training exercises after delivery;
   o pelvic floor exercises can be initiated immediately postpartum;

o advise to gradually resume exercise routine after delivery once it is medically safe depending on mode of delivery and presence of medical or surgical complications; and

o advise breastfeeding mothers to feed their infants before exercising to avoid breast discomfort.

44.5   Weight loss management strategy

44.5.1   Postpartum weight loss

- Postpartum weight loss[17] at a rate of 1 kg/month can occur without affecting lactation
  o women will typically retain 2 lbs above the prepregnancy weight at 1 year postpartum;
  o encourage exercise and healthy eating habits, particularly in women with residual postpartum retention of weight gained during pregnancy:
    ■ diet alone for 11 days was associated with increased weight loss compared to usual care;
    ■ diet plus exercise was associated with decreased body weight with increase in return to prepregnancy weight;
    ■ none of the interventions adversely affected breastfeeding performance.
  o Addition of Internet-based weight-loss programme to Special Supplemental Nutrition Program for Women, Infants, and Children (WIC programme) may increase postpartum weight loss in low-income women at 12 months:
    ■ Internet-based programme provided calorie and physical activity goals and included weekly lessons, web diary, instructional videos, computerised feedback, text messages, and monthly face-to-face group visits at clinics.
  o Dietary modification (with or without exercise regimen) may be more effective for sustainable weight loss than exercise alone during postpartum period.

44.5.2   Breastfeeding and obesity

- Provide breastfeeding support and counsel mothers that breastfeeding is associated with improved weight loss and decreased risk of subsequent diabetes.[18]
- Education and social support interventions might increase any breastfeeding at 6 months compared to usual care in women with overweight or obesity.
- High BMI before or during pregnancy is associated with increased risk of postpartum anaemia and early termination of breastfeeding.
- Increasing prepregnancy BMI is associated with decreased initiation of breastfeeding, any breastfeeding, and exclusive breastfeeding at 12 weeks postpartum.
- Obesity class II may be associated with increased risk of failing to exclusively breastfeed at hospital discharge compared to overweight.[19]
- Obesity during pregnancy may be associated with increased risk of discontinuing breastfeeding.
- Breastfeeding exclusively for ≥ 4 months and continuation of any breastfeeding for ≥ 12 months associated with less retained pregnancy weight in obese women.
- Obese mothers may receive less exposure to pro-breastfeeding hospital practices.

44.6   Obesity, weight loss, and future fertility planning

44.6.1   Management overview

- Weight loss of 5%−15% may greatly reduce complications in persons with overweight and obesity.[20]
- General treatment strategy[21] as per the American Association of Clinical Endocrinologists/ American College of Endocrinology (AACE/ACE):
  - therapy should be aimed at reducing weight-related complications through weight loss rather than weight loss as the only goal (AACE/ACE grade D);
  - make available a structured lifestyle intervention programme consisting of a healthy meal plan, physical activity, and behavioural interventions (AACE/ACE Grade A, Best evidence level 1);
  - base strategy on concepts of prevention for chronic diseases (AACE/ACE Grade C, Best evidence level 4):
    - for preventing additional weight gain and weight-related complications (secondary prevention)
      - use BMI to screen weight;
      - use BMI and waist circumference (as applicable) to assess for risk of complications;
      - use lifestyle and behavioural interventions with or without weight loss medication.
    - if weight-related complications are present (tertiary prevention)
      - treat overweight and obesity with lifestyle or behavioural interventions plus weight loss medication;
      - consider bariatric surgery.
  - if considering weight-loss therapy, screen for binge-eating disorder and night-eating syndrome (AACE/ACE Grade B, Best evidence level 3).
- Weight loss diets require caloric expenditure (including the resting metabolic rate, thermic effect of feeding, and physical activity) to exceed the caloric intake to be effective; consider specific types of diet or dietary patterns to optimise adherence, eating patterns, metabolic profiles, risk factor reduction, and/or other clinical outcomes:
  - different dietary approaches are effective for weight loss and weight maintenance as long as target calorie reduction is achieved; recommended options include
    - energy deficit of 500−750 calories below estimated energy needs;
    - consumption of 1200−1500 calories/day for women or 1500−1800 calories/day for men (calories adjusted for individual's body weight);
    - adherence to restrictive diet (such as low-carbohydrate diet or high-fibre diet) to create energy deficit through reduced food intake.
  - dietary changes that may be associated with increased weight loss include increasing intake of vegetables and fruit, high-fibre and whole-grain foods, and water while reducing dietary sugar.
  - guideline recommended diets include Mediterranean diet, portion control plates, and Weight Watchers, Jenny Craig, or Nutrisystem.
- Exercise may promote weight loss, especially when combined with dietary change:[1]
  - multiple short-bout exercise (four 10-minute sessions/day 5 days/week) might be equivalent to long-bout exercise (40-minute sessions 5 days/week);
  - regular exercise (brisk walking) slightly reduces body weight and body fat in postmenopausal women;

- o low amount of exercise (walking 30 minutes/day) appears adequate to avoid weight gain and higher amounts promote weight loss.
- Behavioural and cognitive-behavioural strategies associated with small reduction in body weight in patients with overweight or obesity
- Acupuncture associated with significant weight loss in patients with obesity compared with placebo and conventional treatments.
- Weight loss medication may be used as adjunct to diet, behaviour therapy, and physical activity:
  - o medications may be useful for patients with BMI $\geq 27 \, \text{kg/m}^2$ with concomitant obesity-related risk factors or diseases (diabetes, hypertension or dyslipidaemia), or those with BMI $\geq 30 \, \text{kg/m}^2$;
  - o FDA-approved medications that are recommended as adjuncts to chronic management (> 6 months) of weight loss in adults include
    - orlistat (Xenical);
    - liraglutide (Saxenda);
    - phentermine plus extended-release topiramate combination (Qsymia);
    - naltrexone/bupropion extended release (Contrave).
- Multicomponent weight-loss programs might be effective for weight loss in adults; maintenance interventions with behavioural and/or lifestyle components may promote further weight loss for up to 18 months in adults with overweight or obesity.
- Surgery:
  - o bariatric surgery
    - consider bariatric surgery for patients with BMI $\geq 40 \, \text{kg/m}^2$ or BMI $\geq 35 \, \text{kg/m}^2$ with comorbidities (BMI $30-34 \, \text{kg/m}^2$ with comorbidities in some cases);
    - compared to nonsurgical management bariatric surgery associated with[1]
      - reduced mortality;
      - long-term weight loss (maintained up to 10 years);
      - reductions in comorbidities such as diabetes, hyperlipidaemia, hypertension, and obstructive sleep apnoea;
      - improved fertility in women with fewer maternal and neonatal complications.
  - o liposuction reported to reduce BMI but not most risk factors for coronary heart disease or mediators of inflammation within 12 weeks in adult women with obesity with or without type 2 diabetes.

44.7 Planning for future pregnancies

44.7.1 Obesity and polycystic ovarian syndrome

Obese ladies will often have polycystic ovarian syndrome (PCOS) that need to be managed to achieve future fertility.

44.7.1.1 Benefits of weight loss

- o Improved endocrine profile, such as decreased insulin resistance, hyperandrogenism, and glucose.
- o Increased likelihood of ovulation (naturally and in response to ovulation induction therapy).
- o Increased pregnancy rates.
- o Greater likelihood of healthy pregnancy.

44.7.1.2   Lifestyle management recommendations for women with polycystic ovarian syndrome seeking fertility

- First-line therapy for all women with PCOS seeking fertility is lifestyle modification targeting weight loss in women with overweight and prevention of weight gain in women who are lean and should include reduced energy (caloric) intake and physical activity (WHO strong recommendation, moderate-quality evidence).[22]
- For women with BMI >40 kg/m², consider discouraging ovulation induction with medications for first-line therapy until weight loss has occurred with diet, exercise, bariatric surgery, or other appropriate means (WHO weak recommendation, low-quality evidence).[22]

44.7.1.3   Lifestyle management recommendations for all women with polycystic ovarian syndrome

- Dietary interventions
  - o In women with PCOS, principles of healthy eating should be followed (NHMRC/ASRM/ESHRE Clinical consensus, Strong recommendation).[23]
  - o For patients with overweight or obesity with PCOS:
    - ■ consider beginning weight loss with reduced-calorie diet (with no evidence that 1 particular diet is superior) (Endocrine Society weak recommendation, low-quality evidence);[24]
    - ■ recommend a variety of balanced dietary approached to decrease energy intake and lead to weight loss (NHMRC/ASRM/ESHRE clinical consensus recommendation, strong recommendation);
    - ■ energy deficit of 30% or 500−750 kcal/day (total intake of 1200−1500 kcal/day) while considering body weight, physical activity levels, and individual energy requirements suggested.
  - o Limited or no evidence to suggest that one energy equivalent diet is better than another or that women with PCOS respond differently to weight management interventions compared to women without PCOS (NHMRC/ASRM/ESHRE Clinical practice point).
  - o Tailoring dietary changes to individual food preferences, allowing for a personalised and flexible approach for decreasing energy intake and avoiding restrictive and nutritionally unbalanced diets suggested (NHMRC/ASRM/ESHRE clinical practice point).

44.7.1.4   Antiobesity medications recommendations for women with polycystic ovarian syndrome seeking fertility

- Antiobesity medications for all women with PCOS[23]
  - o Consider antiobesity medications an experimental therapy in women with PCOS for the goal of improving fertility; current risk to benefit ratios are too uncertain to recommend antiobesity medications for fertility therapy (NHMRC/ASRM/ESHRE clinical consensus recommendations).
  - o When considering antiobesity medications, consider avoidance of pregnancy while taking antiobesity medications (NHMRC/ASRM/ESHRE clinical practice point).

44.7.1.5 Bariatric surgery recommendations for women with polycystic ovarian syndrome seeking fertility

- Bariatric surgery may be considered to improve fertility outcomes in women.
- With PCOS who are anovulatory, have a BMI ≥ 35 kg/m^2, and who remain infertile despite intensive lifestyle management involving reduced caloric intake, exercise, and behavioural interventions for ≥ 6 months (WHO weak recommendation, low-quality evidence).[22]
- Risk to benefit ratios are unclear to advocate bariatric surgery as fertility therapy in women with PCOS (NHMRC/ASRM/ESHRE clinical consensus recommendation, conditional recommendation against the option)[22]
- Consider bariatric surgery an experimental therapy in women with PCOS for the goal of having a healthy baby (NHMRC/ASRM/ESHRE clinical consensus recommendation, conditional recommendation against the option).
- Considerations for patients choosing bariatric surgery:
  - Women with PCOS having bariatric surgery should be managed in an interdisciplinary setting that includes a dietitian and other staff members trained to work with patients who have had bariatric surgery (WHO strong recommendation, low-quality evidence).
  - Consider availability of structured weight management programme involving physical activity, diet, and interventions to continue to improve psychological, cardiovascular, and musculoskeletal health postoperatively (NHMRC/ASRM/ESHRE clinical practice point).
  - Counsel women on
    - risks of surgery (WHO Strong recommendation, Low-quality evidence), including risks to long-term health and pregnancy:
      - pregnancy risks persist in women who remain with obesity after bariatric surgery;[25]
      - neonatal risks include premature delivery, small for gestational age, and possibly increased infant mortality (NHMRC/ASRM/ESHRE clinical practice point).
    - pre and postoperative nutritional deficiencies (WHO strong recommendation, low-quality evidence).
  - Consider possible benefits, including reduced incidence of gestational diabetes and large for gestational age foetus (NHMRC/ASRM/ESHRE clinical practice point).
  - Pregnancy should be avoided during periods of rapid weight loss (WHO strong recommendation, low-quality evidence; NHMRC/ASRM/ESHRE clinical practice point).
  - Recommendations for delaying pregnancy after bariatric surgery vary:
    - avoid pregnancy for 6−12 months postsurgery (WHO strong recommendation, low-quality evidence);
    - delay for minimum of 12 months after bariatric surgery and suggest use of appropriate contraception (NHMRC/ASRM/ESHRE clinical practice point).
  - if pregnancy occurs after bariatric surgery:
    - clinicians and patients should be aware of and prevent nutritional deficiencies (both pre and postoperative), ideally in a specialist interdisciplinary setting (NHMRC/ASRM/ESHRE clinical practice point);
    - monitor foetal growth during pregnancy (WHO strong recommendation, low-quality evidence; NHMRC/ASRM/ESHRE clinical practice point).

44.7.1.6   Screening for glucose intolerance

44.7.1.6.1   Gestational diabetes mellitus

- For women with gestational diabetes mellitus (GDM), screen for glucose intolerance at 6–12 weeks postpartum and manage as indicated (strong recommendation). Repeat testing at least every 3 years in women with GDM and a normal initial postpartum screen (strong recommendation).[26]

44.7.1.6.2   Pregestational diabetes mellitus

- Affects 1% of all pregnancies (diabetes before pregnancy or who meet the diagnostic criteria at initial prenatal visit).
- Poor glycaemic control in early pregnancy is associated with
  - adverse maternal outcomes including preeclampsia, preterm delivery, caesarean section, and maternal mortality;
  - adverse foetal outcomes including congenital malformations, perinatal mortality, and macrosomia (which affects 45% of affected infants).
- All women with preexisting type 1 or type 2 diabetes should receive preconception care to optimise glycaemic control and assess for complications; evaluations include:[26]
  - blood tests to monitor for glycaemic control
    - serum glucose for continued monitoring;
    - HbA1c for surveillance of blood glucose control.
  - evaluation for underlying vasculopathy, including
    - dilated eye exam by an ophthalmologist to evaluate for retinopathy;
    - 24-hour urine collection for protein excretion and creatinine clearance;
    - lipid profile (total cholesterol, low-density lipoprotein cholesterol, high-density lipoprotein cholesterol, and triglycerides) for risk assessment in all patients with diabetes at initial medical evaluation (ADA Grade E);
    - baseline echocardiogram, particularly in women with longstanding diabetes or any vascular complications;
    - patients with type 1 diabetes should have evaluation of thyroid function;
    - patients with type 1 diabetes should have evaluation of eyes for retinopathy;
    - gradual deterioration of visual or evidence of progressive diabetic retinopathy is a relative contraindication for pregnancy.
- Develop or adjust a management plan to achieve near-normal glycaemia, while minimising significant hypoglycaemia.

44.7.1.6.3   Obesity

- Screen obese pregnant women for glucose intolerance and obstructive sleep apnoea during first prenatal visit with history, physical exam, and laboratory investigations as needed.
- Glucose tolerance screening:
  - need for early pregnancy screening for glucose intolerance should be based on risk factors, such as (ACOG Level C):[26]
    - maternal BMI ≥ 30 kg/m^2;
    - known impaired glucose metabolism;
    - previous gestational diabetes.
  - consider alternative testing for gestational diabetes in women with previous malabsorptive-type bariatric surgery (ACOG Level C).[27]

References

1. DynaMed [Internet], Ipswich (MA): EBSCO Information Services. Record No. T115009, Obesity in adults. <https://www-dynamed-com.knowledge.idm.oclc.org/topics/dmp~AN~T115009>; 1995 Accessed February 2021.
2. Global BMIMC Di, Angelantonio E, Bhupathiraju Sh N, Wormser D, Gao P, Kaptoge S, et al. Body-mass index and all-cause mortality: individual-participant-data meta-analysis of 239 prospective studies in four continents. *Lancet (London, England)*. 2016;388 (10046):776–786.
3. Visscher TL, Rissanen A, Seidell JC, Heliövaara M, Knekt P, Reunanen A, et al. Obesity and unhealthy life-years in adult Finns: an empirical approach. *Arch Int Med*. 2004;164(13):1413–1420.
4. Murphy NF, MacIntyre K, Stewart S, Hart CL, Hole D, McMurray JJ. Long-term cardiovascular consequences of obesity: 20-year follow-up of more than 15000 middle-aged men and women (the Renfrew-Paisley study). *Eur Heart J*. 2006;27(1):96–106.
5. Hayashi T, Boyko EJ, Leonetti DL, McNeely MJ, Newell-Morris L, Kahn SE, et al. Visceral adiposity is an independent predictor of incident hypertension in Japanese Americans. *Ann. Intern Med*. 2004;140(12):992–1000.
6. Rubino F, Puhl RM, Cummings DE, Eckel RH, Ryan DH, Mechanick JI, et al. Joint international consensus statement for ending stigma of obesity. *Nat. Med*. 2020;26 (4):485–497.
7. Yaniv-Salem S, Shoham-Vardi I, Kessous R, Pariente G, Sergienko R, Sheiner E. Obesity in pregnancy: what's next? Long-term cardiovascular morbidity in a follow-up period of more than a decade. *J Maternal-fetal Meonatal Med*. 2016;29(4):619–623.
8. Aune D, Saugstad OD, Henriksen T, Tonstad S. Maternal body mass index and the risk of fetal death, stillbirth, and infant death: a systematic review and *meta*-analysis. *Jama*. 2014;311(15):1536–1546.
9. Meehan S, Beck CR, Mair-Jenkins J, Leonardi-Bee J, Puleston R. Maternal obesity and infant mortality: a *meta*-analysis. *Pediatrics*. 2014;133(5):863–871.
10. McCloskey K, Ponsonby AL, Collier F, Allen K, Tang MLK, Carlin JB, et al. The association between higher maternal pre-pregnancy body mass index and increased birth weight, adiposity and inflammation in the newborn. *Pediatr Obes*. 2018;13(1):46–53.
11. Hussen HI, Persson M, Moradi T. Maternal overweight and obesity are associated with increased risk of type 1 diabetes in offspring of parents without diabetes regardless of ethnicity. *Diabetologia*. 2015;58(7):1464–1473.
12. Reynolds RM, Allan KM, Raja EA, Bhattacharya S, McNeill G, Hannaford PC, et al. Maternal obesity during pregnancy and premature mortality from cardiovascular event in adult offspring: follow-up of 1 323 275 person years. *BMJ (Clin Res Ed)*. 2013;347:f4539.
13. Tepper NK, Boulet SL, Whiteman MK, Monsour M, Marchbanks PA, Hooper WC, et al. Postpartum venous thromboembolism: incidence and risk factors. *Obstetr Gynecol*. 2014;123(5):987–996.
14. Xu H, Wade JA, Peipert JF, Zhao Q, Madden T, Secura GM. Contraceptive failure rates of etonogestrel subdermal implants in overweight and obese women. *Obstetr Gynecol*. 2012;120(1):21–26.
15. Practice Bulletin No. 186. Long-acting reversible contraception: implants and intrauterine devices. *Obstetr Gynecol*. 2017;130(5):e251–e69.
16. ACOG Committee. Opinion No. 650: Physical activity and exercise during pregnancy and the postpartum period. *Obstetr Gynecol*. 2015;126(6):e135–e42.

17. Kilpatrick S.J., Papile L.A., Macones, G.A. (eds), (2017). Guidelines for perinatal care (8th ed.). *American Academy of Pediatrics.*
18. Garrison AW. Obesity in pregnancy. *Am Fam Phys.* 2013;87(9):606. 8.
19. Martinez JL, Chapman DJ, Pérez-Escamilla R. Prepregnancy obesity class is a risk factor for failure to exclusively breastfeed at hospital discharge among latinas. *J Human Lactat.* 2016;32(2):258–268.
20. Kushner RF, Ryan DH. Assessment and lifestyle management of patients with obesity: clinical recommendations from systematic reviews. *Jama.* 2014;312(9):943–952.
21. Garvey WT, Mechanick JI, Brett EM, Garber AJ, Hurley DL, Jastreboff AM, et al. American Association of Clinical Endocrinologists and American College of Endocrinology Comprehensive Clinical Practice guidelines for medical care of patients with obesity. *Endocr Pract.* 2016;22(suppl 3):1–203.
22. Balen AH, Morley LC, Misso M, Franks S, Legro RS, Wijeyaratne CN, et al. The management of anovulatory infertility in women with polycystic ovary syndrome: an analysis of the evidence to support the development of global WHO guidance. *Hum Reprod Update.* 2016;22(6):687–708.
23. Teede HJ, Misso ML, Costello MF, Dokras A, Laven J, Moran L, et al. Recommendations from the international evidence-based guideline for the assessment and management of polycystic ovary syndrome. *Fertil Steril.* 2018;110(3):364–379.
24. Legro RS, Arslanian SA, Ehrmann DA, Hoeger KM, Murad MH, Pasquali R, et al. Diagnosis and treatment of polycystic ovary syndrome: an endocrine society clinical practice guideline. *J Clin Endocrinol Metab.* 2013;98(12):4565–4592.
25. Vitek W, Hoeger K, Legro RS. Treatment strategies for infertile women with polycystic ovary syndrome. *Minerva Ginecol.* 2016;68(4):450–457.
26. ACOG practice bulletin no 156. Obesity in pregnancy. *Obstet Gynecol.* 2015;126(6): e112–e26.
27. ACOG practice bulletin no. 105. Bariatric surgery and pregnancy. *Obstetr Gynecol.* 2009;113(6):1405–13.

Index

Note: Page numbers followed by "*f*," "*t*," and "*b*" refer to figures, tables, and boxes, respectively.

Printed in the United States
by Baker & Taylor Publisher Services